CONTROL SYSTEMS ENGINEERING AND DESIGN

CONTROL SYSTEMS ENGINEERING AND DESIGN

S. THOMPSON

Department of Mechanical and Manufacturing Engineering
The Queen's University of Belfast

Longman Scientific & Technical

Copublished in the United States with
John Wiley & Sons, Inc., New York

Longman Scientific & Technical,
Longman Group UK Limited,
Longman House, Burnt Mill, Harlow,
Essex CM20 2JE, England
and Associated Companies throughout the world.

Copublished in the United States with
John Wiley & Sons, Inc., 605 Third Avenue, New York, NY 10158

First published 1989

British Library Cataloguing in Publication Data
Thompson, S. (Steve)
    Control systems engineering and design.
    1. Control systems. Design
    I. Title
    629.8′312

ISBN 0-582-99468-3

Library of Congress Cataloging-in-Publication Data
Thompson, S., 1949–
    Control systems engineering and design.
    Bibliography: p.
    Includes index.
    1. Automatic control.   I. Title.
TJ213.T493   1989   629.8   88-24582
ISBN 0-470-21322-1 (USA only)

Set in Monotype Lasercomp Times

Produced by Longman Group (FE) Limited
Printed in Hong Kong

CONTENTS

PREFACE

As an undergraduate studying mechanical engineering, I found the study of control fascinating. This early interest was no doubt due to the skills of the lecturers who presented the material in such a vital way. After graduation, my studies of control continued with equal enthusiasm, to encompass an M.Sc. and then, much later, a Ph.D. However, despite having (academically speaking) discovered my first love, I was left feeling disillusioned. How could the techniques of control theory be applied to the objects of reality? All the books I read, or lecturers I listened to, seemed to assume the existence of a mathematical model from which all else followed. But how was such a model to be established? Clearly, this was the corner stone at which theory and practice met, and yet – it is a topic which is ignored in most textbooks on control.

In writing this book my main aim has therefore been to address the complete control system design problem: from development of a model, through the design of a controller, to the application and implementation of techniques on equipment accessible to students. The book is not intended to be an introduction to the various disciplines of control theory; there are many excellent books that do just that. Rather, it concentrates on linear, single input/single output, analogue control systems, or classical control, and attempts to provide a thorough practical and theoretical treatment of the subject.

The book is organized as three main parts: theory, case studies and computer programs. The first and largest part consists of Chapters 1 to 9. Chapter 1 introduces the various aspects of control system design, and the lumped parameter technique of control system modelling. Chapter 2 deals with differential equations and Laplace transforms, and Chapter 3 with transfer functions and block diagrams. These three chapters establish the mathematical and conceptual framework on which subsequent chapters are built. The graphical nature of the subject is emphasized in Chapter 4, and provides the link between the various stability and design techniques developed in Chapters 5, 6 and 7. Chapter 8 covers controller and compensator design, and Chapter 9 presents some simple identification procedures. Throughout, emphasis is placed on understanding the techniques, their relationship to each other and their practical applications.

Four case studies are presented in Chapter 10. Each is based on a standard piece of laboratory equipment and is used to illustrate various aspects of the theory. These studies, together with the programs in Chapter 11, are intended to be used in laboratory work and class demonstrations.

The final part, Chapter 11, contains a number of documented BASIC programs suitable for implementation on most computers. These programs mirror the material of Chapters 1 to 9, and are referred to throughout the text.

A more detailed study of the contents will reveal such topics as root loci and PID controllers, inverse polar plots, time delays and conditional stability which are not normally covered in an introductory text on control. Their inclusion is in keeping with the stated objectives of providing a sound practical and theoretical knowledge of the subject. Also, the treatment of the various topics reflects the change in emphasis brought about by the widespread use of computers. Inverse polar plots, for example, are a necessary prerequisite for the study of multivariable systems in the frequency domain, which is invariably computer based. The ubiquitous microcomputer, with its data logging and control facilities, makes identification a practical and attractive alternative to modelling.

Throughout the book the needs of the student and lecturer have been borne in mind. Each chapter contains worked examples, and concludes with a representative set of problems for which partial solutions are provided.

Steve Thompson, May 1988

turn on investment. Such a statement may seem self-evident, but it is
sarily an easy matter to identify the key parts of a complex system,
ly if there are many elements, all of which could be controlled. However,
tified, most control problems are solved by adopting a strategy similar
own in Fig. 1.1.

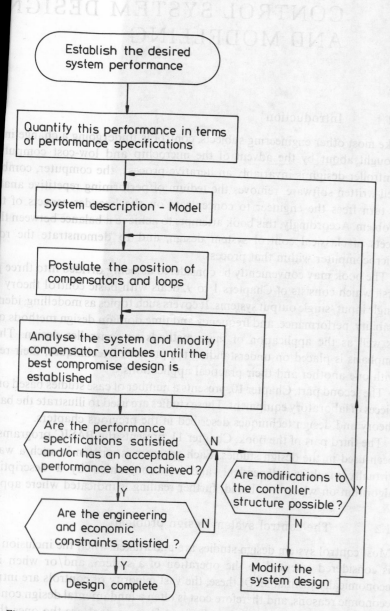

Figure 1.1 Typical control system design procedure

Acknowledgements

It is a pleasure to thank all those who, in their various ways, have made the writing
of this text possible. My parents, Joe and Elsie Thompson, who never thought
they would see their names in print, and have proved a constant source of
encouragement throughout my undergraduate and postgraduate years. Those
teachers, lecturers and colleagues to whom, through formal lectures, written works
and discussions, I am greatly indebted. In particular I would like to thank
Dr George Irwin for his active interest and contributions, especially to Chapters 2
and 3, the representatives of Armfield, Feedback and TecQuipment for their
assistance, the typists, especially Rhona Mitchell, for their excellent work, and
Diane Gordon, who produced all the artwork. On a lighter note, I must mention
daddy's two wee helpers, Andrea and Philip, aged 3 and 1 respectively, who always
ensured an early start. Also, I acknowledge the additional burden placed on my
dear wife Stella, without whose support this work would not have been completed.

I can think of no better conclusion than to quote from the Bible. To my
mind, this verse eloquently summarizes the learning and writing process, and
acknowledges the efforts of those who have gone before:

> For it is precept upon precept, precept upon precept, line upon line, line
> upon line, here a little, there a little.

Isaiah, chapter 28 verse 10.

We are indebted to the following for permission to reproduce copyright material:

Armfield Technical Education Co Ltd for fig 10.5 and **fig** 10.22; Feedback
Instruments Ltd for figs 10.1 (a) & (b); TecQuipment Ltd for fig 10.15.

CHAPTER 1

CONTROL SYSTE
AND MODELLING

1.1   Introduction

Like most other engineering subjects, control is und
brought about by the advent of the microchip an
controller design is invariably an iterative process, t
well-written software, removes the tedium of perfor
in turn frees the engineer to concentrate on the b
problem. Accordingly, this book attempts to achieve a
facets of classical control system design, and to d
microcomputer within that process.

The book may conveniently be considered as being
first, which consists of Chapters 1 to 9, deals with ba
single input/single output systems. It covers such topics
stability, performance, and frequency and time domain
as well as the application of these methods to cont
emphasis is placed on understanding the presented te
with one another and their practical applications.

The second part, Chapter 10, presents a number of cas
pieces of laboratory equipment. These studies are used to
theory and design techniques described in the previous

The third part of the book, Chapter 11, presents the
been used in the design studies. Each program is writt
virtually machine-independent, and is accompanied
algorithm on which it is based; further reading is indica

1.2   The control system design problem

Most control system design studies are undertaken when
is considered essential to the operation of a system, a
economically justifiable. Of these, the vast majority of co
economic reasons, and therefore cost is often a fundament
elements of a system which benefit most from control a

1.2.1   System performance and performance specifications

The initial phase of the design process depicted in Fig. 1.1 is that of establishing and then quantifying the desired response characteristics of the system. It is presumed that the equipment is capable of achieving the required performance, and that the current system could be improved or is inadequate, for only under these conditions can there be a control problem to be solved.

Consider the level control system shown in Fig. 1.2. The operator controls the level of fluid in the tank by manipulating the valve. Variations in the inlet flow rate, temperature changes (these would alter the fluids viscosity and consequently the flow rate through the valve) and similar disturbances must be compensated for by the operator. Without control the tank is likely to flood, or run dry. Using an operator may produce an adequate control system but, social arguments apart, such a situation is not likely to be economically justifiable. Clearly this system is a candidate for control; the equipment is capable of maintaining a level within prescribed limits, and the current setup could be improved.

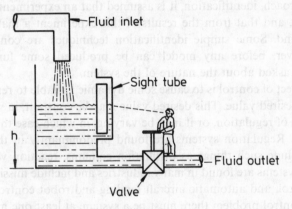

Figure 1.2 Level control system

For the system shown in Fig. 1.2 the designer's objective for the controller may be to maintain the fluid level constant at, or near, some desired value. Furthermore, following a disturbance the controller should respond quickly and return the fluid level to its desired value. It may also be required that any fluctuation in level following a disturbance should respond in some acceptable manner. These desired response characteristics may be quantified, together with any additional system requirements.

Maintenance of a level implies that the system should be stable. If the control system is to respond to a change in level, then it must be supplied with information about those changes, so the tank will have to incorporate a level sensor. Also, there will have to be a transmission system which collects information from the sensor and relays it in a form that the controller can accept.

The requirement for response to a disturbance suggests that design constraints are imposed on the controller. These are easily quantified in terms of rise time, peak overshoot, decay ratio and so on (these terms are discussed in subsequent chapters). After observing some change in level detected by the sensor, the controller will require some action to be taken which will manipulate the valve and thus counteract the effect. The device which manipulates the valve is called an actuator. A second transmission path, from the controller to the actuator, is required.

A control problem now exists and, as indicated in Fig. 1.1, the next stage in the design procedure is to produce a model.

1.2.2   System description – modelling

Control system modelling is a subject in its own right. Essentially there are two approaches to finding a model. In the first the system is broken down into elements for which a mathematical description is then established. The simplest such technique available is lumped parameter modelling; considered later in this chapter. In the second approach, identification, it is assumed that an experiment is carried out on the system, and that from the results of this experiment a mathematical model can be found. Some simple identification techniques are considered in Chapter 9. However, before any model can be produced some fundamental questions must be asked about the nature of the system.

The primary object of control is to cause some dynamic variable to remain fixed at, or near, some desired value. This desired value may be constant, in which case the problem is one of regulation, or it may be varying, in which case the problem is one of tracking. Regulation systems are found predominantly in the process industries, and include such things as distillation columns, reaction vessels and boilers. Tracking systems are found in many industries and include missile control, machine tool control, and automatic aircraft landing and robot control systems. In both types of control problem there must be a system, at least one measurable output, and at least one manipulable input which modifies the measurable outputs in some quantifiable manner. There are likely to be other inputs which cannot be manipulated, but which nevertheless affect the measurable outputs; these are called disturbance inputs. Establishing the system, together with its various inputs and outputs, constitutes the first stage in any modelling exercise.

Consider again the level control system shown in Fig. 1.2. A system may be defined as an arrangement of physical components connected or related in such a manner as to form and/or act as an entire unit. The level control system shown is, by this definition, clearly a system. However, it is made up of a number of subsystems. Not counting the operator, the system consists of the ensemble of fluid, tank, fluid inlet and fluid outlet. Such a system is often referred to as the process, or the plant. The dynamic variable to be controlled is the fluid level in the tank, the measurable output is in the fluid level in the sight tube, and the manipulable input is the valve on the fluid outlet pipe. Changes in temperature, inlet flow rate and so on will all affect the level, but since they cannot be manipulated they are what are known as disturbance inputs. The plant is an open-loop system since (without the operator) the measured output has no effect on the valve, the manipulable input. When the operator is included, the system is said to be a closed-loop system, since the measurable output quantity will affect the manipulable input. This is achieved by a property known as feedback, which permits the output to be compared with the input so that an appropriate controlling action may be formed. The required control system design model is the one that describes the dynamic relationship between the open-loop plant's manipulable input and the measurable output. Ideally, the model would include the effect of disturbance inputs, although often an exact dynamic description of these is not necessary.

Before considering the next stage in the design process, it is useful to have another look at the system shown in Fig. 1.2. The control requirement is that the tank level h remains constant, but the measurable output is the sight tube level. Only under steady-state conditions (when fluid inlet flow equals fluid outlet flow, and when all the transients within the system have subsided) are the tank and sight tube levels likely to be identical. The sight tube is itself a subsystem with its own forcing input (the tank level), which will be related dynamically to the sight tube's indicated level. What the operator is doing is controlling the sight tube level, not the tank level. With regulation systems this is not a major problem, provided there is a one-to-one steady-state relationship between the variable to be controlled and the measurable output. However, for tracking systems a one-to-one dynamic relationship is required. In practice this means that the dynamics (response time) of the sensor must be much faster than the dynamics of the variable to be controlled. Furthermore, any delay between a change occurring in the variable to be controlled and its detection should be eliminated, or minimized. The proper selection and location of sensors and, to a lesser extent, actuators, is an important aspect of control system design, but one that is beyond the scope of this book. Suffice it to say that poor measurement implies poor control.

1.2.3   Loops and compensators

Once a model is produced, the structure of the controlling system is established. Typically, for a single input/single output (SISO) plant there will be a feedback loop which compares the measured output with some desired reference value. An error signal, which is the difference between the reference and measurement signals, is passed to a forward path compensator, the output of which is used to drive an actuator. The actuator, together with the inevitable power amplification device, acts directly on the manipulable input variable. A block diagram of a typical SISO control system is shown in Fig. 1.3. Such a closed-loop structure forms the basis of almost all control system design work. Other structures are possible, including such things as feedback compensation, cascade controllers and feedforward

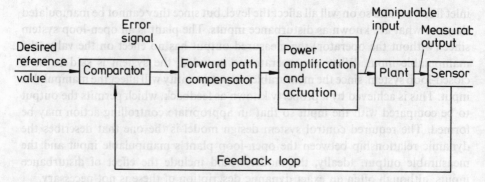

Figure 1.3 Block diagram of a simple SISO control system

compensators, but in almost every case they are additional to the basic structure. (These alternatives are considered in Chapter 8.)

Given the structure of the closed-loop control system, together with a model of the plant, including actuator and sensor dynamics, the design problem reduces to one of establishing the compensation device and its settings. Again, there are only a few compensator types that need to be considered. These are the PID (proportional–integral–derivative) and lead–lag compensators and their derivatives. The choice of a device depends on the nature or dynamic behaviour of the plant, and also on the required performance specifications. Flow loops tend to be noisy, temperature loops tend to be slow and pressure loops tend to be fast; experience indicates that in each case certain types of compensation are either mandatory, or precluded. Also, before any analysis of the system is carried out, certain performance specifications will indicate the type of compensation to be included. A regulation system requiring zero steady-state error would suggest the inclusion of integral action, fast rise times suggest derivable action, or lead compensation, and in all cases proportional action is mandatory.

1.2.4   Control system analysis

Following the procedure laid down in Fig. 1.1, the open-loop plant model is then analysed for closed-loop stability, together with its closed-loop steady-state and dynamic performance. A number of techniques are available. Routh's array (Chapter 5) tests for stability, whilst Evans' root loci method (Chapter 6) clearly indicates closed-loop performance. The various frequency domain techniques (Chapter 7) have a greater range of application than either the Routh or the root loci method, but give a less clear indication of closed-loop response. These techniques enable the final form of the compensator and its settings to be established.

Analysis procedures, as indicated in Fig. 1.1, tend to be iterative. Furthermore, the various techniques are used to support each other rather than being used in isolation. It is therefore this aspect of control system design that will most obviously benefit from computer assistance. And once the use of a computer has been justified, it can also assist in other aspects of the design process, in particular in identification (Chapter 9), where it is used to establish control models.

1.2.5   Assessing a proposed design

To complete the design procedure depicted in Fig. 1.1, the response of the compensated closed-loop system must be assessed. If its performance is not acceptable then either the structure of the closed-loop plant, including its sensors and actuators, or the plant itself will have to be modified. Finally, the design must be checked to ensure that the economic and engineering constraints have not been violated.

With the possible exception of some military projects such as the USA's Strategic Defense Initiative, the economic constraints will have played a large part throughout the design process. Unless the project was considered economically justifiable it would not have been started. Each stage of the design is likely to have had a budget, and the final design will undoubtedly be a compromise between that which is desirable, in the engineering sense, and that which is justifiable on the grounds of cost. The final check is necessary to determine that no part of the system was over-designed in relation to another, which would have incurred unnecessary capital costs, and that the operational requirements are within budget.

Engineering constraints, which can have many forms, are included to take account of those aspects of the design not catered for by adopting a linear plant model. They include such things as signal noise, effects of non-linearities, disturbance inputs, and physical constraints imposed by the plant and its associated equipment.

Briefly, noise (spurious unwanted signals) will affect most system signals and is almost impossible to eliminate completely. High-frequency noise is usually the most troublesome since it precludes the use of lead compensation, or derivative action, and imposes a bandwidth restriction on the closed-loop system. This in turn limits the system's speed of response.

Most practical systems contain some non-linearity. This could include such things as backlash (the free play in a mechanism), dead zones (regions of insensitivity) or non-linearities in the physical behaviour of the plant. If the non-linearities are small they may be neglected, but if they are significant a linear design will be valid for only a small region around a specified operating point.

Disturbance inputs are often unknown or inadequately specified. Ideally the effect of these inputs on the measured output is compensated for by the control system. However, this compensation is additional to that anticipated and may result in a degradation of system performance as a result of excessive wear of components. Unexpected changes in the environment can also have a dramatic effect on hardware and control system performance.

Physical constraints include such things as the limiting values of current, voltage, temperature, pressure, stress and actuator motion. For example, linear analysis may indicate that a 180° deflection in an aircraft control surface may be desirable, but in practice this is not achievable.

1.3   Control system modelling

Control system modelling is a specialization of the more general area of mathematical modelling. Like all mathematical models, a control system model provides insight into the operation of the system and defines the cause-and-effect relationships between variables. With a control system model the important relationship is that between the manipulable input and the measurable output. Ideally, this relationship should be linear and capable of being described by an expression of low order. Most commonly, a low-order, linear differential equation model is used.

The basis for the development of any mathematical model is provided by the fundamental physical laws governing the behaviour of the system. For control system modelling it is usual to analyse an idealized equivalent of the physical system. In this equivalent system each element has a single property or function. For example, an actual mass becomes a concentrated mass having no stiffness or damping, and a resistor has perfect resistance with no inductance or capacitance. The advantage of considering such an idealized or 'lumped parameter' system is that each element has one independent variable – time – so that the system can then be described using a differential equation model.

Probably the most important task in lumped parameter modelling is determining which assumptions can be validly made. Obviously an extremely rigorous model that includes every phenomenon down to microscopic detail would take a long time to develop, and might be impossible to solve. On the other hand, an oversimplified model could bear no dynamic resemblance to the original system. In practice, the amount of detail incorporated tends to be a function of the available solution techniques. Nevertheless, any assumption that is made should be carefully considered and listed, since it will impose limitations on the model which should be borne in mind when evaluating the system's predicted behaviour.

A mathematical model is obtained by applying the fundamental physical laws to the lumped parameter model. The resulting equations may be non-linear, in which case further assumptions may have to be made in order to produce an ordinary linear differential equation model. In such cases it is not unusual to assume that small perturbations occur about a given operating condition. If the assumed operating region is small enough, most non-linear plants may be adequately described by a set of linear equations.

Once all the equations of the mathematical model have been written, it is usually a good idea, particularly with complex systems of equations, to make sure that the number of variables equals the number of equations. If it does not, the system is either underspecified or overspecified and something is wrong with the formulation of the problem. This kind of consistency check may seem trivial, but it can save many hours of frustration and confusion. Checking that the units of all terms in all equations are consistent is perhaps another trivial and obvious step, but one that is often overlooked.

Finally, one of the more important parts of model development is that of model validation. In this context, validation is the art of proving that the mathematical model does indeed describe the real-world situation. Sometimes this cannot be done at the design stage because the system has not yet been built. However, even in this situation there is usually either a similar existing system or a pilot plant from which some experimental data can be obtained.

Example 1.1

Produce a lumped parameter model for a simple pendulum.

Solution

A simple pendulum may be modelled as a massless, rigid rod hinged at one end to a frictionless bearing which is fixed in space, as shown in Fig. 1.4. The other end of the rod is attached to a point mass.

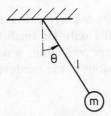

Figure 1.4 A simple pendulum model

It is assumed that the pendulum can vibrate only in the plane of the paper. With this constraint, only one coordinate – the angular displacement of the rod, θ – is required for a complete specification of the pendulums geometric location. Such a mechanical system is said to have one degree of freedom. A further assumption is that the only force acting on the system is gravitational.

For linear systems, any measurable output will be a linear combination of the specified coordinates. Therefore, in this case there is only one independent output to specify since the system has only one degree of freedom. There is no input that can be manipulated, but there are two possible disturbance inputs: the initial angular displacement of the rod and the initial angular velocity.

A mathematical model can now be derived from the lumped parameter model of the pendulum. By applying Newton's second law of motion it can be shown that

$$m \frac{d^2\theta}{dt^2} + \frac{mg}{l} \sin \theta = 0$$

where l is the pendulum's length and θ is the angle it makes with the vertical, and g is the acceleration due to gravity.

The above equation is non-linear, but it may be linearized by assuming small perturbations; that is, by restricting the model to small angles of swing. Replacing $\sin \theta$ by the first term of its Taylor series expansion, namely θ, gives

$$\frac{d^2\theta}{dt^2} + \frac{g}{l}\theta = 0$$

Checking the units for consistency indicates that each term should have units of rad/s^2.

If the lumped parameter model contained all the pertinent information, the above equation would describe the motion of the pendulum. At this stage an experiment would normally be required to validate the model. However, most students have at some stage performed a simple pendulum experiment, and it is well known that the results obtained have proved to have a good correlation with the physical system. It is worth pointing out that the design of experiments to test the validity of a model can sometimes present a real challenge and should be carefully thought out.

The modelling exercise is now complete, and the required linear equation of motion established. However, this equation is valid only for small oscillations about the vertical, and any attempt to use it for large oscillations would result in completely erroneous information about the behaviour of the pendulum.

Example 1.2

Example 1.1, although trivial, does illustrate the various aspects of modelling. A more realistic modelling exercise will now be attempted: to illustrate the development of a lumped parameter model which will describe the dynamic characteristics of a car engine.

Solution

The engine is assumed to consist of four interacting subsystems, as shown in Fig. 1.5. This model has two manipulable inputs: the accelerator pedal position and the spark advance. Load torque is a disturbance input. Four outputs are indicated, two of which – the engine speed and the manifold pressure – are easily measured. The other two outputs indicated could be measured, but for reasons of cost they are not normally measured in normal production engines.

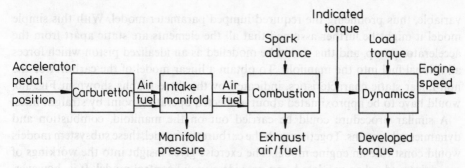

Figure 1.5 Subsystems of an engine model

Each subsystem which goes to make up the engine model may now be considered in turn. The degree of complexity will depend on the information required from the model. A typical carburettor model, for example, may take the form shown in Fig. 1.6. This model has three inputs, the manipulable accelerator pedal position and the disturbance manifold and atmospheric pressures. Manifold pressure, although shown as an input to this system, is a feedback loop from the manifold. (Indeed, as shown in Fig. 1.5, manifold pressure is an output of the engine model.) There are two outputs, air and fuel, which are the inputs driving the manifold model.

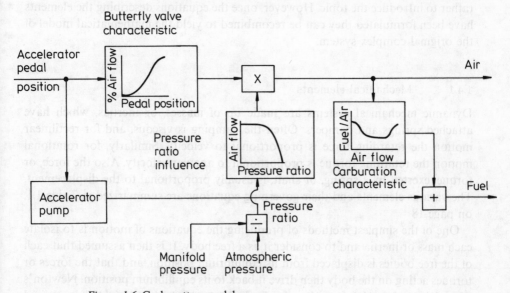

Figure 1.6 Carburettor model

For control purposes, this carburettor model is probably good enough for describing the engine's normal running conditions. Each element could be formulated such that the dynamic characteristics have time as the one independent variable, thus producing the required lumped parameter model. With this simple model it might be further assumed that all the elements are static apart from the accelerator pump, and this would be modelled as an idealized piston which forces additional fuel into the manifold. To obtain a linear model of the carburettor the non-linear static characteristics, indicated by the three graphs shown in Fig. 1.6, would have to be approximated about a selected operating point by straight lines.

A similar procedure could be carried out on the manifold, combustion and dynamics subsystems. Together with the carburettor model, these subsystem models would constitute an engine model. The exercise gives insight into the workings of an engine and, when validated, can provide a useful control model. It is, however, only a functional description of the engine and as such is of limited value for anything other than simple control studies.

1.4   Mathematical models

In Section 1.3 it has been shown how a complex system can be broken down into elements. Since each element is assumed to have time as the one independent variable, the resulting model is referred to as a lumped parameter model. In this section mathematical models are formed by applying physical laws to some simple lumped parameter elements. The treatment is intended not to be exhaustive, but rather to introduce the topic. However, once the equations describing the elements have been formulated they can be recombined to yield the mathematical model of the original complex system.

1.4.1   Mechanical elements

Dynamic mechanical systems are made up of masses, or inertias, which have attached springs and dampers. Often the damping is viscous, and for rectilinear motion the retarding force is proportional to velocity. Similarly, for rotational motion the retarding torque is proportional to angular velocity. Also the force, or torque, exerted by a spring, or shaft, is usually proportional to the displacement. The various elements and their governing equations are summarized in Table 1.1, on page 18.

One of the simplest methods of producing the equations of motion is to isolate each mass or inertia and to consider it as a free body. It is then assumed that each of the free bodies is displaced from its equilibrium position, and that the forces or torques acting on the body then drive it back to its equilibrium position. Newton's second law of motion can then be applied to each body to yield the required equations of motion.

For a rectilinear system Newton's second law indicates that for a consistent system of units:

The sum of forces equals the mass times the acceleration.

In the SI system of units, force is measured in newtons (N), mass in kilograms (kg) and acceleration in metres per second squared (m/s^2).

For a rotational system Newton's second law becomes:

> *The sum of the moments equals the moment of inertia times the angular acceleration.*

The moment, or torque, has units of Newton-metres (N m), the inertia units of kilograms per metre squared (kg m^2) and the angular acceleration units of radians per second squared (rad/s^2).

Example 1.3

A simple mechanical accelerometer is shown in Fig. 1.7. The displacement x of the mass M with respect to the accelerometer case is related to the acceleration of the case; k is the spring stiffness and b the damping. Determine this relationship.

Assume all motions take place in the directions shown.

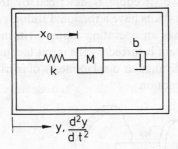

Figure 1.7 Mechanical accelerometer

Solution

The input to the system is the acceleration d^2y/dt^2 of the accelerometer case, and the output is the deflection x_0 of the mass.

Since the accelerometer mass M is constrained to linear motion, the system has one degree of freedom and so one free-body diagram is required. Giving the mass M a displacement x from its equilibrium position produces the free-body diagram shown in Fig. 1.8. Note that x_0 is equal to x plus some fixed displacement.

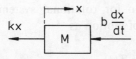

Figure 1.8

Applying Newton's second law of motion gives

$$Mf = -kx - b\frac{\mathrm{d}x}{\mathrm{d}t}$$

where f is the acceleration of the mass relative to the Earth and is given by

$$f = \frac{\mathrm{d}^2x}{\mathrm{d}t^2} - \frac{\mathrm{d}^2y}{\mathrm{d}t^2}$$

Substituting for f gives the required equation of motion as

$$M\frac{\mathrm{d}^2x}{\mathrm{d}t^2} + b\frac{\mathrm{d}x}{\mathrm{d}t} + kx = M\frac{\mathrm{d}^2y}{\mathrm{d}t^2}$$

Note that each term in the above equation has units of newtons (N).

Example 1.4

A rotational system with two degrees of freedom is described by the lumped parameter model shown in Fig. 1.9. The shafts have a torsional stiffness k (in units of N m/rad), and the linear dashpot has an operating radius r (in metres) and damping b (in units of N s/m). The system is forced by the input torque $T(t)$ and has two outputs θ_1 and θ_2, which are the angular displacements of inertias I_1 and I_2, respectively. Find the equations of motion.

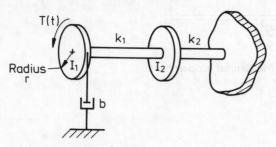

Figure 1.9

Solution

Two free-body diagrams are required as shown in Fig. 1.10: one for inertia I_1, which is given an angular displacement θ_1, and the other for inertia I_2, with its angular displacement θ_2. The dashpot acting on inertia I_1 produces a linear retarding force of $br(\mathrm{d}\theta_1/\mathrm{d}t)$.

Applying Newton's second law of motion for rotational systems to both free-body diagrams yields the two equations of motion as

$$I_1\frac{\mathrm{d}^2\theta_1}{\mathrm{d}t^2} + (br)r\frac{\mathrm{d}\theta_1}{\mathrm{d}t} + k_1(\theta_1 - \theta_2) = T(t)$$

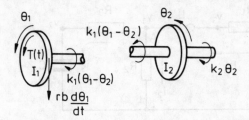

Figure 1.10

and

$$I_2 \frac{d^2\theta_2}{dt^2} + k_2\theta_2 - k_1(\theta_1 - \theta_2) = 0$$

In this example there are two output variables, θ_1 and θ_2, and two equations of motion. Note also that each term in each equation has units of N m.

1.4.2   Electrical elements

For electrical systems the three basic elements are the capacitor, the inductor and the resistor. Again, the governing equations describing the idealized elements are given in Table 1.1, on page 18. For idealized elements, the equations describing the system's behaviour may be obtained from Kirchhoff's laws:

> *Kirchhoff's first law: The total current flowing towards a junction is equal to the total current flowing from that junction, i.e. the algebraic sum of the currents flowing towards a junction is zero.*

> *Kirchhoff's second law: In a closed circuit, the algebraic sum of the voltages across each part of the circuit is equal to the applied e.m.f.*

Example 1.5

Find the equations which describe the dynamic response of the electrical network shown in Fig. 1.11.

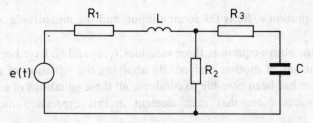

Figure 1.11

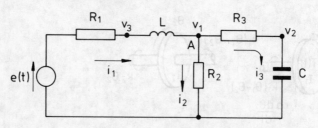

Figure 1.12

Solution A, using Kirchhoff's current law
For the node at A (see Fig. 1.12),

$$i_1 - i_2 - i_3 = 0$$

Now, for each element

$$i_1 = \frac{e - v_3}{R_1} = \frac{1}{L} \int (v_3 - v_1)\, dt$$

$$i_2 = \frac{v_1 - v_2}{R_3} = C \frac{dv_2}{dt}$$

$$i_3 = \frac{v_1}{R_2}$$

The current law for the node at A may be written as

$$2i_1 - 2i_2 - 2i_3 = 0$$

When the above values of i_1, i_2 and i_3 are substituted into this expression, the required equation becomes

$$\frac{e - v_3}{R_1} + \frac{1}{L} \int (v_3 - v_1)\, dt - \frac{v_1 - v_2}{R_3} - C \frac{dv_2}{dt} - 2 \frac{v_1}{R_2} = 0$$

or

$$\frac{v_3}{R_1} + \frac{v_1 - v_2}{R_3} + 2 \frac{v_1}{R_2} + C \frac{dv_2}{dt} - \frac{1}{L} \int (v_3 - v_1)\, dt = \frac{e}{R_1}$$

In the above equation e/R_1 is the forcing input, and the measurable outputs are v_1, v_2 and v_3.

To produce the above equation, three variables, i_1, i_2 and i_3, have been specified and three equations of motion derived. By applying the equation of constraint, $i_1 - i_2 - i_3 = 0$, it has been possible to condense all three equations of motion into a single expression. Note that each element in this expression has units of amperes (A).

Solution B, using Kirchhoff's voltage law
Two loops could be identified with the circulating currents i_1 and i_2, as shown in Fig. 1.13. For loop 1,

$$R_1 i_1 + L\frac{di_1}{dt} + R_2(i_1 - i_2) = e$$

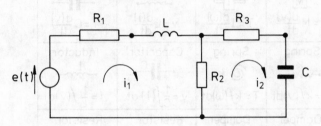

Figure 1.13

For loop 2,

$$R_3 i_2 + \frac{1}{C}\int i_2 \, dt + R_2(i_2 - i_1) = 0$$

In the above equation e is the forcing input, and the measurable outputs are i_1 and i_2.

Again, consistency checks would indicate the correct number of equations, and that each term in those equations has units of volts (V).

1.4.3   Electrical and mechanical analogies

So far, mechanical and electrical systems have been treated separately. However, the dynamic behaviour of these systems is analogous. It is possible, for example, to take a rotational mechanical system, convert it into an equivalent electrical circuit and use Kirchhoff's laws to find the equations of motion. Similarly, an electrical circuit can be transformed into an equivalent mechanical system and the governing equations deduced using Newton's second law.

Table 1.1 gives the various lumped parameters for mechanical and electrical circuits, together with their governing equations. In the first column are the linear mechanical elements and their governing equations in terms of force. In the second column the equivalent rotational mechanical elements are given; not surprisingly, the governing equations expressed in terms of torque are similar to those of the linear elements. This could be thought of as the torque–force analogy between linear and torsional mechanical systems. What *are* surprising are the voltage–force and the current–force analogies shown in columns three and four. The following example illustrates how this table could be used to transform a mechanical system

Table 1.1 Electrical and mechanical analogies

MECHANICAL		ELECTRICAL	
Linear	Torsional Torque/Force	Voltage/Force	Current/Force
Mass $\boxed{M}$ $f = M\dfrac{d(v)}{dt}$	Inertia I $T = I\dfrac{d(\omega)}{dt}$	Inductor L $v = L\dfrac{d(i)}{dt}$	Capacitor C $i = C\dfrac{d(v)}{dt}$
Spring k $f = k\!\int\!(v)dt$	Spring k $T = k\!\int\!(\omega)dt$	Capacitor C $v = \dfrac{1}{C}\!\int\!(i)\,dt$	Inductor L $i = \dfrac{1}{L}\!\int\!(v)dt$
Damper b $f = bv$	Damper b $T = b\omega$	Resistor R $v = Ri$	Resistor R $i = \dfrac{1}{R}\,v$

into an electrical analogue, although clearly an electrical circuit could just as easily be transformed into a mechanical analogue.

Example 1.6

Draw the equivalent electrical voltage–force and current–force analogies for the mechanical system with two degrees of freedom shown in Fig. 1.14.

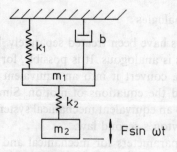

Figure 1.14

Solution

In the voltage–force system the masses become inductors, the springs capacitors and the damper a resistor. The force on one side of m_1 is made up of a spring force from the spring with stiffness k_1 and the damper force. In addition to these forces, there is a force from the spring with stiffness k_2 which acts on the other

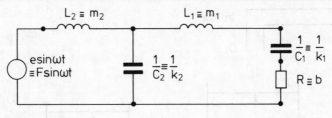

Figure 1.15

side of m_1 and also on mass m_2. Also, a sinusoidal force acts directly on m_2. The required analogy therefore takes the form shown in Fig. 1.15.

In the current–force analogy the masses become capacitors, the springs inductors and the damper a resistor. Here the velocity of m_1 is the same as the velocity across both k_1 and b. The velocity across k_2 is the difference between the velocities of m_2 and m_1. This gives the analogous electrical circuit shown in Fig. 1.16. Note that in this analogy the forcing function is indicated as a current source.

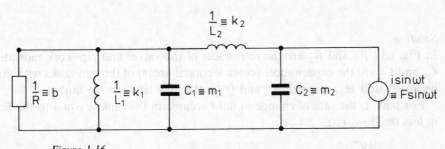

Figure 1.16

1.4.4   Fluid elements

Fluid systems are made up of tanks connected together by means of pipes and valves. The equations describing such systems are found by applying the fundamental laws of fluid mechanics.

In a simplified analysis the valves are assumed to be linear – that is, the flow through the valve is proportional to the liquid head, or

$$Q = H/R \tag{1.1}$$

where

Q = volumetric flow rate (m³/s)
H = head (m)
R = valve resistance (s/m³)

Also the tanks are assumed to be open, and to have a constant capacitance or cross-sectional area. Equations describing the behaviour of the system are derived from the continuity equation – rate of change of volume equals flow in less flow out.

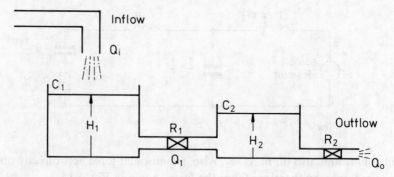

Figure 1.17

Example 1.7

Determine the equations describing the response of the fluid system shown in Fig. 1.17 between the inflow Q_i and outflow Q_o.

Solution

In Fig. 1.17 R_1 and R_2 are the resistances of the valves and pipework indicated, C_1 and C_2 are the capacitances (cross-sectional areas) of the two tanks with fluid heads H_1 and H_2, respectively, and Q_1 is the flow between the tanks.

For tank 1, the rate of change of fluid volume in the tank is equal to the flow in less the flow out:

$$\frac{dV_1}{dt} = Q_i - Q_1$$

Since $V_1 = C_1 H_1$, then

$$C_1 \frac{dH_1}{dt} = Q_i - Q_1 \tag{1.2}$$

The flow between the tanks is given by

$$Q_1 = \frac{H_1 - H_2}{R_1}$$

which, on substitution into Equation (1.2) yields

$$C_1 R_1 \frac{dH_1}{dt} + H_1 = R_1 Q_i + H_2 \tag{1.3}$$

For tank 2 the continuity equation is

$$C_2 \frac{dH_2}{dt} = Q_1 - Q_o \tag{1.4}$$

Also,

$$Q_o = \frac{H_2}{R_2} \tag{1.5}$$

Substituting into Equation (1.4) for Q_1 and Q_o produces

$$R_1 C_2 R_2 \frac{dH_2}{dt} + R_1 H_2 + R_2 H_1 = R_2 H_1 \tag{1.6}$$

Equations (1.3), (1.5) and (1.6) are the required system equations.

1.4.5   Thermal elements

Thermal elements include such things as mixing of hot and cold streams, heat transfer, combustion and chemical reaction. Most systems have some thermal component, and the various modelling techniques are well covered in the literature.

For simple lumped parameter models, the first law of thermodynamics may be used to determine the temperature changes in a body. The rate of change of a body's internal energy is equal to the flow of heat into the body less the flow of heat out of the body. This may be expressed as

$$C \frac{dT}{dt} = q_i - q_o \tag{1.7}$$

where

> C = thermal capacity of body (J/K)
> T = temperature (K)
> q = heat flow rate (W)

The thermal capacity of a body may be found from the equation

$$C = MS \tag{1.8}$$

where

> M = mass of body (kg)
> S = specific heat capacity of material (J/kg K)

The heat flow rate through a body is a function of the thermal resistance of the body. This is normally assumed to be linear, and therefore

$$q = \frac{T_1 - T_2}{R} \tag{1.9}$$

where

> R = thermal resistance (K/W)
> $T_1 - T_2$ = temperature change (K)

Example 1.8

The insulated tank of water shown in Fig. 1.18 is heated by an electrical element.
Develop an equation for the rise in water temperature.

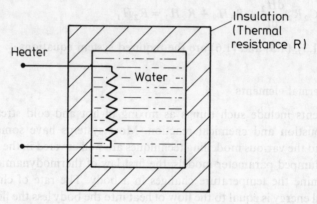

Heater

Insulation
(Thermal
resistance R)

Water

Figure 1.18

Solution

Assume that the water has a uniform temperature and that there is no heat storage
in the insulation. Let:

T_w = temperature of the water
T_a = temperature of air
q = rate of heat supply
q_o = rate of heat loss

From Equation (1.7),

$$C\frac{dT_w}{dt} = q - q_o$$

but, from Equation (1.9),

$$q_o = \frac{T_w - T_a}{R}$$

Hence

$$\frac{dT_w}{dt} + \frac{T_w}{RC} = \frac{q}{C} + \frac{T_a}{RC}$$

which is the required equation.

1.4.6   Lagrange's equations

Lagrange's equations provide a very powerful means of determining the equations of motion of a dynamic system. They are easily shown to be based on Hamilton's principle, which can be loosely interpreted as:

> *For a dynamic system in which the work of all forces is accounted for in the Lagrangian, an admissible motion between specific configurations at times t_1 and t_2 is a natural motion if, and only if, the energy of the system remains constant.*

Lagrange's equation may be written as

$$\frac{d}{dt}\left(\frac{\partial L}{\partial \dot{q}_i}\right) - \frac{\partial L}{\partial q_i} = 0 \tag{1.10}$$

for conservative systems, or more generally as

$$\frac{d}{dt}\left(\frac{\partial L}{\partial \dot{q}_i}\right) - \frac{\partial L}{\partial q_i} + \frac{\partial P}{\partial \dot{q}_i} = Q_i \tag{1.11}$$

where

$L = T - V$ is the Lagrangian
$T =$ kinetic energy
$V =$ potential energy
$P =$ power function describing the dissipation energy of the system
$Q_i =$ generalized external forces acting on the system
$q_i =$ generalized coordinates

As we have seen in Example 1.1, the number of degrees of freedom of a body is the number of independent quantities that must be specified if the position of the body is to be uniquely defined. Any unique set of such quantities is referred to as a set of generalized coordinates for the system. In Example 1.3, the position of the mass of the accelerometer is defined by the displacement x, and therefore x is a generalized coordinate for this system. The system in Example 1.4 has two degrees of freedom, and two generalized coordinates are required; these could be θ_1 and θ_2, the angular displacements of the two inertias. Clearly, from Section 1.4.3, there must be an electrical analogy. Loop currents (or, more usually, the charges in the various loops) would form a set of generalized coordinates.

The various energy forms for linear, mechanical and electrical elements are summarized in Table 1.2. Note that the current i is expressed in terms of charge q as

$$i = \frac{dq}{dt} = \dot{q} \tag{1.12}$$

Table 1.2 Energy expressions for mechanical and electrical elements

ENERGY TYPE	MECHANICAL	ELECTRICAL
Kinetic energy T	Mass $\boxed{m}$ $T = \frac{1}{2} m \dot{x}^2$	Inductor $\underset{L}{\text{—www—}}$ $T = \frac{1}{2} L \dot{q}^2$
Potential energy V	Spring k —www— $V = \frac{1}{2} k x^2$	Capacitor $\dashv\vdash^C$ $V = \frac{1}{2} C v^2 = \frac{1}{2C} q^2$
	Gravitational $\boxed{m}$ $V = mgh$	–
Dissipative energy P	Damper $\overset{b}{\text{—⊏Ⴇ—}}$ $P = \frac{1}{2} b \dot{x}^2$	Resistor $\overset{R}{\text{—▭—}}$ $P = \frac{1}{2} R \dot{q}^2$

and that linear velocity v is expressed in terms of displacement x as

$$v = \frac{dx}{dt} = \dot{x} \tag{1.13}$$

Lagrange's equations have many modelling applications, although the intention in this section is to demonstrate their effectiveness in developing mathematical models for electromechanical systems.

Example 1.9

The lumped parameter model of a capacitor microphone is shown in Fig. 1.19. In the equilibrium position and with no external force applied to the moving plate, there is a charge q_0 on the capacitor. This charge results in a force of attraction

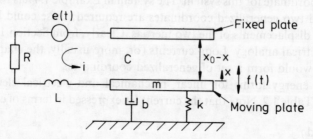

Figure 1.19

between the plates which pre-tensions the spring. A sound wave applies a force to the moving plate, resulting in displacement x from the equilibrium position. This motion alters the capacitance C of the capacitor and results in a change in charge.

Produce a mathematical model for this lumped parameter system.

Solution

The system has two degrees of freedom, one mechanical and the other electrical. Two coordinates are therefore required, and the charge q and displacement x from equilibrium are selected.

The kinetic energy function for the system is

$$T = \frac{1}{2}L\dot{q}^2 + \frac{1}{2}M\dot{x}^2$$

The potential energy function is

$$V = \frac{1}{2C}q^2 + \frac{1}{2}kx^2$$

Since C is a function of x, given by

$$C = \frac{\varepsilon A}{x_0 - x}$$

where

A = area of plates
ε = dielectric constant of air

then

$$V = \frac{1}{2\varepsilon A}(x_0 - x)q^2 + \frac{1}{2}kx^2$$

The dissipation function of the system is

$$P = \frac{1}{2}R\dot{q}^2 + \frac{1}{2}b\dot{x}^2$$

Applying Lagrange's equation of the coordinate x gives

$$\frac{d}{dt}\left(\frac{\partial L}{\partial \dot{x}}\right) - \frac{\partial L}{\partial x} + \frac{\partial P}{\partial \dot{x}} = f(t)$$

where $L = T - V$ and $f(t)$ is the external force at x, whence

$$M\ddot{x} + b\dot{x} - \frac{1}{2\varepsilon A}q^2 + kx = f(t) \tag{1.14}$$

Similarly, applying Lagrange's equation to coordinate q yields

$$L\ddot{q} + R\dot{q} + \frac{1}{\varepsilon A}(x_0 - x)q = e(t) \tag{1.15}$$

Equations (1.14) and (1.15) form the required mathematical model.

1.5   System models

With complex systems, the proposed modelling technique is to reduce the system to a number of manageable lumped parameter elements. The mathematical model for each element is then determined, and a system model developed by combining all the equations for all the elements. However, when adopting this approach care should be taken to ensure that one element does not interact with, or load, another. In Example 1.7 the two tanks interact with each other since the head of fluid in tank 1 is partially dependent on the head of fluid in tank 2. The mathematical model describing the changing head in tank 1 requires a feedback of information on the head of fluid in tank 2. If the tanks were non-interacting (that is, if the outlet flow from tank 1 poured into tank 2, as in Fig. 1.20), then both tanks could be modelled independently.

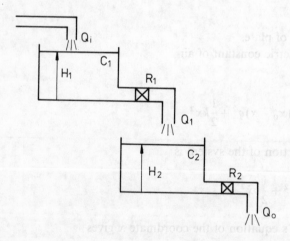

Figure 1.20 Non-interacting tanks

In general, then, if elements are non-interacting the output from one stage becomes the input to the next. When interactions occur they must be accounted for in the model, which usually requires the feedback of information from a later stage to an earlier stage.

Example 1.10

Demonstrate that the electrical network shown in Fig. 1.21(a) cannot in general be modelled by the independent networks shown in Figs. 1.21(b) and (c). Under what conditions could independent modelling be carried out? Assume that the line L carries no current.

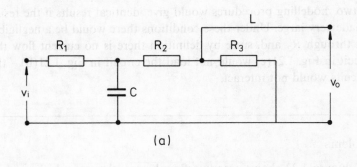

(a)

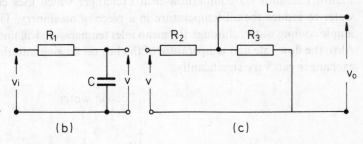

(b)   (c)

Figure 1.21

Solution

The equation relating the input voltage v_i to the output voltage v_o in Fig. 1.21(a), for no current flow along L, is given by

$$\frac{R_1}{R_3}(R_2 + R_3)C\frac{dv_o}{dt} + \frac{R_1 + R_2 + R_3}{R_3}v_o = v_i$$

For Fig. 1.21(b) the equation relating v and v_i is

$$R_1 C\frac{dv}{dt} + v = v_i \tag{1.16}$$

For Fig. 1.21(c) the equation relating v_o and v, again for no current flow in L, is

$$\frac{R_2 + R_3}{R_3}v_o = v \tag{1.17}$$

Eliminating v from Equations (1.16) and (1.17) yields

$$\frac{R_1}{R_3}(R_2 + R_3)C\frac{dv_o}{dt} + \frac{R_2 + R_3}{R_3}v_o = v_i$$

which demonstrates the inadequacy of independent modelling for interacting systems.

The two modelling procedures would give identical results if the resistance R_3 were made very large. Under these conditions there would be a negligible flow of current through R_3 and, since by definition there is no current flow through L, the circuit in Fig. 1.21(c) would not load the circuit in Fig. 1.21(b) – that is, the two circuits would not interact.

Problems

1.1. Figure 1.22 shows a counterflow heat exchanger which uses cooling water to reduce the oil temperature in a piece of machinery. There is ample cooling water, although the mean inlet temperature will fluctuate. Also, the flow rate and temperature of the hot oil flowing into the heat exchanger can vary significantly.

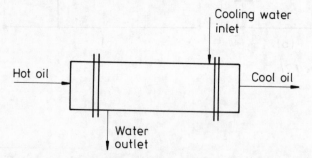

Figure 1.22 A counterflow heat exchanger

What are the disturbance and manipulable inputs? What are the measurable and controllable outputs? Describe a control scheme which could be used to maintain a constant outlet oil temperature.

1.2. A typical industrial flow control scheme is shown in Fig. 1.23. Draw a block diagram of the system, using a schematic notation similar to that used in Fig. 1.3.

1.3. Produce a lumped parameter model, consisting of springs, masses and dampers, which describes the vertical motion of a car's suspension system.

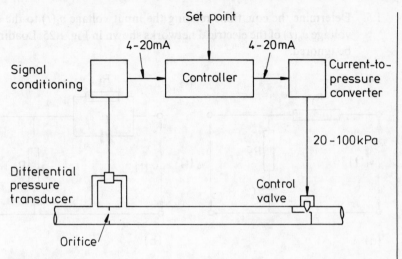

Figure 1.23

1.4. Determine the equations of motion for the mechanical systems shown in Fig. 1.24. In each case the input displacement is $u(t)$, and the output displacement is $y(t)$.

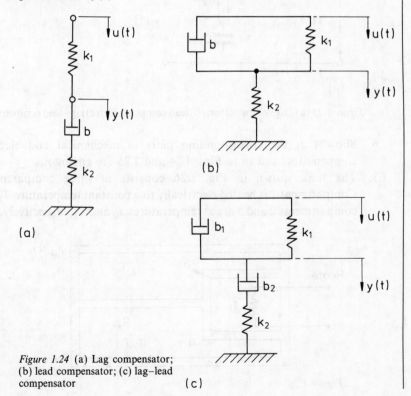

Figure 1.24 (a) Lag compensator; (b) lead compensator; (c) lag–lead compensator

1.5. Determine the equations relating the input voltage $v_i(t)$ to the output voltage $v_o(t)$ of the electrical networks shown in Fig. 1.25. Loading may be ignored.

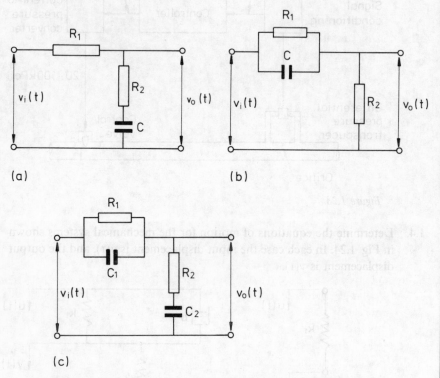

(a)

(b)

(c)

Figure 1.25 (a) Lag compensator; (b) lead compensator; (c) lag–lead compensator

1.6. Show that the corresponding pairs of mechanical and electrical compensators shown in Figs. 1.24 and 1.25 are analogous.

1.7. The tank shown in Fig. 1.26 consists of three compartments. Compartment 1 is heated electrically to a constant temperature T_1, and compartments 2 and 3 are at temperatures T_2 and T_3, respectively. Heat

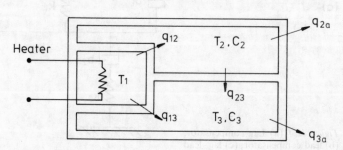

Figure 1.26

can flow only from compartment 1 to compartments 2 and 3, as shown. There will be an exchange of heat between compartments 2 and 3, both of which can lose heat to the environment, which is at temperature T_a.

Develop equations for the change in temperature of fluid in tanks 2 and 3, given that their respective thermal capacities are C_2 and C_3.

1.8. Using the current–force analogy, find the mechanical system equivalent to the electrical circuit shown in Fig. 1.27.

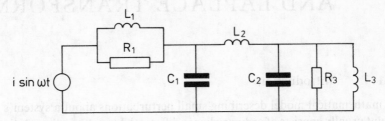

Figure 1.27

1.9. The loudspeaker model shown in Fig. 1.28 is based on an ideal moving coil transducer, a device which converts electrical energy into mechanical energy without loss. A change in charge on the electrical side produces a change in force on the mechanical side. Given that the magnetic energy for the transducer is

$$T = \tfrac{1}{2}L(x)\dot{q}^2$$

produce a mathematical model for this system relating the current i to the displacement x.

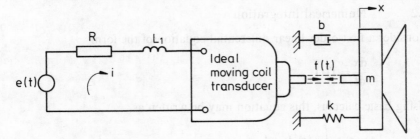

Figure 1.28

CHAPTER 2

DIFFERENTIAL EQUATIONS
AND LAPLACE TRANSFORMS

2.1 Introduction

A mathematical model describing small perturbations about a system's operating point usually consists of ordinary linear differential equations, the solution of which yields the system's dynamic and steady-state characteristics. Unfortunately it is not easy to deal with differential equations directly, and any change in the system's forcing input or control structure can significantly change the method of solution. For these reasons, if for no other, alternative, indirect methods of solution have been sought. This chapter briefly examines numerical solutions which are based on integration routines for computer implementation and, in more detail, algebraic solutions based on Laplace transforms. It should be emphasized that Laplace transforms provide the mathematical foundation for all the classical control theory developed in this book.

2.2 Numerical integration

Consider a first-order linear differential equation of the form

$$\frac{dx}{dt} = f(t) \tag{2.1}$$

Using basic calculus, this equation may be written as

$$x(t) = \int f(t)\, dt + k \tag{2.2}$$

where k is the constant of integration. This equation gives a direct relationship between the dependent variable $x(t)$ and the independent variable t, so Equation (2.2) is a solution of Equation (2.1). The conversion of differential equations into integral equations is a fundamental part of digital computer simulation (see Program 1 in Chapter 11).

If the differential equation is higher than first order, it may be converted into a set of first-order equations. Consider the equation

$$\frac{d^3x}{dt^3} + \frac{5d^2x}{dt^2} + \frac{6dx}{dt} + 3x = f(t) \tag{2.3}$$

Let

$$x(t) = u_1(t) \tag{2.4}$$

and, on differentiation of $u_1(t)$,

$$\frac{du_1}{dt} = u_2 \tag{2.5}$$

$$\frac{du_2}{dt} = u_3 \tag{2.6}$$

Differentiation of $u_3(t)$ is expressed in terms of $u_3(t)$, $u_2(t)$ and $u_1(t)$, which are all functions of $x(t)$ and its derivatives, so from Equations (2.3)–(2.6)

$$\frac{du_3}{dt} = -5u_3 - 6u_3 - 3u_1 + f(t) \tag{2.7}$$

Equations (2.5)–(2.7) are equivalent to Equation (2.3) – that is, a third-order differential equation has been converted into three first-order differential equations. Clearly, the method will be effective for any linear nth-order differential equation, and may be extended to many non-linear equations.

Typically, a mathematical model of a dynamic system consists of a number of coupled high-order differential equations. The method outlined is easily extended, and the problem again reduces to one of selecting an appropriate integration routine.

Example 2.1

The roll stabilization system of an aircraft consists of a closed-loop control system which automatically develops a restoring torque if the aircraft rolls from a desired angular position ϕ. The dynamics of the aircraft may be approximately described by the equation

$$I \frac{d^2\theta}{dt^2} = k\alpha$$

where

θ = angle of roll
I = moment of inertia
α = deflection of ailerons
$k\alpha$ = torque produced by deflection

The servomotor which operates the ailerons is governed by an equation of the form

$$T\frac{d\alpha}{dt} + \alpha = u$$

and the controller which generates the signal u behaves in accordance with the equation

$$u = \varepsilon + aT\frac{d\varepsilon}{dt}$$

where $\varepsilon = \phi - \theta$, and a and T are positive parameters.

Produce a set of first-order differential equations for this system.

Solution

The closed-loop system takes the form shown in Fig. 2.1. Converting the equations into a set of first-order linear differential equations is straightforward, apart, that is, from the controller, which has input ε and output u.

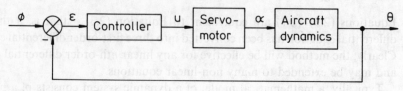

Figure 2.1

Let

$$x_3 = \theta \quad \text{and} \quad x_5 = \alpha$$

Furthermore, let

$$x_1 = \phi_1 \tag{2.8}$$

and

$$x_2 = \frac{d\phi_1}{dt} \tag{2.9}$$

The controller equation is then

$$u = (x_1 - x_3) + aT(x_2 - x_4) \tag{2.10}$$

the servomotor equation is

$$\frac{dx_5}{dt} = -\frac{x_5}{T} + \frac{u}{T} \tag{2.11}$$

and the aircraft roll dynamics are given by

$$\frac{\mathrm{d}x_3}{\mathrm{d}t} = x_4 \tag{2.12}$$

and

$$\frac{\mathrm{d}x_4}{\mathrm{d}t} = \frac{k}{I}x_5 \tag{2.13}$$

Equations (2.10)–(2.13) are the equations of the closed-loop system; Equations (2.8) and (2.10) are the changes in the demand input position and velocity, respectively.

2.3   Laplace transforms

Laplace transforms provide the mathematical foundation for much of classical control. The method is a substitutional one in which the linear differential equation model of the system is transformed into the complex frequency, or Laplace, domain. This greatly simplifies the mathematics since the operation of integration associated with the time domain solution is now replaced by algebraic manipulation of the transformed equations. After manipulation, the required time domain solution is obtained by making the inverse Laplace transformation. This process, shown schematically in Fig. 2.2, is similar to that of using logarithmic tables to simplify the problem of multiplying a series of numbers together.

The analogy between logarithms and Laplace transforms is worth pursuing further. When using logarithms, the requirement is a good set of log and antilog

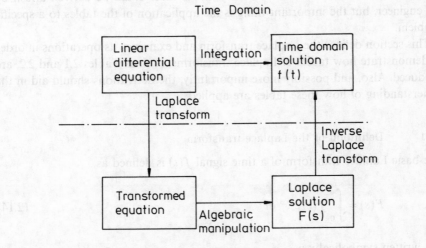

Figure 2.2 Laplace transformation

Table 2.1 Laplace transform operations

Operation		$f(t)$	$F(s)$
1.	Transform integral	$f(t)$	$\int_0^\infty f(t)e^{-st}\,dt$ or $\mathscr{L}\{f(t)\}$
2.	Linearity	$f_1(t) \pm f_2(t)$	$F_1(s) \pm F_2(s)$
3.	Constant multiplication	$af(t)$	$aF(s)$
4.	Complex shift theorem	$e^{\pm at}f(t)$	$F(s \mp a)$
5.	Real shift theorem	$f(t-T)$	$e^{-Ts}F(s)\,(T \geqslant 0)$
6.	Scaling theorem	$f(t/a)$	$aF(as)$
7.	First derivative	$\dfrac{d}{dt}f(t)$	$sF(s) - f(0)$
8.	nth derivative	$\dfrac{d^n}{dt^n}f(t) \equiv f^n(t)$	$s^nF(s) - \sum\limits_{r=1}^{n} \dfrac{d^{r-1}}{dt^{r-1}}f(0)s^{n-r}$
9.	First integral	$\int_0^t f(t)\,dt$	$\dfrac{1}{s}F(s)$
10.	Convolution integral	$\int_0^t f_1(\tau)f_2(t-\tau)\,d\tau$	$F_1(s)F_2(s)$

tables. It is useful to know how these tables are formulated, but such knowledge is not essential. Similarly, with Laplace transforms the requirement is a good set of Laplace transform tables. Each entry in the table could be derived afresh by the engineer, but the important thing is the application of the tables to a specific problem.

This section defines the Laplace transform and examines its operations in order to demonstrate how tables of Laplace transforms, such as Tables 2.1 and 2.2, are produced. Also, and possibly more importantly, the derivations should aid in the understanding of how these tables are applied.

2.3.1   Definition of the Laplace transform

The basic Laplace transform of a time signal $f(t)$ is defined as

$$F(s) = \int_0^\infty f(t)\,e^{-st}\,dt \tag{2.14}$$

and written symbolically as

$$F(s) = \mathscr{L}[f(t)] \tag{2.15}$$

Table 2.2 Laplace transforms of common functions

Time function $f(t)$	Laplace transform $\mathcal{L}\{f(t)\} = F(s)$
1. $\delta(t)$: unit impulse	1
2. $u(t)$: unit step	$\dfrac{1}{s}$
3. t	$\dfrac{1}{s^2}$
4. t^n	$\dfrac{n!}{s^{n+1}}$
5. e^{at}	$\dfrac{1}{s-a}$
6. $\cos \omega t$	$\dfrac{s}{s^2+\omega^2}$
7. $\sin \omega t$	$\dfrac{\omega}{s^2+\omega^2}$
8. $e^{-at}\cos \omega t$	$\dfrac{s+a}{(s+a)^2+\omega^2}$
9. $e^{-at}\sin \omega t$	$\dfrac{\omega}{(s+a)^2+\omega^2}$
10. $\dfrac{e^{-at}-e^{-bt}}{b-a}$	$\dfrac{1}{(s+a)(s+b)}$
11. $\dfrac{\omega}{\sqrt{(1-\zeta^2)}}e^{-\zeta\omega t}\sin[\omega\sqrt{(1-\zeta^2)}t]$	$\dfrac{\omega^2}{s^2+2\zeta\omega s+\omega^2}$
12. $\dfrac{1}{T^n(n-1)!}t^{n-1}e^{-t/T}$	$\dfrac{1}{(1+sT)^n}$
13. $1-\cos \omega t$	$\dfrac{\omega^2}{s(s^2+\omega^2)}$
14. $1-e^{-t/T}$	$\dfrac{1}{s(1+Ts)}$
15. $1-\dfrac{t+T}{T}e^{-t/T}$	$\dfrac{1}{s(1+Ts)^2}$
16. $\dfrac{\omega^2}{\sqrt{(1-\zeta^2)}}e^{-\zeta\omega t}\sin[\omega\sqrt{(1-\zeta^2)}t+\phi]$	$\dfrac{s\omega^2}{s^2+2\zeta\omega s+\omega^2}$
where $\phi = \tan^{-1}\dfrac{\sqrt{(1-\zeta^2)}}{-\zeta}$	

It is common practice to use a capital letter F for the transform of the time signal $f(t)$; F is a function of the variable s. Also, it is assumed that $f(t)$ is zero for all times before $t = 0$.

In Equation (2.14) the exponent st must be dimensionless, else the expression e^{-st} is meaningless. Thus the variable s has the dimension $1/\text{time}$, which is the dimension of frequency; and since s is a complex quantity it is often referred to as the complex frequency. More is said on the complex nature of s in Chapter 4.

Two useful properties follow directly from the definition of Equation (2.14). These are the properties of linearity and constant multiplication, respectively

$$\mathscr{L}[f_1(t) + f_2(t)] = \mathscr{L}[f_1(t)] + \mathscr{L}[f_2(t)] \tag{2.16}$$

and

$$\mathscr{L}[af(t)] = a\mathscr{L}[f(t)] \tag{2.17}$$

Such properties are referred to as Laplace transform operations. A list of such operations is given in Table 2.1.

2.3.2   Laplace transforms of some common functions

The Laplace transforms of a number of common functions are given in Table 2.2. For example, Equation (2.14) may be used to find the Laplace transform of

$$f(t) = e^{at} \tag{2.18}$$

as

$$\mathscr{L}[e^{at}] = \int_0^\infty e^{at} e^{-st} \, dt \tag{2.19}$$

or

$$F(s) = \int_0^\infty e^{-(s-a)t} \, dt \tag{2.20}$$

which becomes

$$F(s) = -\frac{1}{s-a} [e^{-(s-a)t}]_0^\infty \tag{2.21}$$

Since s is complex, the term within the square brackets will approach zero for increasing t provided the real part of $s - a$ is positive, leaving

$$F(s) = \frac{1}{s-a} \tag{2.22}$$

which, from entry 5 of Table 2.2, is the required solution.

Equation (2.22) may be used to find other standard transforms. If the constant a

equals zero, then

$$\mathcal{L}[e^{0t}] = \frac{1}{s} \qquad (2.23)$$

Since it is assumed that $f(t)$ is zero for all time before $t = 0$, Equation (2.23) defines a unit step (see entry 2 of Table 2.2).

Setting $a = j\omega$ in Equation (2.22) gives

$$\mathcal{L}[e^{j\omega t}] = \frac{1}{s - j\omega} \qquad (2.24)$$

Since

$$e^{j\omega t} = \cos \omega t + j \sin \omega t \qquad (2.25)$$

then

$$\mathcal{L}[\cos \omega t + j \sin \omega t] = \frac{1}{s - j\omega} \qquad (2.26)$$

Applying the property of linearity to the left-hand side of Equation (2.26) and rationalizing the right-hand side produces

$$\mathcal{L}[\cos \omega t] + j\mathcal{L}[\sin \omega t] = \frac{s}{s^2 + \omega^2} + j\frac{\omega}{s^2 + \omega^2} \qquad (2.27)$$

Equating the real and imaginary parts of Equation (2.27) gives entries 6 and 7 of Table 2.2.

Further useful results are obtained if a is set equal to unity in Equation (2.22). Then

$$\mathcal{L}[e^t] = \frac{1}{s - 1} \qquad (2.28)$$

Now, both sides of Equation (2.28) have power series expansions, namely

$$e^t = 1 + t + \frac{1}{2!}t^2 + \frac{1}{3!}t^3 + \cdots \qquad (2.29)$$

and

$$\frac{1}{s - 1} = \frac{1}{s} + \frac{1}{s^2} + \frac{1}{s^3} + \cdots \qquad (2.30)$$

Using the property of linearity and equating corresponding terms in the two series produces entries 3 and 4 in Table 2.2.

To generate further entries for Table 2.2, the best procedure is to build on previous results in preference to applying the defining integral. To this end, further Laplace transform operations are required.

2.3.3   Laplace transform operations

Apart from linearity and constant multiplication, there are a number of other useful Laplace transform operations (see Table 2.1) which can be used to establish other entries in Table 2.2. Each of the operations listed in Table 2.1 follows from the Laplace transform definition. Rather than proving each entry, this section concentrates on a few of their applications.

By the complex shift theorem, for example, the Laplace transform of $e^{-3t} \sin 4t$ becomes

$$\mathscr{L}[e^{-3t} \sin 4t] = \frac{4}{(s+3)^2 + 16} \tag{2.31}$$

which agrees with entry 9 of Table 2.2.

The real shift theorem is useful for systems containing a time delay. Time delays, or dead times, are frequently encountered in chemical engineering systems in which a process stream is flowing through a pipe in essentially plug flow. If an individual element of fluid takes time T to flow from the entrance to the exit of a pipe, then the pipe acts as a time delay. More is said on time delays in subsequent chapters.

For control purposes, entries 7 and 8 of Table 2.1 are probably the most frequently used operations. They are best illustrated by means of an example. Assume that the mathematical model of an unforced lumped parameter system is given by

$$\frac{d^2 x}{dt^2} + 5\frac{dx}{dt} + 6x = 0 \tag{2.32}$$

The free response of the system will depend on the initial conditions, and for the sake of this illustration let

$$x(0) = 0 \tag{2.33a}$$

and

$$\frac{dx}{dt}(0) = 7 \tag{2.33b}$$

Using linearity to find the Laplace transform of each term in Equation (2.32), together with the constant multiplication and derivative operations, yields

$$\left(s^2 X(s) - sx(0) - \frac{dx}{dt}(0)\right) + 5(sX(s) - x(0)) + 6X(s) = 0 \tag{2.34}$$

which, with some algebraic manipulation, becomes

$$X(s) = \frac{s+5}{s^2 + 5s + 6}x(0) + \frac{1}{s^2 + 5s + 6}\frac{dx}{dt}(0) \tag{2.35}$$

Inserting the initial conditions gives

$$X(s) = \frac{7}{(s+2)(s+3)} \qquad (2.36)$$

The corresponding time response can, in this case, be obtained directly from Table 2.2 (entry 10) as

$$x(t) = 7\frac{e^{-2t} - e^{-3t}}{3 - 2}$$

or

$$x(t) = 7e^{-2t} - 7e^{-3t} \qquad (2.37)$$

Clearly the Laplace transform technique is very powerful. The ability to convert a constant-parameter, ordinary linear differential equation in x into an algebraic equation in s simplifies the solution procedure considerably. For control system design, most of the work is performed using the Laplace transform model. Only when a design is deemed acceptable in the Laplace (or frequency) domain can the solution of the equations be sought in the time domain. However, rather than follow a chronological design sequence, the subsequent sections consider the various techniques that are used to find time domain solutions.

2.4   Inverse Laplace transforms

The formal definition of the inverse Laplace transform, which converts $F(s)$ to the corresponding time function $f(t)$, is

$$f(t) = \frac{1}{2\pi j} \int_{\sigma - j\omega}^{\sigma + j\omega} F(s)\, e^{st}\, ds \qquad (2.38)$$

which is written symbolically as

$$f(t) = \mathscr{L}^{-1}[F(s)] \qquad (2.39)$$

This integral is difficult to evaluate directly, and the normal procedure is to manipulate $F(s)$ into one of the standard forms for which the inverse Laplace transform is known. Table 2.2, for example, provides a number of Laplace transforms for common functions. However, this table can be used in either direction: just as, given $f(t)$, it is possible to find $F(s)$, so, given $F(s)$, it is possible to find $f(t)$. The amount and type of manipulation applied to $F(s)$ depends on the proposed method of solution – graphical, hand calculation or by computer. Heaviside's partial-fraction expansion method provides the basis for the development of all three approaches.

Example 2.2

Use Tables 2.1 and 2.2 to find the time functions corresponding to the transforms

$$\frac{10}{s^2+9} \quad \text{and} \quad \frac{2e^{-2s}}{s+3}$$

Solution

The first function is the transform of sin 3t with the multiplying factor 10/3. The corresponding time function is therefore

$$\mathscr{L}^{-1}\left[\frac{10}{3}\frac{3}{s^2+3^2}\right] = \frac{10}{3}\sin 3t$$

The second transform is that for $2e^{-3t}$ multiplied by e^{-2s}, and the inverse Laplace transform then follows from the real shift theorem as

$$2e^{-3(t+2)}$$

Example 2.3

Find the inverse Laplace transform of

$$F(s) = \frac{s+5}{s^2+2s+5}$$

Solution

This example demonstrates that manipulation is often required to change the transform into a standard form. In this transform the denominator can be rewritten to give

$$F(s) = \frac{s+5}{(s+1)^2+4}$$

or

$$F(s) = \frac{s+1}{(s+1)^2+4} + \frac{4}{(s+1)^2+4}$$

The complex shift theorem and the standard forms for sine and cosine functions in Table 2.2 then allow the inverse transform to be found by inspection:

$$\mathscr{L}^{-1}[F(s)] = e^{-t}\cos 2t + 2e^{-t}\sin 2t$$

2.4.1   Partial fractions

The Laplace transform solution of an ordinary linear differential equation often takes the form of a rational polynomial:

$$F(s) = \frac{N(s)}{D(s)} \tag{2.40}$$

where $N(s)$ and $D(s)$ are polynomials. Hence Equation (2.40) could be rewritten as

$$F(s) = \frac{b_m s^m + b_{m-1} s^{m-1} + \cdots + b_0}{s^n + a_{n-1} s^{n-1} + \cdots + a_0} \tag{2.41}$$

For rationality, the degree m of the numerator must be less than the degree n of the denominator. If m is greater than or equal to n, then the numerator can always be made of lower degree by dividing the denominator into the numerator. This process will produce a rational polynomial, plus a remainder polynomial. However, for engineering systems the Laplace transform solution is invariably a rational polynomial, and the problem is one of finding the inverse transform. To do so, a partial fraction decomposition of $F(s)$ into a sum of simpler terms is required. This can be illustrated by considering the terms

$$F(s) = -\frac{1}{s+1} - \frac{3}{s-2} + \frac{4}{s+3} \tag{2.42}$$

The inverse transform of Equation (2.42) may be obtained directly from Tables 2.1 and 2.2. However, if Equation (2.42) is written with a common denominator – that is, in rational polynomial form – then

$$F(s) = \frac{7s - 5}{s^3 - 4s^2 + s + 6} \tag{2.43}$$

and the inverse transform is not immediately obvious.

The normal procedure for dealing with a rational polynomial is first to factorize the denominator polynomial $D(s)$. (A computer program for finding the roots of a polynomial is given in Chapter 11.) The particular type of expansion then employed is determined by the types of pole involved.

For hand calculation only two cases need be considered:

(i)   An unrepeated, real or complex root of $D(s)$

If $-p_i$ is an unrepeated pole of $F(s)$, there is a corresponding partial fraction

$$\frac{C_i}{s + p_i} \tag{2.44}$$

for which it is evident that

$$\mathscr{L}^{-1}\left[\frac{C_i}{s + p_i}\right] = C_i e^{-p_i t} \tag{2.45}$$

Note that a root of $D(s)$ is a pole of $F(s)$ since, at $s = -p_i$, $F(s)$ becomes infinite.

(ii)   *A repeated, real or complex root of* D(s)

If $-p_i$ is a pole of order n (that is, if there are n poles all having value $-p_i$), there is a corresponding partial-fraction expansion of the form

$$\frac{C_{i,n}}{(s+p_i)^n}+\frac{C_{i,n-1}}{(s+p_i)^{n-1}}+\cdots+\frac{C_{i,1}}{s+p_i} \tag{2.46}$$

and the inverse transform is, from Tables 2.1 and 2.2,

$$\mathscr{L}^{-1}\left[\frac{C_i(s)}{(s+p_i)^n}\right]=e^{-p_it}\sum_{k=1}^{n}C_{i,k}\frac{t^{k-1}}{(k-1)!} \tag{2.47}$$

Equations (2.45) and (2.47) may be combined so that, if $F(s)$ has a total of N distinct poles, the complete inverse transform is given by

$$f(t)=\sum_{i=1}^{N}\left(e^{-p_it}\sum_{k=1}^{n}C_{i,k}\frac{t^{k-1}}{(k-1)!}\right) \tag{2.48}$$

where i denotes the pole, and k the order of the pole in the partial fraction.

For machine calculation, Equation (2.48) would normally be used only to find the inverse transform of real poles. If a pole is complex, it is more convenient to combine it with its complex conjugate, which will always exist if the coefficients of $D(s)$ are real. Therefore, for a pair of complex conjugate poles, $-a_i+jb_i$ and $-a_i-jb_i$, it is usual to write the corresponding partial function in the form

$$\frac{As+B}{(s+a)^2+b^2} \tag{2.49}$$

where A and B are real constants.

In this form two further expansions need to be considered:

(iii)   *An unrepeated pair of complex conjugate root of* D(s)

A quadratic, unrepeated factor of $D(s)$ with poles at $s=-a\pm jb$ contributes a term in the time response $f(t)$ defined by

$$f_1(t)=\mathscr{L}^{-1}\left[\frac{As+B}{(s+a)^2+b^2}\right] \tag{2.50}$$

In partial-fraction form this is

$$f_1(t)=\mathscr{L}^{-1}\left[\frac{C_R+jC_I}{s+a+jb}+\frac{C_R-jC_I}{s+a-jb}\right] \tag{2.51}$$

and the inverse Laplace transform is

$$f_1(t)=(C_R+jC_I)e^{-(a+jb)t}+(C_R-jC_I)e^{-(a-jb)t} \tag{2.52}$$

Equation (2.52) may be rearranged, using Equation (2.25), to give

$$f_1(t) = e^{-at}(2C_R \cos bt + 2C_I \sin bt) \tag{2.53}$$

which is the required inverse transform.

(iv)   *A repeated pair of complex conjugate roots of* D(s)

If $D(s)$ has n repeated quadratic factors of the form $[(s+a)^2 + b^2]^n$, there corresponds a partial-fraction expansion having an inverse Laplace transform given by

$$e^{-at} \sum_{k=1}^{n} \left(\frac{t^{k-1}}{(k-1)!} (2C_{R_k} \cos bt + 2C_{I_k} \sin bt) \right) \tag{2.54}$$

This result follows directly from the previous three cases considered.

In each of the above four cases the partial-fraction expansion leads directly to the inverse Laplace transform, but it remains to find the coefficients C of the various partial-fraction expansions. This is done by using the procedure which has been credited to the electrical engineer Oliver Heaviside (1850–1925) and called the Heaviside technique. The method consists of four formulae for calculating the inverse Laplace transforms of components in the decomposed form of $F(s)$. Each formula corresponds to one of four types of factor, already considered, which can arise in the denominator of $F(s)$. They are as follows:

(i)   *A distinct linear factor of the form* $s + p_i$

If the denominator of $F(s)$ contains a linear factor $s + p_i$ (that is, if there is a first-order real pole at $s = -p_i$), there will be a term in $f(t)$ corresponding to this factor which can be computed from the equation

$$f(t)_{s+p_i} = [(s+p_i)F(s)e^{st}]|_{s=-p_i} \tag{2.55}$$

Thus, the coefficient C_i in Equation (2.45) may be written as

$$C_i = \lim_{s \to -p_i} [(s+p_i)F(s)] \tag{2.56}$$

or

$$C_i = \lim_{s \to -p_i} \left[(s+p_i)\frac{N(s)}{D(s)} \right] \tag{2.57}$$

or

$$C_i = \frac{N(s)}{D'(s)}\bigg|_{s=-p_i} \tag{2.58}$$

where $D'(s)$ denotes the differential of $D(s)$ with respect to s.

Coefficients of simple, non-repeated linear factors are called residues; Equations (2.57) and (2.58) may be used to find the residue C_i.

(ii)   *Repeated linear factors of the form* $(s + p_i)^n$

If there is a factor $(s + p_i)^n$ in the denominator of $F(s)$, there will be corresponding terms in $f(t)$ which can be computed from the equation

$$f(t)_{(s+p_i)^n} = \frac{1}{(n-1)!} \left(\frac{d^{n-1}}{ds^{n-1}} [(s+p_i)^n F(s) e^{st}] \right) \Bigg|_{s=-p_i} \qquad (2.59)$$

Alternatively, Equation (2.59) may be written in the form

$$f(t)_{(s+p_i)^n} = e^{-p_i t} \sum_{k=1}^{n} \left(\frac{t^{k-1}}{(n-k)!(k-1)!} \frac{d^{n-k}}{ds^{n-k}} [(s+p_i)^n F(s)] \right) \Bigg|_{s=-p_i} \qquad (2.60)$$

Thus, $C_{i,k}$ in Equation (2.47) is given by

$$C_{i,k} = \frac{1}{(n-k)!} \left(\frac{d^{n-k}}{ds^{n-k}} (s+p_i)^n F(s) \right) \Bigg|_{s=-p_i} \qquad (2.61)$$

The above two methods of finding coefficients are valid for both real and complex poles. However, as has already been indicated, for machine computation it is preferable to use other formulae when the poles are complex.

(iii)   *A distinct quadratic factor of the form* $(s + a)^2 + b^2$

A quadratic unrepeated factor representing a pair of complex conjugate poles at $s = -a \pm jb$ contributes a term in $f(t)$ defined by

$$f(t)|_{(s+a)^2+b^2} = e^{-at} (2C_R \cos bt + 2C_I \sin bt) \qquad (2.62)$$

where C_R and C_I are the real and imaginary parts of

$$[(s+a-jb)F(s)]|_{s=-(a+jb)} \qquad (2.63)$$

(iv)   *Repeated quadratic factors of the form* $[(s + a)^2 + b^2]^n$

Finally, for each repeated quadratic factor there will be a term of the form

$$f(t)|_{[(s+a)^2+b^2]^n} = e^{-at} \sum_{k=1}^{n} \frac{t^{k-1}}{(k-1)!} (2C_{R_k} \cos bt + 2C_{I_k} \sin bt) \qquad (2.64)$$

where

$$C_{R_k} + jC_{I_k} = \frac{1}{(n-k)!} \left(\frac{d^{n-k}}{ds^{n-k}} [(s+a-jb)^n F(s)] \right) \Bigg|_{s=-(a+jb)} \qquad (2.65)$$

Equations (2.55), (2.59) and (2.62)–(2.65) are the required Heaviside formulae.

The remainder of this section illustrates with examples some techniques of partial-fraction expansion. An examination of these examples, together with the Laplace transform. Tables 2.1 and 2.2, should enable each of the Heaviside formulae to be derived. Heaviside formulae for inverse Laplace transforms are considered further in Chapters 4 and 11.

Example 2.4

Prove that equation (2.57) gives the partial-fraction coefficient associated with a linear unrepeated factor of the denominator of $F(s)$.

Solution
Let

$$F(s) = \frac{N(s)}{D(s)} = \frac{C_i}{s + p_i} + \text{(all other terms arising from the factors of } D(s)\text{)}$$

To find the partial-fraction coefficient C_i, multiply throughout by $s + p_i$ to give

$$(s + p_i)\frac{N(s)}{D(s)} = C_i + (s + p_i) \times \text{(all other terms)}$$

If s is set equal to $-p_i$, then

$$C_i = \left((s + p_i)\frac{N(s)}{D(s)} \right)\Bigg|_{s = -p_i}$$

which is the required solution. In practice, the above equation amounts to what is sometimes called the cover-up rule. The factor $s + p_i$ in the denominator of the rational polynomial $F(s)$ is covered up, and to find the coefficient C_i all the remaining values of s are set equal to $-p_i$.

Example 2.5

Prove that Equation (2.61) gives the partial-fraction coefficients associated with a linear repeated factor in the denominator polynomial $D(s)$.

Solution
Let

$$F(s) = \frac{C_{i,n}}{(s + p_i)^n} + \frac{C_{i,n-1}}{(s + p_i)^{n-1}} + \cdots + \frac{C_{i,1}}{s + p_i} + \text{(all remaining terms)}$$

Multiplying throughout by $(s + p_i)^n$ and setting $s = -p_i$ will yield the sum $C_{i,n}$. This is exactly the same procedure as used in the previous example, Example 2.4.

To find $C_{i,n-1}$ first multiply throughout by $(s+p_i)^n$ to give

$$(s+p_i)^n F(s) = C_{i,n} + C_{i,n-1}(s+p_i) + (s+p_i)^n \times \text{(all other terms)}$$

Differentiating both sides of the above equation with respect to s gives

$$\frac{d}{ds}[(s+p_i)^n F(s)] = C_{i,n-1} + n(s+p_i)^{n-1} \times \text{(all other terms)}$$

Evaluating at $s = -p_i$ gives $C_{i,n-1}$ and, by induction,

$$C_{i,k} = \frac{1}{(n-k)!}\left(\frac{d^{n-k}}{ds^{n-k}}(s+p_i)^n F(s)\right)\bigg|_{s=-p_i}$$

which is the required proof (see equation (2.61)).

Example 2.6

Find the inverse Laplace transform of

$$F(s) = \frac{s+3}{(s+1)^2(s-2)}$$

Solution

The partial-fraction expansion of $F(s)$ is

$$\frac{s+3}{(s+1)^2(s-2)} = \frac{C_1}{s-2} + \frac{C_{2,2}}{(s+1)^2} + \frac{C_{2,1}}{s+1}$$

The coefficients C_1 and $C_{2,2}$ are easily calculated using the cover-up rule to be $5/9$ and $-2/3$, respectively; $C_{2,1}$ is obtained by differentiation:

$$C_{2,1} = \left(\frac{d}{ds}\frac{s+3}{s-2}\right)\bigg|_{s=-1} = -\frac{5}{9}$$

The inverse Laplace transform then follows as

$$f(t) = -\frac{2}{3}te^{-t} - \frac{5}{9}e^{-t} + \frac{5}{9}e^{2t}$$

An alternative way to find the partial-fraction expansion would be to let

$$F(s) = \frac{1}{s+1}\frac{s+3}{(s+1)(s+2)}$$

Taking a partial-fraction expansion of the right-hand side gives

$$F(s) = \frac{1}{s+1}\left(\frac{-2/3}{s+1} + \frac{5/3}{s-2}\right)$$

$$= \frac{-2/3}{(s+1)^2} + \frac{5/3}{(s+1)(s+2)}$$

A further partial-fraction expansion yields the required result:

$$F(s) = \frac{-2/3}{(s+1)^2} - \frac{5/9}{s+1} + \frac{5/9}{s+2}$$

Example 2.7

Find the inverse Laplace transform of

$$F(s) = \frac{s+2}{(s-1)^2 s^3}$$

Solution

This example is included since it requires repeated differentiation. The expansion of $F(s)$ takes the form

$$F(s) = \frac{C_{1,2}}{(s-1)^2} + \frac{C_{1,1}}{s-1} + \frac{C_{2,3}}{s^3} + \frac{C_{2,2}}{s^2} + \frac{C_{2,1}}{s}$$

The coefficients $C_{1,2}$ and $C_{2,3}$ are readily evaluated using the cover-up rule as 3 and 2, respectively; $C_{1,1}$ and $C_{2,2}$ require a single differentiation:

$$C_{1,1} = \left(\frac{d}{ds} \frac{s+2}{s^3} \right)\bigg|_{s=1} = -8$$

and

$$C_{2,2} = \left(\frac{d}{ds} \frac{s+2}{(s-1)^2} \right)\bigg|_{s=0} = 5$$

The final coefficient is obtained by double differentiation. From equation (2.61),

$$C_{2,1} = \frac{1}{2!} \left(\frac{d^2}{ds^2} \frac{s+2}{(s-1)^2} \right)\bigg|_{s=0} = 8$$

from which it follows that the time function is

$$f(t) = 3te^t - 8e^t + t^2 + 5t + 8$$

Before leaving this section it is worth noting that if a factor p_i is complex, then the coefficient C_i is complex too. This poses no real problem for hand calculation, but it does make machine computation awkward.

2.4.2   The convolution integral

The convolution integral (entry 10 of Table 2.1) is defined by

$$\mathscr{L}^{-1}[F_1(s)F_2(s)] = \int_0^t f_1(\tau) f_2(t - \tau)\, d\tau \qquad (2.66)$$

where $F_1(s)$ and $F_2(s)$ are the transforms of $f_1(t)$ and $f_2(t)$, respectively. Proof of this equation may be obtained from the definition of the Laplace transform (Equation (2.14)), and will not be given here. It is, however, instructive to examine a graphical interpretation of the convolution process.

Suppose, for the sake of illustration, that $f_1(\tau)$ and $f_2(\tau)$ are the time signals shown in Fig. 2.3. The term $f_2(t-\tau)$ in Equation (2.66) is obtained by delaying $f_2(\tau)$ by a time t to give $f_2(\tau-t)$, followed by reflection about a vertical axis at $\tau = t$, as shown in Fig. 2.4. The convolution integral involves the product of $f_1(\tau)$ and $f_2(t-\tau)$. Figure 2.5 shows this product for various values of t. In this figure

Figure 2.3 Time signals $f_1(\tau)$ and $f_2(\tau)$

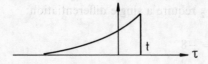

Figure 2.4 The signal $f_2(t-\tau)$

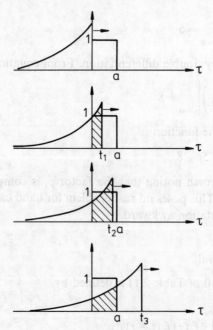

Figure 2.5 The product $f_1(\tau)f_2(t-\tau)$ for various values of t

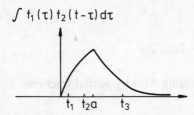

$$\int f_1(\tau)\, f_2(t-\tau)\, d\tau$$

$t_1 \quad t_2 a \qquad t_3$

Figure 2.6 The convolution integral $\int f_1(\tau) f_2(t-\tau)\, d\tau$

the shaded areas represent the values of the convolution integral for the two signals at the specified times. Combining all these values for each time instant allows the convolution integral to be generated, as shown in Fig. 2.6.

Convolution and the related technique of deconvolution are particularly useful in signal processing and in some forms of identification. However, it has been introduced here for two reasons: to illustrate that the inverse transform of a Laplace transform product is not the product of the individual time functions, and to demonstrate the use of the integral in finding inverse transforms.

Example 2.8

Find the inverse Laplace transform of

$$F(s) = \frac{2}{(s+3)(s^2+4)}$$

using the convolution integral.

Solution

Let $F(s) = F_1(s) F_2(s)$, where

$$F_1(s) = \frac{1}{s+3} \quad \text{with inverse } f_1(t) = e^{-3t}$$

and

$$F_2(s) = \frac{2}{s^2+4} \quad \text{with inverse } f_2(t) = \sin 2t$$

The convolution integral indicates that

$$\mathcal{L}^{-1}[F_1(s)F_2(s)] = \int_0^t \sin 2\tau\, e^{-3(t-\tau)}\, d\tau$$

$$= e^{-3t} \int_0^t \sin 2\tau\, e^{3\tau}\, d\tau$$

Integrating by parts twice gives

$$f(t) = \frac{1}{13}(2e^{-3t} + 3\sin 2t - 2\cos 2t)$$

This answer may be checked by expanding $F(s)$ by partial fractions.

2.5   Initial and final value theorems

These two further properties of Laplace transforms are useful for checking the solution of a differential equation, and also for extracting information about, for example, the steady-state response of a system without having to perform the inverse Laplace transform operation.

2.5.1   Initial value theorem

Suppose $f(t)$ and $F(s)$ are a Laplace transform pair. Then, provided the limit of $sF(s)$ as $s \to \infty$ exists, the initial value of the time function is given by

$$f(0) = \lim_{t \to 0}[f(t)] = \lim_{s \to \infty}[sF(s)] \qquad (2.67)$$

2.5.2   Final value theorem

Suppose $f(t)$ and $F(s)$ are a Laplace transform pair and that $F(s)$ is stable. (Stability is dealt with in Chapter 5, but essentially the restriction implies that $f(t)$ is bounded – that is, it does not become infinite as t tends to infinity.) Under these conditions

$$f(\infty) = \lim_{t \to \infty}[f(t)] = \lim_{s \to 0}[sF(s)] \qquad (2.68)$$

Example 2.9

Find the initial and final values of the time function whose Laplace transform is

$$Y(s) = \frac{2s^2 + 7s + 4}{s(s+2)(s+1)}$$

Solution
From the initial value theorem (Equation (2.67)),

$$y(0) = \lim_{s \to \infty}[sY(s)]$$

Hence

$$y(0) = \lim_{s \to \infty}\left[\frac{2s^2 + 7s + 4}{(s+2)(s+1)}\right]$$

which, on dividing the numerator and denominator by s^2 and letting $s \to \infty$, gives

$$y(0) = 2$$

The final value theorem applies, since the poles of $Y(s)$ will give rise to bounded terms in $y(t)$. Hence, from Equation (2.68),

$$y(\infty) = \lim_{s \to 0} [sY(s)] = 2$$

If a partial-fraction expansion is performed on $Y(s)$, the resulting inverse transform is found to be

$$y(t) = 2 - e^{-2t} + e^{-t}$$

Letting $t = 0$, and then $t = \infty$, produces results that agree with those already obtained.

Problems

2.1. A system is represented by the first-order equations

$$\dot{y}_1(t) = y_3(t)$$

$$\dot{y}_2(t) = y_4(t)$$

$$\dot{y}_3(t) = -\frac{K}{M}y_1(t) - \frac{B}{M}y_3(t) + \frac{U}{M}y_4(t)$$

$$\dot{y}_4(t) = -\frac{U}{L}y_3(t) - \frac{R}{L}y_4(t) + \frac{1}{L}e(t)$$

Find the two second-order differential equations relating the outputs $y_1(t)$ and $y_2(t)$ to the input $e(t)$.

2.2. Using the Laplace transform integral (Equation (2.14)), prove the following:

$$\mathcal{L}[a_1 f_1(t) + a_2 f_2(t)] = a_1 F_1(s) + a_2 F_2(s)$$

$$\mathcal{L}[u(t - \tau)f(t - \tau)] = e^{-s\tau}F(s)$$

$$\mathcal{L}\left[\int_0^\infty f_1(t - \tau)f_2(\tau)\,d\tau\right] = F_1(s)F_2(s)$$

2.3. Find the inverse Laplace transforms of the following functions:

$$Y_1(s) = \frac{1}{s(s+2)}$$

$$Y_2(s) = \frac{10}{(s+4)(s+10)^3}$$

$$Y_3(s) = \frac{1}{s^2 + 2s + 3}$$

2.4. Calculate the initial and final values of the functions having the following Laplace transforms:

$$F_1(s) = \frac{2(s+1)}{s(s+3)(s+5)^2}$$

$$F_2(s) = \frac{4}{s^3 + 5s^2 + 12s + 8}$$

Check that these values are correct by finding the inverse Laplace transforms and evaluating at $t = 0$ and $t = \infty$.

2.5. Use Program 2 of Chapter 11 to find the inverse Laplace transforms of the functions in Problems 2.3 and 2.4.

2.6. Use Laplace transforms to find the responses of the systems governed by the following equations:

(a) $$\frac{d^2 x}{dt^2} + 3\frac{dx}{dt} + 2x = 1$$

given that $x(0) = 1$ and $\dot{x}(0) = 1$, and

(b) $$\frac{d^2 x}{dt^2} + 4\frac{dx}{dt} + 4x = 3\frac{dy}{dt} + 2y$$

given that $x(0) = \dot{x}(0) = 0$ and $y = e^{-3t}$ for $t \geqslant 0$.

CHAPTER 3

BLOCK DIAGRAMS

3.1   Introduction

The previous two chapters have covered modelling and the mathematical foundation
of the Laplace transform method. In this chapter the concepts of a transfer function
and its related block diagram are introduced.

Block diagrams provide a pictorial representation of a system and its associated
control structure and compensators. The application of block reduction techniques,
or 'block diagram algebra', condenses the Laplace transform system equations
together with any controller equations into a form suitable for either design studies
or inverse transformation. Block diagram algebra may also be used to demonstrate
many of the advantages of the feedback control system.

3.2   Transfer functions

The concept of a transfer function is conveniently introduced by recalling the
modelling techniques developed in Chapter 1. Consider, for example, the electrical
circuit shown in Fig. 3.1. Here the problem is to determine the current response $i(t)$
due to a change in voltage $e(t)$.

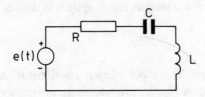

Figure 3.1 An *RLC* electric circuit

Application of Kirchhoff's voltage law around the circuit gives

$$e(t) = Ri + \frac{1}{C}\int_0^t i\,dt + L\frac{di}{dt} \qquad (3.1)$$

Ignoring initial conditions, this equation transforms to

$$E(s) = RI(s) + \frac{I(s)}{sC} + LsI(s) \tag{3.2}$$

and Equation (3.2) may be rearranged to give

$$\left(sL + R + \frac{1}{sC}\right)I(s) = E(s) \tag{3.3}$$

In general, a Laplace transform system model will take the form

$$(\text{system geometry}) \times (\text{response output})$$

$$= (\text{forcing input}) + (\text{initial conditions})$$

Control system studies are concerned more with the forced response than with the transient due to the initial conditions. If the system is stable, the influence of the initial conditions on the output becomes negligible as time progresses. It is therefore common practice to assume all the initial conditions to be zero, in which case any system model could be written as the ratio of output over input. Hence Equation (3.3) may be written as

$$\frac{I(s)}{E(s)} = \frac{Cs}{CLs^2 + CRs + 1} \tag{3.4}$$

and represented pictorially as in Fig. 3.2.

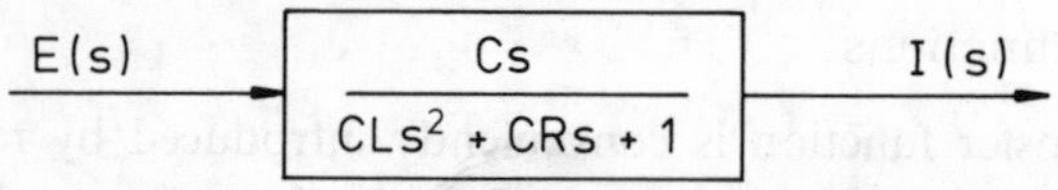

Figure 3.2 A block diagram for the *RLC* circuit in Fig. 3.1

In general, the transfer function of a constant linear system is defined as

$$F(s) = \frac{\mathscr{L}[\text{output}]}{\mathscr{L}[\text{input}]} \tag{3.5}$$

with all initial conditions set to zero. The input/output relationship defined by Equation (3.4) and the block diagram in Fig. 3.2 provide the same transfer function information. In Fig. 3.2 the transfer function contained in the block may be thought of as transforming the input signal $E(s)$ into the output signal $I(s)$. Note that, as in this case, transfer functions of engineering systems are invariably rational polynomials.

Most complex modelling exercises result in a number of subsystems, each of which has a transfer function representation. The use of block diagrams is a convenient method of pictorially grouping these subsystems such that the system is represented in a mathematically meaningful way, and this is best demonstrated by an example.

Example 3.1

Figure 3.3 shows a system of coupled tanks (the same system that was analysed in Example 1.7). Derive an equation for each of the three subsystems – the two tanks and the outflow – and put them into block diagram form. Combine the three diagrams to form a block diagram of the overall system.

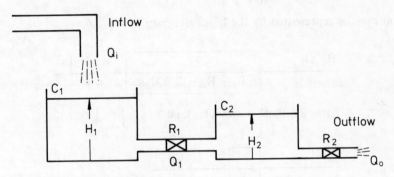

Figure 3.3 A coupled-tanks system

Solution

For tank 1, the governing equation is

$$C_1 R_1 \frac{dH_1}{dt} + H_1 = R_1 Q_i + H_2 \tag{3.6}$$

and with zero initial conditions its Laplace transform is

$$(1 + C_1 R_1 s)H_1(s) = R_1 Q_i(s) + H_2(s) \tag{3.7}$$

Equation (3.7) has one output, $H_1(s)$, and two inputs, $Q_i(s)$ and $H_2(s)$. The block diagram representation of this element is shown in Fig. 3.4. The circle with a cross in it represents a comparator, which takes the sum of the incoming signals. The two plus signs indicate the sign of the signals.

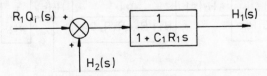

Figure 3.4 Block diagram for tank 1

For tank 2, the governing equation is

$$R_1 C_2 R_2 \frac{dH_2}{dt} + H_2 R_1 + H_2 R_2 = R_2 H_1 \tag{3.8}$$

Its Laplace transform form, with zero initial conditions, is

$$(R_1 + R_2 + C_2 R_1 R_2 s) H_2(s) = R_2 H_1(s) \tag{3.9}$$

The block diagram representation is shown in Fig. 3.5. The equation describing the outflow from tank 2 is

$$R_2 H_2(s) = Q_o(s) \tag{3.10}$$

and may be represented by the block diagram in Fig. 3.6.

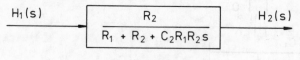

$$H_1(s) \qquad \boxed{\dfrac{R_2}{R_1 + R_2 + C_2 R_1 R_2 s}} \qquad H_2(s)$$

Figure 3.5 Block diagram for tank 2

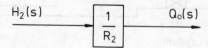

$$H_2(s) \qquad \boxed{\dfrac{1}{R_2}} \qquad Q_o(s)$$

Figure 3.6 Block diagram for outflow equation

All three block diagrams may now be combined into a single diagram for the system, as shown in Fig. 3.7. From Fig. 3.7 the transfer function for tank 1, for example, is obtained as

$$\frac{H_1(s)}{E_1(s)} = \frac{1}{1 + C_1 R_1 s} \tag{3.11}$$

where

$$E_1(s) = R_1 Q_1(s) + H_2(s) \tag{3.12}$$

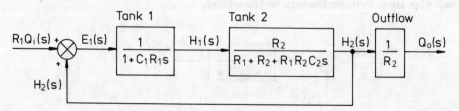

Figure 3.7 Complete block diagram for the coupled-tanks system in Fig. 3.3

This could of course have been obtained directly from Equation (3.7). However, if the transfer function between $Q_o(s)$ and $Q_i(s)$ were required, Equations (3.7), (3.9) and (3.10) would have to be manipulated algebraically. The advantage of the block diagram representation of transfer functions is that it makes possible a structured approach to the manipulation of such system equations. This aspect of block diagrams is considered in Section 3.4.

3.3   Transfer functions and impulse responses

A transfer function provides a Laplace transform description of a system. The link with the corresponding time description may be developed in the following way. Suppose a unit impulse $\delta(t)$ is applied to a system with a transfer function $F(s)$,

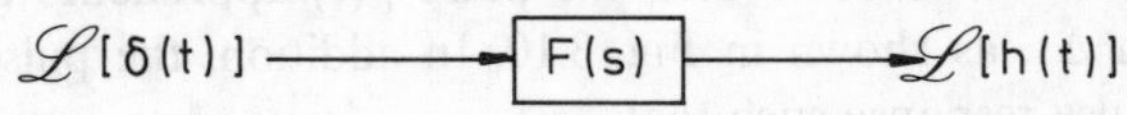

$$\mathscr{L}[\delta(t)] \longrightarrow \boxed{F(s)} \longrightarrow \mathscr{L}[h(t)]$$

Figure 3.8 Input/output representation by a transfer function $F(s)$

as shown in Fig. 3.8. The output signal $h(t)$ is called the impulse response. From Equation (3.5),

$$\mathscr{L}[h(t)] = F(s)\mathscr{L}[\delta(t)] = F(s) \qquad (3.13)$$

since $\mathscr{L}[\delta(t)]$ is unity. This result means that the transfer function is the transform of the system's impulse response. More generally, the response to any input, starting at $t = 0$, follows from Equations (3.5) and (3.13). Hence, for a system with input $U(s)$ and output $Y(s)$,

$$Y(s) = \mathscr{L}[h(t)]U(s) \qquad (3.14)$$

Using the convolution integral, the inverse transform is obtained as

$$y(t) = \int_0^t h(t - \tau)u(\tau)\, d\tau \qquad (3.15)$$

which is the time domain equivalent of Equation (3.14).

It is helpful to interpret Equation (3.15) physically, as follows. The first step is to approximate the system input $u(t)$ in terms of pulses of duration $\Delta\tau$, as in Fig. 3.9. If $p(t)$ is a pulse of unit amplitude starting at time $t = 0$, then $u(n\Delta\tau)p(t - n\Delta\tau)$

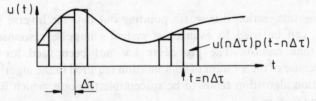

Figure 3.9 Decomposition of a time signal into pulses

represents a pulse starting at time $n\Delta\tau$ whose amplitude is that of $u(t)$ at time $n\Delta\tau$. The mathematical expression for the approximation of the time signal as a superposition of these pulses is

$$u(t) \approx \sum_{n=0}^{\infty} p(t - n\Delta\tau)u(n\Delta\tau) \tag{3.16}$$

Now, if $h_1(t)$ is the response of the system to a single unit pulse $p(t)$, the total response to the signal $u(t)$ may be approximated by the sum of the responses to the individual pulses in Fig. 3.9, and hence

$$y(t) \approx \sum_{n=0}^{\infty} h_1(t - n\Delta\tau)u(n\Delta\tau) \tag{3.17}$$

In Equation (3.17), as $\Delta\tau$ tends to zero the pulse $p(t)$ approaches an impulse function of strength $\Delta\tau$, as shown in Fig. 3.10. In addition, the pulse response approaches an impulse response such that

$$h_1(t - n\Delta\tau) \to \Delta\tau h(t - n\Delta\tau) \tag{3.18}$$

and if $n\Delta\tau$ is written as τ, then $\Delta\tau$ tends to dτ and, in this limit, from Equation (3.17), the output becomes

$$y(t) = \int_{0}^{\infty} h(t - \tau)u(\tau)\, d\tau \tag{3.19}$$

which is the convolution integral of Equation (3.15). The original input $u(t)$ has now been decomposed into a continuum of impulses, rather than pulses, and the response $y(t)$ is made up of a superposition of impulse responses, as expressed mathematically in Equation (3.19).

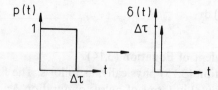

Figure 3.10 An impulse as a limiting form of a pulse

Before leaving this section it is worth pointing out that the inverse procedure, deconvolution, can be used to estimate a system's impulse response from its input/output time records. The procedure has not been used for computer calculations because of the amount of computation required in the algorithm. Also, the deconvolution algorithm tends to be susceptible to noise, which is probably its most limiting feature.

3.4 Block diagrams and block diagram algebra

In the examples given so far, block diagrams have been used to represent the Laplace transform equations for the plant. However, the technique is easily extended to include the control equipment and its associated feedback paths, as shown in Fig. 3.11, which contains all the basic elements associated with a single input/single output control system. This figure uses a standard notation for paths and elements: $R(s)$ denotes the actual reference or demand signal applied to the system; $E(s)$ is an error signal, or the output of a comparator (in this case $E(s)$ is the difference between $R(s)$ and the feedback signal, $B(s)$); $U(s)$ and $Y(s)$ are, respectively, the plant's manipulable input and its measured output. All forward path elements are denoted by G, with subscripts to distinguish between the various blocks, for example the control element and the plant are denoted by $G_c(s)$ and $G_p(s)$ in Fig. 3.11. Similarly, all feedback blocks are denoted by H and distinguished by means of subscripts. Typically, all the Gs and Hs are rational polynomials.

For control purposes, a representation with only one block between some input and output is all that is normally required. So, for example, if the object were to design the compensating elements $G_c(s)$ in Fig. 3.11, an open-loop transfer function between $U(s)$ and $B(s)$ would be required. If, however, a closed-loop time response were required, a single block would give the relationship between the command input and the output of the indirectly controlled variable. The technique of condensing a number of blocks into a single block is called block reduction.

Block reduction techniques are based on three simple rules. The first deals with the combination of two blocks on the same path, the second with the combination of two blocks, one of which is on a forward path, and the third with the combination of two blocks, one of which is on a feedback path. These rules are summarized in Table 3.1, where for convenience the variable s has been omitted and for clarity signals are given lower-case letters. Note that, for each rule, the governing equation must hold for both the original and the equivalent diagram.

Block reduction occasionally requires some block manipulation. Table 3.2 shows most of the manipulations that are likely to be encountered in practice. Again, the variable s has been omitted and, contrary to normal convention, the Laplace domain signals are given lower-case letters.

An alternative to block reduction is block diagram algebra. With block diagram algebra the procedure involves the following steps:

1. Label the output from each comparator: $E_1(s)$, $E_2(s)$, ...
2. Establish the transfer function equations for the system and comparator outputs.
3. Eliminate those variables which are not required.

In step 2 the transfer function equations are obtained by tracing the signal, or signals, back to either a comparator or an output position. This and the block reduction method are best illustrated by examples.

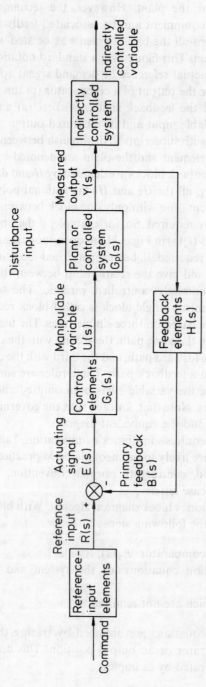

Figure 3.11 Block diagrams of a feedback control system

Table 3.1 Block reduction rules

	RULE	ORIGINAL DIAGRAM	EQUIVALENT DIAGRAM	EQUATION
1	Blocks in series	$u \to G_1 \to G_2 \to y$	$u \to \boxed{G_1\,G_2} \to y$	$y = G_1 G_2 u$
2	Blocks in parallel	$u \to G_1$ (+), G_2 (±) $\to y$	$u \to \boxed{G_1 \pm G_2} \to y$	$y = (G_1 \pm G_2)\,u$
3	Blocks in a feedback loop	$u\ (+) \to \otimes \to G_1 \to y$, feedback G_2 $(\pm)$	$u \to \dfrac{G_1}{1 \mp G_1\,G_2} \to y$	$y = G_1(u \pm G_2 y)$

Example 3.2

The block diagram of a multiple-loop feedback control system is shown in Fig. 3.12. Use block diagram reduction to simplify this to a single block connecting $Y(s)$ to $R(s)$. Note that, for convenience, the variable s has been omitted from the transfer functions within the blocks.

Solution

This system contains a feedforward loop G_5, as well as a positive feedback loop $G_4 G_6 H_1$. The first step is to combine G_1 and G_2 using entry 1 in Table 3.1, and to move the summing junction of the feedforward loop in front of the G_4 block, as shown in Fig. 3.13. (The latter manipulation may not be immediately obvious: it is the result of combining entries 1 and 10 from Table 3.2.)

At this stage blocks G_4 and G_6 may be combined and the feedback loop $G_4 G_6 H_1$ reduced using entry 3 in Table 3.1. Also, from entry 2 of Table 3.1, the parallel blocks G_5/G_4 and G_3 may be combined, as shown in Fig. 3.14. The forward path blocks may now be combined and the feedback loop H_2 eliminated to produce the required solution, which is shown in Fig. 3.15.

Example 3.3

Use block diagram algebra to solve Example 3.2.

Table 3.2 Block manipulation rules

	MANIPULATION	ORIGINAL DIAGRAM	EQUIVALENT DIAGRAM	EQUATION
1	Moving a summing point ahead of an element			$y = Gu_1 - u_2$
2	Moving a summing point beyond an element			$y = G(u_1 - u_2)$
3	Moving a take-off point ahead of an element			$y = Gu$
4	Moving a take-off point beyond an element			$y = Gu$ $u = y/G$
5	Removing an element from a forward path			$y = (G_1 - G_2)u$
6	Inserting an element in a forward path			$y = G_1 u - u$
7	Removing an element from a feedback path			$y = \dfrac{Gu}{1 + GH}$
8	Inserting an element in a feedback path			$y = \dfrac{G_1 u}{1 + G_1}$
9	Rearrangement of summing points			$y = u_1 - u_2 - u_3$
10	Interchange of summing points			$y = u_1 - u_2 + u_3$
11	Moving a take-off point ahead of a summing point			$y = u_1 - u_2$
12	Moving a take-off point beyond a summing point			$y = u_1 - u_2$ $u_1 = y + u_2$

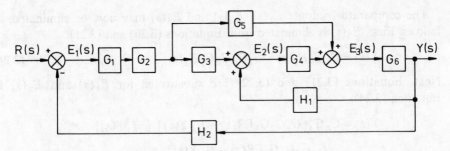

Figure 3.12

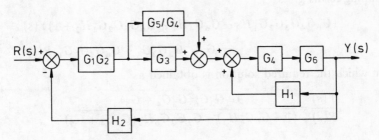

Figure 3.13

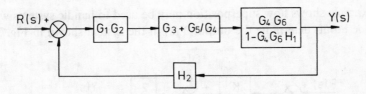

Figure 3.14

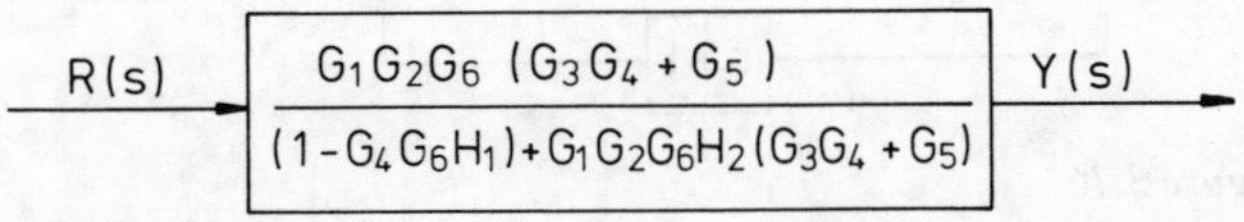

$$R(s) \longrightarrow \boxed{\dfrac{G_1G_2G_6\,(G_3G_4+G_5)}{(1-G_4G_6H_1)+G_1G_2G_6H_2(G_3G_4+G_5)}} \longrightarrow Y(s)$$

Figure 3.15

Solution

There are three comparators, with outputs $E_1(s)$, $E_2(s)$ and $E_3(s)$, and one system output, $Y(s)$. The four transfer function equations are therefore

$$Y(s) = G_6 E_3(s) \tag{3.20}$$

$$E_1(s) = R(s) - H_2 Y(s) \tag{3.21}$$

$$E_2(s) = G_3 G_2 G_1 E_1(s) + H_1 Y(s) \tag{3.22}$$

$$E_3(s) = G_4 E_2(s) + G_5 G_2 G_1 E_1(s) \tag{3.23}$$

The comparator outputs $E_1(s)$, $E_2(s)$ and $E_3(s)$ may now be eliminated, as follows. First, $E_3(s)$ is eliminated from Equations (3.20) and (3.21):

$$Y(s) = G_6G_4E_2(s) + G_6G_5G_2G_1E_1(s) \qquad (3.24)$$

Next, Equations (3.21) and (3.22) are substituted for $E_1(s)$ and $E_2(s)$ in Equation (3.24):

$$Y(s) = G_6G_4\{G_3G_2G_1[R(s) - H_2Y(s)] + H_1Y(s)\}$$
$$+ G_6G_5G_2G_1[R(s) - H_2Y(s)]$$

Grouping terms gives

$$(G_6G_4G_3G_2G_1H_2 - G_6G_4H_1 + G_6G_5G_2G_1H_2 + 1)Y(s)$$
$$= (G_6G_4G_3G_2G_1 + G_6G_5G_2G_1)R(s)$$

from which the required solution is obtained as

$$\frac{Y(s)}{R(s)} = \frac{G_1G_2G_6(G_3G_4 + G_5)}{(1 - G_6G_4H_1) - G_1G_2G_6H_2(G_3G_4 + G_5)}$$

Example 3.4

This example shows how superposition may be used to handle systems with more than one input: determine the output $Y(s)$ in the system shown in Fig. 3.16.

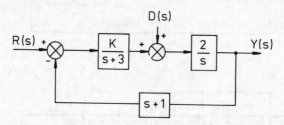

Figure 3.16

Solution

Setting $D(s) = 0$ gives the transfer function between $Y(s)$ and $R(s)$ as

$$\frac{Y(s)}{R(s)} = \frac{2K}{s(s+3) + 2K(s+1)}$$

Setting $R(s) = 0$ gives the transfer function between $Y(s)$ and $D(s)$ as

$$\frac{Y(s)}{D(s)} = \frac{2(s+3)}{s(s+3) + 2K(s+1)}$$

With both inputs present, the output is given by

$$Y(s) = \frac{2KR(s)}{s(s+3) + 2K(s+1)} + \frac{2(s+3)D(s)}{s(s+3) + 2K(s+1)}$$

or

$$Y(s) = \frac{2KR(s) + 2(s+3)D(s)}{s(s+3) + 2K(s+1)}$$

3.5   Feedback

There are essentially two types of feedback that are likely to be encountered: naturally occurring feedback and feedback which has been deliberately introduced to achieve some control objective. In both cases, feedback is that property of the system which permits a change in an output variable to adjust some input variable and hence the system's response. If the adjustment is additional to the input variable it is referred to as positive feedback, and if subtracts from the input variable it is called negative feedback. Most engineering control systems employ negative feedback since, in general, positive feedback will produce stability problems. By using techniques developed in this chapter, this section demonstrates some of the advantages of the feedback structure.

3.5.1   Natural feedback

Consider the two tanks shown in Fig. 3.17. The tank in Fig. 3.17(a) has a metering pump at its outlet. Provided there is fluid in the tank the outflow f_o is constant, so any change in inflow f_i will cause the tank to flood or to run dry. Such a system will not regulate itself and must include control. Figure 3.17(b) shows the same tank with a valve on its outlet. For turbulent flow, the discharge is found from

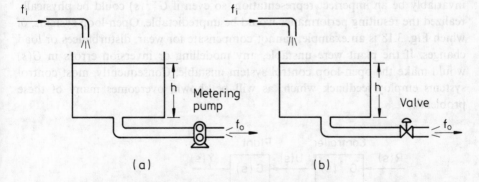

Figure 3.17 Identical tanks with different outflow characteristics: (a) without and (b) with self-regulation

Bernoulli's law, which may be stated as

$$f_o = C_d \sqrt{(2gh)} \tag{3.25}$$

where C_d is the coefficient of discharge and g the acceleration due to gravity.

Equation (3.25) is non-linear, but for small perturbations it may be approximated by finding the change in head required to cause a change in flow. The result is called the hydraulic resistance, R:

$$R = \frac{dh}{df_o} = \frac{2f_o}{2C_d^2 g} = \frac{2h}{q} \tag{3.26}$$

Laminar and turbulent flow are now both dealt with by the equation

$$f_o = h/R \tag{3.27}$$

Consequently, irrespective of the type of valve flow, there is a natural feedback since, for any given condition, there will be an associated head at which the inflow f_i will equal the outflow f_o. Such a system is referred to as being self-regulating, and control is required only if the fluctuations in head are excessive.

Naturally occurring feedback can produce problems in modelling and in the block diagram representation of models. The coupled-tanks system shown in Fig. 3.3, for example, has natural feedback, as demonstrated by its block diagram representation given in Fig. 3.7. The second tank loads the first, and the loading is represented by means of a feedback loop. Such problems occur in all types of systems; the reader may find it helpful to have another look at Section 1.5, on system models.

3.5.2   Deliberately introduced feedback

In theory, perfect control may be obtained by using a controller which is the mathematical inverse of the plant – see Fig. 3.18. However, there are many reasons why such a system should not be used in practice. The plant model $G(s)$ will invariably be an imperfect representation, so even if $G^{-1}(s)$ could be physically realized the resulting performance would be unpredictable. Open-loop systems, of which Fig. 3.18 is an example, cannot compensate for wear, disturbances or load changes. If the plant were unstable, any modelling or inversion errors in $G(s)$ would make the open-loop control system unstable. Consequently, most control systems employ feedback which, as will be shown, overcomes many of these problems.

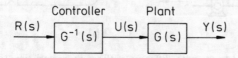

Figure 3.18 Ideal open-loop control

Feedback, or more precisely negative feedback, has many advantages but one main disadvantage. Its disadvantage is that it can change an open-loop stable system into a closed-loop unstable one. Stability considerations therefore act as a constraint which tends to limit the advantages of using feedback.

Consider the open- and closed-loop systems shown in Fig. 3.19, in which $R(s)$ is the desired value and $Y(s)$ the measured value. The input/output relationships for the two systems are, for Fig. 3.19(a)

$$\frac{Y(s)}{R(s)} = G(s) \tag{3.28}$$

and for Fig. 3.19(b)

$$\frac{Y(s)}{R(s)} = \frac{G(s)}{1 + G(s)} \tag{3.29}$$

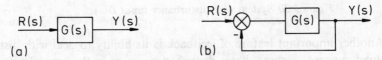

(a)   (b)

Figure 3.19 (a) Open- and (b) closed-loop systems

Assume for the sake of argument that both systems are stable. For a fixed input $R(s)$, a small change $\Delta G(s)$ in $G(s)$ will produce a change $\Delta Y(s)$ in the output $Y(s)$ or, in the open-loop case,

$$Y(s) + \Delta Y(s) = (G(s) + \Delta G(s))R(s) \tag{3.30}$$

Subtracting Equation (3.28) and substituting for $R(s)$, again from Equation (3.28), produces

$$\Delta Y(s) = \frac{\Delta G(s)}{G(s)} Y(s) \tag{3.31}$$

In general, Equation (3.31) indicates that any change in plant or controller characteristics is passed directly to the output. Note also that with only output measurements it is impossible to distinguish between a change in $G(s)$ and a change in $R(s)$.

For the closed-loop system a similar analysis produces (ignoring small quantities)

$$Y(s) = \frac{1}{1 + G(s)} \frac{\Delta G(s)}{G(s)} Y(s) \tag{3.32}$$

The feedback structure has introduced the factor

$$\frac{1}{1 + G(s)} \tag{3.33}$$

which tends to reduce the change in output.

In general, under steady-state conditions, high loop gains are desirable since they reduce the sensitivity to variations in plant and controller characteristics (see Equation (3.32)). Also, Equation (3.29) shows that a high loop gain increases accuracy. These results are important since it is sometimes possible to achieve a desired design by using less accurate – and hence less expensive – control components.

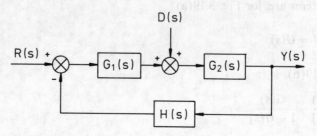

Figure 3.20 System with disturbance input $D(s)$

Another important feature of feedback is its ability to deal with disturbances, or load changes. Assume that a disturbance occurs in the control system shown in Fig. 3.20. (This type of system has already been examined in Example 3.3.) The change in output due to the disturbance $D(s)$ is given by

$$\Delta Y(s) = \frac{G_2(s)D(s)}{1 + G_1 G_2 H(s)} \tag{3.34}$$

which suggests that the effect at the output has been reduced by the factor

$$\frac{G_2(s)}{1 + G_1 G_2 H(s)} \tag{3.35}$$

If $D(s)$ is harmonic, the effective reduction will depend on the dynamics of Equation (3.35). In some circumstances the amplitude of $D(s)$ could increase. The harmonic analysis of systems is considered in Chapter 7.

From the points made in this section, it should be evident that deliberately introduced feedback is the cornerstone on which control system design is based. In the following chapters the design problem is therefore considered as one of ensuring closed-loop stability and performance, given the system's open-loop transfer function.

Problems

3.1. The following equations represent various systems, each having an input $r(t)$ and output $y(t)$. Put these equations in block diagram form.

(a) $\qquad \dfrac{d^2y}{dt} + 6\dfrac{dy}{dt} + 3y = r$

(b) $\qquad \dfrac{d^2y}{dt^2} + 6\dfrac{dy}{dt} + 3y = 5\dfrac{dr}{dt} + 4r$

(c) $\qquad y = \dfrac{d^2r}{dt^2} + \dfrac{dr}{dt} + r$

(d) $\qquad \dfrac{dy}{dt} + Ay = v \quad \text{and} \quad \dfrac{d^2v}{dt^2} + B\dfrac{dv}{dt} = C\dfrac{dr}{dt} + Dr$

3.2. A simplified model of an aircraft's pitch control system is shown in Fig. 3.21. In this model the pitch angle is $\theta_o(s)$, $\theta_i(s)$ is the pilot's input signal and $V_v(s)$ is the vertical velocity. Determine the differential equation relating $\theta_o(t)$ to $\theta_i(t)$.

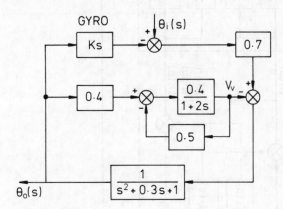

Figure 3.21

3.3. A two input/two output multivariable plant together with its controller and feedback loops is shown in Fig. 3.22. The controller consists of two dynamic compensators, $C_1(s)$ and $C_2(s)$, and two forward path gains, K_1 and K_2. In this arrangement the compensators are used to decouple the open-loop plant – that is, for the decoupled plant a disturbance at $E_i(s)$ produces a response at output $Y_i(s)$ ($i = 1$ or 2), and the other output is unaffected. In this way the multivariable plant behaves as though it were two independent single input/single output plants. The forward path gains act as simple proportional controllers which, together with their corresponding feedback loops, are used to control the decoupled plant.

For this system, find the transfer function equations $C_1(s)$ and $C_2(s)$ which decouple the plant. Also determine the closed-loop transfer

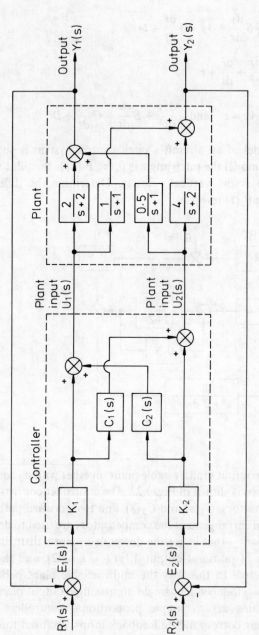

Figure 3.22

functions for both decoupled loops, and reduce each transfer function to its simplest form.

3.4. Determine the overall transfer function of the multiloop control system shown in Fig. 3.23.

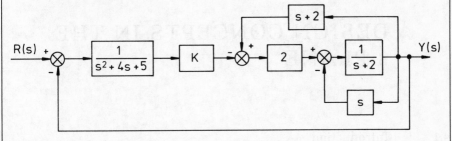

Figure 3.23

3.5. Show that for zero initial charge, the Laplace transform of the loop equations of the network shown in Fig. 3.24 are

$$V_i(s) = R_1 I_1(s) + \frac{1}{C_1 s}(I_1(s) - I_2(s))$$

$$0 = \frac{1}{C_1 s}(I_2(s) - I_1(s)) + \left(R_2 + \frac{1}{C_2 s}\right)I_2(s)$$

$$V_o(s) = \frac{I_2}{C_2 s}$$

Put these equations into block diagram form, and hence determine the overall transfer function $V_o(s)/V_i(s)$.

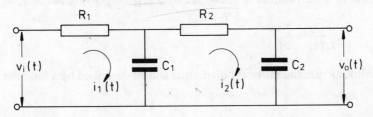

Figure 3.24

3.6. Determine the transfer function models for the mechanical and electrical compensators in Problems 1.4 and 1.5 (Chapter 1).

3.7. Find the transfer function $H_2(s)/Q_i(s)$ for the non-interacting tanks shown in Fig. 1.20 (Chapter 1).

CHAPTER 4

DESIGN CONCEPTS IN THE s-PLANE

4.1 Introduction

This chapter provides the foundations for the graphical study of single input/single output systems. After a brief return to Laplace transform systems (covered in Chapter 2), the concept of a pole–zero s-plane plot is introduced. Emphasis is placed on the graphical interpretation of Heaviside's partial-fraction expansion technique, since this provides the link between the s-plane plot and inverse Laplace transforms. Once this relationship is established, performance criteria are developed and the root loci plot is introduced. Material from this chapter is used in the chapters on stablity and frequency design methods (Chapters 5 and 7).

In the previous chapters it has been shown that most systems may be represented by means of ordinary linear differential equations which, in turn, are Laplace-transformable into algebraic equations in the s-domain. A Laplace-transformed system equation is normally presented in the form of a rational polynomial in s. That is to say, the transformed system output $Y(s)$ divided by the input $U(s)$ is in the form of a polynomial in s, divided by a second polynomial in s;

$$\frac{Y(s)}{U(s)} = \frac{N(s)}{D(s)} \tag{4.1}$$

For convenience, the rational polynomial will be described by a function $F(s)$ such that

$$F(s) = \frac{N(s)}{D(s)} \tag{4.2}$$

If the divisor polynomial is set equal to zero,

$$D(s) = 0 \tag{4.3}$$

the resulting equation is called the system's characteristic equation, since it can be shown to characterize the system dynamics. Its roots (or zeros) are the poles of $F(s)$, the values of s that make the rational polynomial infinite. The roots of the numerator polynomial $N(s)$ are the zeros of $F(s)$, and are the values of s which make the rational polynomial zero.

Both the poles and zeros of $F(s)$ could be complex values of s, and as such will have real and imaginary parts. Since only physical engineering systems are being considered, any complex pole or zero must have a complex conjugate. In general, any root will be of the form

$$s = \sigma + j\omega \tag{4.4}$$

and may be plotted on an Argand diagram. Such a plot, with σ on its real axis and ω on its imaginary axis, is referred to as an s-plane plot. It is common practice to identify the poles of $F(s)$ by a superimposed cross and the zeros by an enclosing circle. All the pole–zero plots in this book follow this convention.

4.2   Graphical interpretation of Heaviside's partial-fraction expansion

In Chapter 2 it was shown that the time response corresponding to a particular Laplace transform may be found by means of a partial-fraction expansion of the transform. A set of inverse transform tables is then used to evaluate the various terms as functions of time. However, it is also possible to obtain the coefficients of the partial-fraction expansion graphically, from an s-plane pole–zero map. Often this procedure may be carried out by inspection, and in such cases a table of Laplace transforms is not needed. Even when the time response cannot be written down directly, the pole–zero map provides valuable information on the system's stability and performance.

If equation (4.2) is in factored form, then

$$F(s) = \frac{K(s+z_1)(s+z_2)\cdots(s+z_m)}{(s+p_1)(s+p_2)\cdots(s+p_n)} \tag{4.5}$$

or

$$F(s) = K \prod_{k=1}^{m} (s+z_k) \Big/ \prod_{l=1}^{n} (s+p_l) \tag{4.6}$$

and may be expressed in partial-fraction form as

$$F(s) = \frac{C_1}{s+p_1} + \frac{C_2}{s+p_2} + \cdots + \frac{C_n}{s+p_n} \tag{4.7}$$

To find C_1, both sides of Equation (4.7) are multiplied by $s + p_1$ to give

$$(s+p_1)F(s) = C_1 + \frac{C_2(s+p_1)}{s+p_2} + \cdots + \frac{C_n(s+p_1)}{s+p_n} \tag{4.8}$$

If s is evaluated at $s = -p_1$, then from Equation (4.8)

$$C_1 = (s + p_1)F(s)|_{s = -p_1} \tag{4.9}$$

By referring to the s-plane, Equation (4.9) may be interpreted graphically. Thus, any coefficient C_i is found to be representable as

$$C_i = \frac{K \times (\text{product of direct distance from each zero to pole } -p_i)}{\text{product of direct distance from all other poles to } -p_i}$$

or, if polar coordinates are used, then

$$C_i = M e^{j\theta} \tag{4.10a}$$

where M is the magnitude of the coefficient C_i, given by

$$M = K \frac{\prod_{k=1}^{m} |(-p_i + z_k)|}{\prod_{\substack{l=1 \\ l \neq i}}^{n} |(-p_i + p_l)|} \tag{4.10b}$$

and θ is the phase angle, given by

$$\theta = \sum_{k=1}^{m} \arg(-p_i - p_k) - \sum_{\substack{l=1 \\ l \neq i}}^{n} \arg(-p_i + p_l) \tag{4.10c}$$

where the arguments, in degrees, are measured counterclockwise with respect to the positive direction of the real axis from the particular pole, or zero, to the pole $-p_i$. Recall that the ps and zs are defined by Equation (4.5), and therefore $-p_l$ (or $-z_k$) indicates that lth pole (or kth zero), and $+p_l$ (or $+z_k$) the negative of the lth pole (or kth zero).

For real poles C_i is real, but its sign could be either positive or negative. If the total phase contribution is a multiple of $360°$, then C_i is positive; if the phase contribution is a multiple of $180°$, then it is negative. In practice there is no need to calculate the phase angle for real poles since the following simple rule may be applied:

> *If the total number of real poles and zeroes to the right of $-p_i$ is even then C_i is positive; if the total is odd then it is negative.*

Complex poles occur in complex conjugate pairs, and have coefficients which are also complex conjugate pairs. From Equations (4.10), the coefficient C_i and C_{i+1}, corresponding to a pair of complex poles, are given by

$$C_i = M e^{j\theta} \tag{4.11a}$$

$$C_{i+1} = M e^{-j\theta} \tag{4.11b}$$

Once all the coefficients have been found, then the inverse Laplace transforms of the terms in Equation (4.7) are evaluated in turn.

For a real pole the corresponding time response is

$$\mathcal{L}^{-1}\left[\frac{C_i}{s+p_i}\right] = C_i e^{-p_i t} \qquad (4.12)$$

If p_i is zero, the time response reduces to the constant C_i.

The time response for a pair of complex conjugate poles is given by

$$\mathcal{L}^{-1}\left[\frac{Me^{+j\theta}}{s+a-jb} + \frac{Me^{-j\theta}}{s+a+jb}\right] = 2Me^{-at}\cos(bt+\theta) \qquad (4.13)$$

An alternative formulation may be established in rectangular coordinates. Again, the measured distance is from each zero, or from all other poles, to the pole for which the coefficient is being evaluated. In rectangular coordinates, the complex coefficients corresponding to Equations (4.11) become

$$C_i = g + jh \qquad (4.14a)$$

$$C_{i+1} = g - jh \qquad (4.14b)$$

In this case the inverse Laplace transform of the complex conjugate pair of poles is

$$\mathcal{L}^{-1}\left[\frac{g+jh}{s+a-jb} + \frac{g-jh}{s+a+jb}\right] = (g+jh)e^{-(a-jb)t} + (g-jh)e^{-(a+jb)t}$$

$$= 2e^{-at}[g\cos(bt) - h\sin(bt)] \qquad (4.15)$$

Example 4.1

Any system having a transfer function which may be written in the form

$$\frac{Y(s)}{U(s)} = \frac{\omega_n^2}{s^2 + 2\zeta\omega_n s + \omega_n^2}$$

is known as a second-order system. In this form, ζ is the dimensionless damping ratio and ω_n the system's natural, undamped frequency of vibration.

Find the time response of a second-order system subjected to a unit step input on $U(s)$. Assume that $\zeta < 1.0$.

Solution

When $U(s)$ is a unit step input, the system's characteristic equation is

$$s(s^2 + 2\zeta\omega_n s + \omega_n^2) = 0 \qquad (4.16)$$

Although the system has no zeros, Equation (4.16) indicates that there are three poles, at

$$s = 0, s = -\zeta\omega_n + j\omega_n\sqrt{(1-\zeta^2)}, \qquad s = -\zeta\omega_n - j\omega_n\sqrt{(1-\zeta^2)}$$

The corresponding pole–zero map is shown in Fig. 4.1.

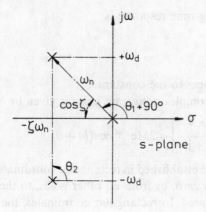

Figure 4.1 Pole–zero map for a second-order system having a step input

To find the time response, a partial-fraction expansion of the system and its forcing input is required. This takes the form

$$Y(s) = \frac{C_1}{s} + \frac{C_2}{s + \zeta\omega_n - j\omega_d} + \frac{C_3}{s + \zeta\omega_n + j\omega_d}$$

where ω_d is the damped frequency and equals $\omega_n\sqrt{(1-\zeta^2)}$.

The graphical technique yields the coefficient C_1 as

$$C_1 = \frac{\omega_n^2}{(\omega_n)(\omega_n)} = +1$$

The sign is positive since there are no real poles or zeros to the right of the pole at $s = 0$.

In polar coordinates (see Equations (4.10)), C_2 is given by

$$C_2 = Me^{j\theta}$$

where

$$M = \frac{\omega_n^2}{|(-\zeta\omega_n + j\omega_d) - (0 + j0)||(-\zeta\omega_n + j\omega_d) - (-\zeta\omega_n - j\omega_d)|}$$

$$= \frac{\omega_n^2}{|(-\zeta\omega_n + j\omega_d)||(2j\omega_d)|} = \frac{\omega_n^2}{(\omega_n)(2\omega_d)} = \frac{1}{2\sqrt{(1-\zeta^2)}}$$

and

$$\theta = -\arg[(-\zeta\omega_n + j\omega_d) - (0 + j0)] - \arg[(-\zeta\omega_n + j\omega_d)$$

$$- (-\zeta\omega_n - j\omega_d)] = -\arg(-\zeta\omega_n + j\omega_d) - \arg(2j\omega_d)$$

$$= -\tan^{-1}(-\zeta\omega_n/\omega_d) - 90°$$

which, from Fig. 4.1, becomes

$$\theta = -(\theta_1 + 90°) - \theta_2 = -180° - \theta_1$$

and, since there is a $180°$ phase shift, the expression for C_2 becomes

$$C_2 = -\frac{1}{2\sqrt{(1-\zeta^2)}}e^{-j\theta_1}$$

Inspection of Fig. 4.1 indicates that M and θ could have been found graphically from the pole–zero map by measuring the appropriate vectors and angles.

By symmetry, the complex conjugate pole at $s = -\zeta\omega_n - \zeta\omega_d$ will have a coefficient C_3 which is the complex conjugate of C_2. From Equations (4.12) and (4.13), the required time response is

$$y(t) = 1 - \frac{2e^{-\zeta\omega_n t}}{2\sqrt{(1-\zeta^2)}}\cos(\omega_d t - \theta_1) \qquad (4.17)$$

However, the addition formulae of the cosine function indicate that

$$\cos(\omega_d t - \theta_1) = \cos(\omega_d t)\cos\theta_1 + \sin(\omega_d t)\sin\theta_1$$

and, from Fig. 4.1,

$$\cos\theta_1 = \omega_d/\omega_n = \sqrt{(1-\zeta^2)} \quad \text{and} \quad \sin\theta_1 = \zeta\omega_n/\omega_n = \zeta$$

which, on substitution Equation (4.17), yields

$$y(t) = 1 - \frac{e^{-\zeta\omega_n t}}{\sqrt{(1-\zeta^2)}}[\zeta\sin\omega_d + \sqrt{(1-\zeta^2)}\cos\omega_d t]$$

This result could have been obtained directly from Equation (4.15) if the coefficients C_1 and C_2 had been calculated in, or transformed into, rectangular form.

A typical time response for this system is shown in Fig. 4.2. It may be noted that the response is contained within two exponential envelopes, at $1 + e^{-\zeta\omega_n t}$ and

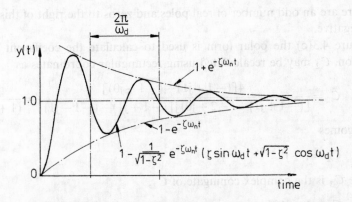

Figure 4.2 Response of a second-order system to a unit step input

$1 - e^{-\zeta\omega_n t}$. From the pole–zero map in Fig. 4.1, the complex poles have their real parts at $-\zeta\omega_n$. Also, the periodic time of the response is given by $2\pi/\omega_d$ (that is, the frequency of the oscillation is ω_d). Again from the pole–zero map, the imaginary parts of the complex poles are at $-\omega_d$ and $+\omega_d$.

Example 4.2

By using a pole–zero map, find the inverse Laplace transform of the transfer function

$$F(s) = \frac{s+1}{(0.5s+1)(0.5s+s+1)}$$

Solution

Since the inverse Laplace transform is to be found by using a pole–zero map, $F(s)$ must first be put into the appropriate form. This change will not alter the pole–zero positions, but it will affect the gain term associated with the transfer function, and hence the magnitude of the coefficients:

$$F(s) = \frac{4(s+1)}{(s+2)(s^2+2s+2)}$$

The pole–zero map for this transfer function is shown in Fig. 4.3. Here $K(=4)$ is the gain associated with the transfer function.

A partial-fraction expansion gives

$$F(s) = \frac{C_1}{s+2} + \frac{C_2}{s+1-j} + \frac{C_3}{s+1+j}$$

From Fig. 4.3, the coefficient of the pole at -2 is

$$C_1 = \frac{4}{\sqrt{2}\sqrt{2}} = -2$$

Since there are an odd number of real poles and zeros to the right of this pole, its sign is negative.

In Figure 4.3(c) the polar form is used to calculate the coefficient C_2. For comparison, C_2 may be recalculated using rectangular coordinates as

$$C_2 = \frac{4[(-1+j1)-(-1+j0)]}{[(-1+j1)-(-2+j0)][(-1+j1)-(-1-j1)]} = \frac{4j}{(1+j1)(2j)}$$

which becomes

$$C_2 = 1 - j1$$

and, since C_3 is the complex conjugate of C_2,

$$C_3 = 1 + j1$$

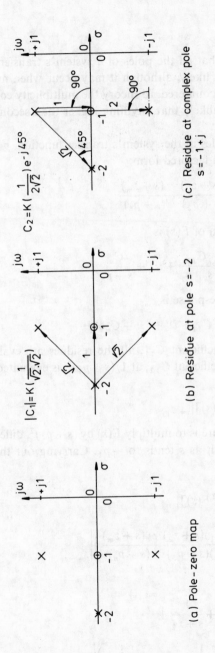

(a) Pole-zero map

(b) Residue at pole $s = -2$

(c) Residue at complex pole $s = -1+j$

$C_2 = K\left(\dfrac{1}{2\sqrt{2}}\right)e^{-j\,45°}$

$|C_1| = K\left(\dfrac{1}{\sqrt{2}\,\sqrt{2}}\right)$

Figure 4.3 (a) Pole–zero map for the function $F(s) = (s+1)/(0.5s+1) \times (0.5s^2 + s + 1)$, and the calculation of residues at (b) pole $s = -2$ and (c) complex pole $s = -1+j$

Hence from Equations (4.12) and (4.15), the inverse Laplace transform is

$$f(t) = -2e^{-2t} + 2e^{-t}(\cos t + \sin t)$$

4.2.1   Repeated poles

So far it has been assumed that all the poles of a system's transfer function are distinct. In most cases this is likely, although it may occur when multiple poles, either real or complex, are encountered. In theory the multiplicity could be of any order, but in practice it is unlikely that anything higher than second order need be considered.

Assume that one of the poles in the system's transfer function, p_1, is a double pole and that $F(s)$ is given in factored form:

$$F(s) = \frac{K(s+z_1)(s+z_2)\cdots(s+z_m)}{(s+p_1)^2(s+p_2)\cdots(s+p_n)} \tag{4.18}$$

The partial-fraction expansion of $F(s)$ is

$$F(s) = \frac{C_{12}}{(s+p_1)^2} + \frac{C_{11}}{s+p_1} + \cdots + \frac{C_n}{s+p_n} \tag{4.19}$$

and the corresponding time response is

$$f(t) = C_{12}t\,e^{-p_1 t} + C_{11}e^{-p_1 t} + \cdots + C_n e^{-p_n t} \tag{4.20}$$

With the exception of the coefficient C_{11}, all the residues are evaluated in the normal way. Note that the coefficient C_{12}, at $1/(s+p_1)^2$, is evaluated in the same way as a simple pole:

$$C_{12} = [(s+p_1)^2 F(s)]|_{s=-p_1} \tag{4.21}$$

To find C_{11} the usual procedure is to multiply $F(s)$ by $(s+p_1)^2$, differentiate with respect to s and take the limit as s tends to $-p_1$. Carrying out this procedure on Equation (4.18) yields

$$C_{11} = \frac{\mathrm{d}}{\mathrm{d}s}[(s+p_1)^2 F(s)]|_{s=-p_1}$$

$$= \left(\frac{\mathrm{d}}{\mathrm{d}s} \frac{K(s+z_1)(s+z_2)\cdots(s+z_m)}{(s+p_2)(s+p_3)\cdots(s+p_n)} \right)\Bigg|_{s=-p_1} \tag{4.22}$$

and thus

$$C_{11} = C_{12}\left(\frac{1}{s+z_1} + \frac{1}{s+z_2} + \cdots \right.$$

$$\left. + \frac{1}{s+z_m} - \frac{1}{s+p_2} - \cdots - \frac{1}{s+p_n} \right)\Bigg|_{s=-p_1} \tag{4.23}$$

With this relationship the coefficient C_{11} can be computed graphically. If $F(s)$ does not contain any zeros, the terms $1/(s + z_i)$, $i = 1, \ldots, m$, are all taken as zero.

A useful check is to apply the initial and final value theorems to $F(s)$, and compare the results with those obtained from the time response function $f(t)$. Indeed, it is very often possible to use these theorems to evaluate some of the coefficients in the partial-fraction expansion.

Example 4.3

Find the inverse Laplace transform of the transfer function

$$F(s) = \frac{8(s + 1)}{(s + 2)(s^2 + 2s + 2)^2}$$

Solution
The pole–zero plot for this transfer function is identical to the one shown in Fig. 4.3, with the exception of the complex poles at $s = -1 + j$ and $s = -1 - j$, which are now double poles. A partial-fraction expansion of $F(s)$ yields

$$F(s) = \frac{C_1}{s + 2} + \frac{C_{22}}{(s + 1 - j)^2} + \frac{C_{22}^*}{(s + 1 + j)^2} + \frac{C_{21}}{s + 1 - j} + \frac{C_{21}^*}{s + 1 + j}$$

where C_{22} and C_{21} are complex numbers of the form $a + jb$, and C_{22}^* and C_{21}^* are the complex conjugates of C_{22} and C_{21}.

It must be remembered that there is a double pole at $s = -1 + j$; the coefficients C_1 and C_{22} are found in the usual way and are

$$C_1 = \frac{8(1)}{(\sqrt{2})^2(\sqrt{2})^2} = -2$$

(since there are an odd number of real poles and zeros to the right of the pole at $s = -2$) and, using rectangular coordinates,

$$C_{22} = -(1 + j)$$

The coefficient C_{21} is found from Equation (4.21), as

$$C_{21} = C_{22}\left(\frac{1}{(-1 + j) + 1} - \frac{1}{(-1 + j) + (1 + j)} - \frac{1}{(-1 + j) + (1 + j)} \right.$$

$$\left. - \frac{1}{(-1 + j) + 2} \right)$$

$$= C_{22}[-j + 0.5j + 0.5j - (0.5 - 0.5j)] = +(1 + j)(0.5 - 0.5j)$$

$$= +1$$

These values may now be substituted into the partial-fraction expansion. Equations (4.12) and (4.15), together with the inverse Laplace transform for a double pole (for a double pole, the inverse Laplace transform will be that for the single pole, multiplied by t), give

$$f(t) = -2e^{-2t} + 2te^{-t}(\sin t - \cos t) + 2e^{-2t}\cos t$$

Example 4.4

Find the inverse Laplace transform of

$$F(s) = \frac{K(s+3)}{(s+1)^2(s+2)^3(s^2+4s+8)}$$

Solution

A partial-fraction expansion of $F(s)$ gives

$$F(s) = \frac{C_{12}}{(s+1)^2} + \frac{C_{11}}{s+1} + \frac{C_{23}}{(s+2)^3} + \frac{C_{22}}{(s+2)^2} + \frac{C_{21}}{s+2} + \frac{C_3}{s+2-2j}$$
$$+ \frac{C_3^*}{s+2+2j}$$

and its inverse Laplace transform is

$$f(t) = C_{12}te^{-t} + C_{11}e^{-t} + C_{23}\frac{t^2}{2!}e^{-2t} + C_{22}te^{-2t} + C_{21}e^{-2t}$$
$$+ 2e^{-2t}(g\cos 2t + h\sin 2t)$$

where g and h are the real and imaginary parts of C_3.

The graphical technique will yield all the coefficients with the exception of C_{21}; to find this coefficient the initial value theorem is employed. The pole–zero map for this transfer function is shown in Fig. 4.4. In this figure the numbers 3 and 2 above the poles at $s = -2$ and $s = -1$ indicate the presence of triple and double poles, respectively.

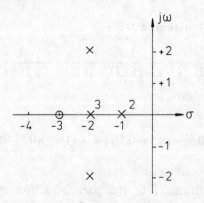

Figure 4.4 Pole–zero map for a system with multiple poles

The coefficients are calculated as follows (note that rectangular coordinates are used where appropriate:

$$C_{12} = \frac{K(2)}{(\sqrt{5})(\sqrt{5})(1)(1)(1)} = 0.4K$$

$$C_{23} = \frac{K(1)}{(2)(2)(1)(1)} = 0.25K$$

$$C_3 = \frac{K(1+2j)}{(2j)(2j)(2j)(4j)(-1+2j)(-1+2j)} = -\frac{11K}{800} - \frac{2Kj}{800}$$

$$C_3^* = -\frac{11K}{800} + \frac{2Kj}{800}$$

These four coefficients have been calculated by the method given in Section 4.2. To find C_{11} and C_{22}, the technique presented in this section is employed:

$$C_{11} = C_{12}\left(\frac{1}{2} - \frac{1}{1} - \frac{1}{1} - \frac{1}{1} - \frac{1}{1-2j} - \frac{1}{1+2j}\right) = -2.9C_{12}$$

$$C_{22} = C_{23}\left(\frac{1}{1} - \frac{1}{-1} - \frac{1}{-1} - \frac{1}{-2j} - \frac{1}{-2j}\right) = 3.0C_{23}$$

The coefficient C_{21} is obtained by using the initial value theorem in conjunction with the time response given by Equation (4.24). From the initial value theorem,

$$\lim_{s \to \infty} [sF(s)] = 0$$

Hence, by setting $t = 0$ in the time response,

$$C_{11} + C_{21} + 2g = 0$$

and therefore

$$C_{21} = -2g - C_{11} = +950K/800$$

Substitution of these coefficients into the time response gives the required inverse Laplace transform.

4.3   The role of poles and zeros

This section deals with the role of poles and zeros in a systems transfer function. Neither the effect on the transfer function of opening or closing a loop nor the way in which the poles and zeros influence this change are considered.

From the preceding sections it is evident that the time response of a linear system depends on its pole–zero locations. Since poles appear only in the dynamic exponential terms, they are said to define the dynamics of the system. Consequently, pole positions indicate stability and speed of response. If all the poles have negative real parts then the system is said to be stable, since it produces an exponential time response which decays to zero; the more negative the real parts, the faster the decay. However, if the real part of any pole is positive, the system becomes unstable since a positive pole gives rise to a term which has a positive exponential in time. Figure 4.5 illustrates the stable and unstable regions in the s-plane, and also shows the effect of pole position on speed of response.

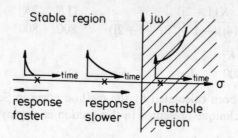

Figure 4.5 Impulse responses for isolated real poles in the s-plane

A transfer function's zeros alter the magnitude of the partial-fraction coefficients associated with each dynamic term. Zeros could therefore be thought of as adjusting the performance of a system. When a zero is close to a pole, the coefficient is smaller and the influence of the pole on the system's response decreases. Pole–zero cancellation occurs if a zero and a pole coincide and the related partial-fraction coefficient becomes zero. It is important to realize that this phenomenon does not indicate that a pole has been lost; clearly this is not possible since the nature of the system, and hence its characteristic equation, remains unchanged. What such a cancellation indicates is that, at the chosen output, the dynamic term associated with the particular pole cannot be observed. For single input/single output systems the distinction between the loss of a pole, through pole–zero cancellation, or its inclusion in the time response, with a magnitude of zero, may seem pedantic as in either case the result is identical. However, that same dynamic term will be affecting other parts of the system, and to ignore it could be a mistake, particularly if the pole happens to be in the right half s-plane. The following example should clarify these points.

Example 4.5

Figure 4.6 shows an idealized lumped parameter model of a machine having a rotating, out-of-balance mass. It has been assumed, not unreasonably, that the imbalance is generating a sinusoidal forcing function which is causing the machine to vibrate, something which often occurs when rotating or reciprocating machinery is sited on a suspended floor. A vibration absorber, consisting of a spring with stiffness k_2 and a mass m_2, has been attached to the machine in order to minimize

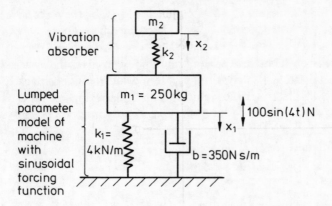

Figure 4.6 A sinusoidally forced machine with a vibration absorber

the amplitude of the displacement at the selected output x_1. The problem is to find the steady-state magnitude of the vibration at x_1 both before and after the absorber is attached.

Solution
Without the vibration absorber the equation of motion for the vibrating machine is found to be

$$250\ddot{x}_1 + 350\dot{x}_1 + 4000x_1 = 100 \sin 4t$$

Note that each term in the above expression has units of force. Taking Laplace transforms and ignoring all initial conditions (which is permissible since only the steady-state response is required) gives

$$(250s^2 + 350s + 4000)X_1(s) = \frac{100(4)}{s^2 + 4^2}$$

Dividing through by 250 and solving for $X_1(s)$ produces

$$X_1(s) = \frac{1.6}{(s^2 + 4^2)(s^2 + 1.4s + 16)}$$

The resulting transfer function has no zeros, and four poles at $s = -4j$, $s = +4j$, $s = -0.7 - 3.94j$ and $s = -0.7 + 3.94j$, as shown in the pole-zero map in Fig. 4.7. Since only the amplitude of the steady-state response is required, only the coefficients associated with the poles at $+4j$ and $-4j$ need be found. The other two poles have negative real parts and produce transient responses which become zero in the steady state. The required amplitude is found to be

$$x_1(t) = 35.7 \text{ mm}$$

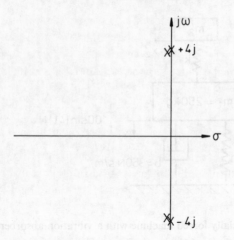

Figure 4.7 Pole–zero map for the sinusoidally forced machine

When the vibration absorber is attached, the equations of motion for the combined system become

$$250\ddot{x}_1 + 350\dot{x}_1 + (4000 + k_2)x_1 - k_2 x_2 = 100 \sin 4t$$

$$m_2 \ddot{x}_2 + k_2 x_2 - 350 x_1 - k_2 x_1 = 0$$

Taking Laplace transforms, assuming zero initial conditions and solving gives the equation for $X_1(s)$ as

$$X_1(s) = \frac{(100)(4)(m_2 s^2 + k_2)}{(s^2 + 4^2)\{[250s^2 + 350s + (4000 + k_2)](m_2 s^2 + k_2) - k_2^2\}}$$

The transfer function now has two zeros and six poles. However, if the absorber mass m_2 and the spring stiffness k_2, both of which are positive, are selected such that $k_2/m_2 = 4^2$, then there is pole–zero cancellation. Since all the remaining poles in the above expression then have negative real parts, the steady-state amplitude of $x(t)$ is zero.

If the analysis were to end at this stage, it might be erroneously assumed that the system is stable in that all the remaining transfer function poles have negative real parts. The system's characteristic equation

$$D(s) = (s^2 + 4^2)\{[250s^2 + 350s + (4000 + k_2)](m_2 s^2 + k_2) - k_2^2\}$$

indicates that this is not so, since there are two poles on the imaginary axis, at $s = -4j$ and $s = +4j$. The influence of these two poles on the system is best demonstrated by finding the displacement of the mass m_2. With zero initial conditions, the transfer function for $X_2(s)$ is

$$X_2(s) = \frac{400k_2}{(s^2 + 4^2)\{[250s^2 + 350s + (4000 + k_2)](m_2 s^2 + k_2) - k_2^2\}}$$

A partial-fraction expansion of $X_2(s)$ results in two complex conjugate poles, at $s = -4j$ and $s = +4j$. On taking the inverse Laplace transform and letting $t \to \infty$ in the resulting time response, $x_2(t)$ performs a continuous steady-state sinusoidal oscillation, the amplitude of which will depend on the values selected for m_2 and k_2.

4.4   Time responses

A mathematical expression for the time response of a specific Laplace transform may be obtained by adding all the terms in the inverse Laplace transform of the partial-fraction expansion. If a range of values of time t are used, say from $t = 0$ to $t = \infty$, then a plot of the time response is obtained. In practice most control systems are designed to be stable (that is, all the poles of the transfer function have negative real parts), so if well-chosen forcing inputs are used the response either settles to some steady-state condition or repeats itself after a finite period of time. The resulting plots are useful design aids in that they give a clear picture of the way in which a system responds, and are often the key to establishing performance criteria.

For design and analysis purposes it is common to use a few standard forcing inputs which are easily manipulated:

(i)   Step input

Figure 4.8 depicts the time response of a step function having magnitude A acting on the system's input u. Mathematically, the step input is defined as

$$u(t) = \begin{cases} 0 & \text{for } t < 0 \\ A & \text{for } t > 0 \end{cases} \tag{4.25}$$

At $t = 0$ the function is undefined, but it is normal practice to assume that $u(t) = A$ when $t = 0$. The Laplace transform of a step input of gain A is A/s. It therefore adds a pole to the system's transfer function, which in turn maps into a point at the origin of the s-plane.

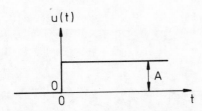

Figure 4.8 A step input of amplitude A

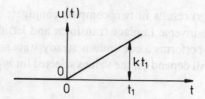

Figure 4.9 A ramp input of gain K

(ii)   Ramp input

The time response of a ramp function is shown in Fig. 4.9. Mathematically, this function is defined as

$$u(t) = \begin{cases} 0 & \text{for } t \leq 0 \\ kt & \text{for } t > 0 \end{cases} \tag{4.26}$$

Its inverse Laplace transform is given by k/s^2, and results in a double pole at the origin of the s-plane. Since the input to the system is unbounded (that is, the input continuously increases with time), the output will also be unbounded and the system's response is said to be unstable. Although ramp inputs are useful for determining the performance of certain systems (e.g. missiles and machine tools), their application to physical plant must be carried out with care.

(iii)   Pulse inputs

There are many kinds and shapes of pulses. Two are particularly useful, one for experimental testing and the other for mathematical convenience. These are the rectangular pulse, which is formed by two successive steps of equal magnitude but opposite sign, shown in Fig. 4.10, and the unit impulse, which is a pulse of unit integral area but zero duration. The Laplace transform of the unit impulse is unity, and consequently it does not affect the system's pole–zero map. This response is often implicitly assumed in order to obtain a time response from the system's transfer function. Without this assumption a stable system would remain undisturbed and the time response solution would be the trivial result that putting nothing into a system makes nothing come out.

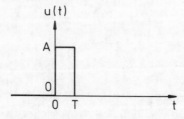

Figure 4.10 A rectangular pulse of amplitude A and duration T

Pulse inputs are particularly useful when dealing with an operating system because there is no prolonged disturbance of the output variable, and also because pulses are readily imposed on the input variable. (Pulse testing is considered in Chapter 9.)

(iv)   Steady-state sinusoidal input

From the point of view of control system design techniques, this is probably the most useful of the forcing inputs. Since the relevant response is the steady-state response, which is observed only after all transient effects have disappeared, the timescale becomes arbitrary. However, it is normal to assume that under steady-state conditions a sine wave

$$u(t) = A \sin \omega t \tag{4.27}$$

is injected into the system (see Fig. 4.11), where A is the amplitude and ω the angular frequency.

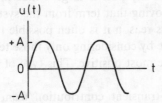

Figure 4.11 A steady-state sinusoidal input

The Laplace transform of this function is

$$U(s) = \frac{A\omega}{s^2 + \omega^2} \tag{4.28}$$

which adds two imaginary poles to the s-plane, one at $s = j\omega$ and the other at $s = -j\omega$. The steady-state sinusoidal or harmonic response of a system and its relationship to the s-plane is considered in Section 4.6.

(v)   Random input

In operating systems, all variables are continually changing. Although for well-controlled systems these changes are small and random, they are the actual forcing inputs. The responses of the operating system and these random inputs may be correlated statistically and used to analyse the system dynamics. To reduce and interpret this data, extensive machine computation is normally required. However, in principle, for operating plants the use of random signals offers a means of dynamic analysis which eliminates the need for significant process disturbance.

As indicated, most of the standard forcing inputs have well-defined Laplace transforms, so a pole–zero map contains all the information necessary for obtaining the associated time domain response. To assist in determining this response, a computer program based on the techniques presented in Section 4.2 is given as Program 2 in Chapter 11. This program can deal with both real and complex poles and zeros, but not with multiple poles.

4.5   Dominant pole analysis

Most stable linear systems could be represented by a transfer function having distinct poles and zeros which are all contained in the left half *s*-plane. The time domain response of such a system may be found from its pole–zero map. This is obtained by performing a partial-fraction expansion, and then taking and summing the inverse Laplace transform of each term. However, each term has a dynamic part consisting of a negative exponential in time and an amplitude which is a function of the relative position of the poles and zeros, both of which influence the response. If the relative amplitude associated with a given term is small, or if the magnitude of the negative real part of any pole is large – thus producing a rapid exponential decay, the effect of removing that term from the system's time response is likely to be negligible. For this reason it is often possible to produce a close approximation to the time response by considering only those terms in the partial-fraction expansion whose poles are most positive. This type of analysis is called a dominant pole analysis.

Dominant pole analysis requires the transient contribution from the non-dominant system poles to be small. This in turn implies either of the following:

1. The non-dominant poles are well to the left of the dominant pole(s), so that the corresponding transients die away rapidly.
2. Any pole near the dominant pole(s) is close to a zero, so that the magnitude of its transient response will be very small.

In general these conditions may be verified by visual inspection of the pole–zero map. Figure 4.12 shows two typical pole–zero maps, one having a dominant single

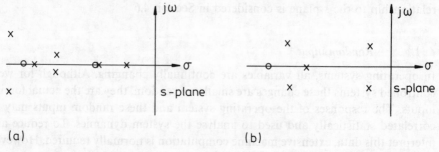

Figure 4.12 Typical pole–zero maps: (a) dominant pole; (b) dominant complex poles

pole and the other a dominant pair of complex conjugate poles. The corresponding time domain responses of systems having similar pole–zero maps appear predominantly first or second order.

For design purposes, it is usually desirable for a system to have an under-damped response. This minimizes the effect of inbuilt non-linearities such as backlash, dead zone and Coulomb friction. An under-damped response requires the existence of dominant complex poles, the examination of which helps in determining performance criteria in the s-plane.

Example 4.6

Find the step response of a system having the transfer function

$$F(s) = \frac{2810.1(s+4)}{(s+3.8)(s+6)(s^2+2s+17)(s^2+10s+29)}$$

and compare it with the response obtained from a dominant pole analysis.

Solution

The pole–zero map for this system, shown in Fig. 4.13, indicates that the dominant poles are complex poles at $s = -1 + 4j$ and $s = -1 - 4j$. Using Program 2 in

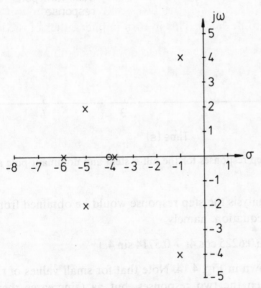

Figure 4.13 Pole–zero map for the system

$$F(s) = \frac{2810.1(s+4)}{(s+3.8)(s+6)(s^2+2s+17)(s^2+10s+29)}$$

Chapter 11, the inverse Laplace transform for the full system is found to be

$$f(t) = 1 + e^{-t}(0.6225 \cos 4t - 0.5714 \sin 4t) - 0.5184 e^{-3.8t}$$
$$+ e^{-5t}(0.9728 \cos 2t - 3.3296 \sin 2t) - 2.0796 e^{-6t}$$

and its step response is shown in Fig. 4.14.

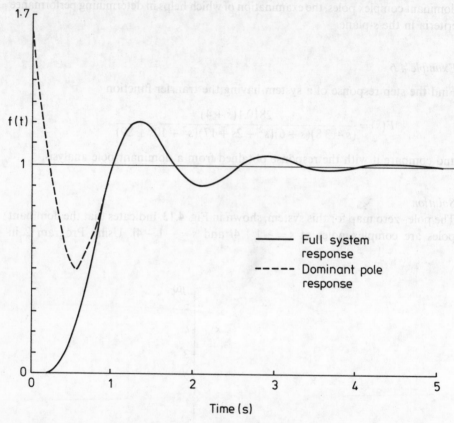

Figure 4.14 Step responses for the full system and dominant pole analysis

In a dominant pole analysis the step response would be obtained from the first two terms in the above equation, namely

$$f(t) = 1 + e^{-t}(0.6225 \cos 4t - 0.5714 \sin 4t)$$

This response is also shown in Fig. 4.14. Note that for small values of t there is a large discrepancy between the two responses, but as t increases they become indistinguishable. Theoretically they would be identical only when t becomes infinite, but for all practical purposes they would, in this case, be considered to have become identical after about 1 time unit.

4.6   Step response dynamics

The dynamic response of a system to a forcing input is made up of a transient response and a steady-state response. For stable systems the transient part of the response is that part which dies away with increasing time. It arises from the negative exponential terms corresponding to the stable system's left half s-plane poles. The steady-state response is that part of the response caused by the forcing input. When the input is a unit step, the steady-state output of a stable system is of constant amplitude. This corresponds to the gain of the coefficient associated with the pole at $s = 0$ and may be found using the final value theorem (see Section 2.5.2).

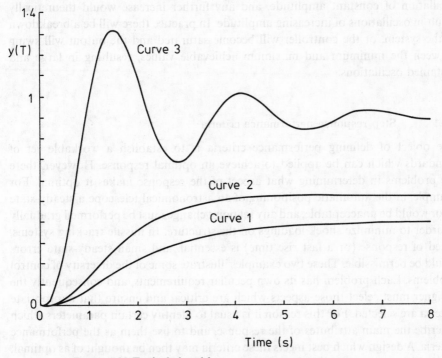

Figure 4.15 Typical closed-loop step responses

Figure 4.15 shows a series of typical closed-loop step response, obtained by increasing the loop gain of a stable open-loop system with a dominant first-order pole. For low values of gain the response is under-damped (curve 1). Also, with low loop gain the steady-state output of the closed-loop system is much smaller than the amplitude of the demand input. In the steady-state, the difference between demand input and measured output is given by

$$e_{ss}(t) = \frac{1}{1 + KG(0)} \tag{4.29}$$

where $KG(0)$ is the gain of the open-loop system. This equation indicates that, as the gain increases, the error decreases and consequently the output signal becomes closer to the demand input. Curve 2 represents a critically damped system in which the loop gain has been increased until a pair of dominant real poles appears. Critical damping is that condition in which the response is the fastest possible without overshooting (that is, without the response exceeding its final steady-state value). An under-damped response (curve 3) is obtained by increasing the loop gain beyond that required to give critical damping. As shown, the response becomes faster, indicated by a more rapid rate of rise, and the steady-state error is reduced, but at the cost of the system becoming more oscillatory. With further increase in gain a point is reached where the response becomes a sustained oscillation of constant amplitude, and any further increase would theoretically result in oscillations of increasing amplitude. In practice there will be a breakdown of the system, or the controller will become saturated and the output will swing between the minimum and maximum achievable values, resulting in large and sustained oscillations.

4.6.1   Step response performance criteria

The object of defining performance criteria is to establish a workable set of standards which can be applied to achieve an optimal response. However, there are problems in determining what aspect of the response makes it optimal. For example, in the automatic positioning of an astronomical telescope a steady-state error would be unacceptable, and any set point change must be performed gradually in order to minimize shock loadings on the structure. In missile tracking systems, speed of response (or a fast rise time) is essential, and small steady-state errors would be permissible. These two examples illustrate some of the diversity of control problems. Each problem has its own peculiar requirements, and consequently the designer must select those aspects which are critical and ensure that appropriate criteria are selected. For this reason it is usual to identify certain parameters which describe the main attributes of the response, and to use them as the performance criteria. A design which best meets these criteria may then be thought of as optimal, although in reality the design will only be as good as the selected criteria.

With step response performance criteria, two approaches are possible. One specifies limits on certain features of the response and the other, which uses integral performance indices, tries to quantify in a single positive measure the whole of the response. Both have advantages and disadvantages, and both have been widely used. Integral performance indices tend to be more appropriate in simulation work and state space analysis, topics not covered in this book.

The step response shown in Fig. 4.16 indicates a number of commonly used performance criteria:

1. *Steady-state error.* The difference between the demand input $r(t)$ and the

steady-state output y_{ss}. Given a stable model of the system, this would normally be found using the final value theorem.

2. *Rise time, t_r.* The shortest time required for the response to achieve some specified percentage of its final value, y_{ss}. Typically, the 100 per cent rise time is used, as shown in Fig. 4.16, although the rise time is sometimes assumed to be the shortest time from, say, 10 per cent of the initial value of the response to when it first achieves, say, 90 per cent of its final value.

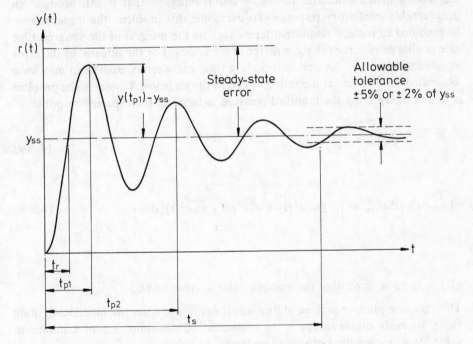

Figure 4.16 Commonly used step response criteria

3. *Peak overshoot, $y(t_{p1}) - y_{ss}$.* Sometimes referred to as the initial or maximum overshoot, peak overshoot is the amplitude of the first peak.
4. *Peak time, t_{p1}.* The time from the initiation of the response to peak overshoot.
5. *Subsidence ratio.* In a decaying oscillation this is the ratio of the amplitudes of successive cycles. A subsidence ratio of 4:1 or 3:1 and a peak overshoot of 30 per cent would provide a practical optimal response for many process control systems.
6. *Settling time, t_s.* The time taken for the response to reach and remain within some specified range of its final value. An allowable tolerance of between 2 and 5 per cent is usual.
7. *Number of oscillations to settling time.* This is self-explanatory.

The formulation of these specifications in the s-plane is considered in Section 4.6.2.

There are many integral performance indices which, if minimized, will optimize the response in some sense. Some of the more common ones are:

(i)   The integral of the squared error (ISE)

The ISE is one of the more popular measures, as it lends itself most readily to mathematical manipulation. Its major disadvantage is that it can produce an unacceptably oscillatory response. To overcome this problem, the measure may be modified to include additional terms such as the integral of the squared time rate of change of error (or the squared error velocity) or the integral of the error squared acceleration. As well as including these extra terms, each term may have associated with it a scalar weighting, w say, as shown below. However, the problem is one of interpreting the modified measure in terms of the expected response.

$$\text{ISE} = \int_0^\infty e^2(t)\, dt \tag{4.30a}$$

$$\text{ISE}_{mod} = \int_0^\infty (w_1 e^2(t) + w_2 e^2(t) + w_3 e^2(t))\, dt \tag{4.30b}$$

(ii)   The integral of the absolute value of error (IAE)

This measure places equal weighting on all deviations from the final steady-state value. Its main disadvantage is the mathematical determination of a minimum value for all except the simplest of systems.

$$\text{IAE} = \int_0^\infty |e(t)|\, dt \tag{4.31}$$

(iii)   The integral of time by absolute error (ITAE)

For the engineer this is probably the most acceptable of the indices since it most nearly matches intuitive expectations. Following a demand step, it is inevitable that there will be a large error in the response, and to penalize this would place an artificial bias on the measure. However, at some later time a smaller error should be heavily penalized. In general, for a system of any order the transfer function minimizing the ITAE will have an acceptable form of transient response.

Again, the main disadvantage is in the analytical determination of a minimum value of the index.

$$ITAE = \int_0^\infty t|e(t)| \, dt \tag{4.32}$$

4.6.2   Performance specifications in the s-plane

In a development of some simple rules for s-plane performance specifications, only the dominant poles of a system need be considered. As has been shown in this chapter, desirable responses tend to be ones which are under-damped, and this occurs if the system has a dominant pair of complex conjugate poles. The corresponding time domain step response may therefore be approximated by a second-order system (see Example 4.1) to produce an equation of the form

$$y(t) \approx Ay_{ss}[1 - 2Be^{-\zeta\omega_n t} \cos(\omega_d t - \theta)] \tag{4.33}$$

where A is the amplitude of the demand step input and y_{ss} is the steady-state output, the system's static gain.

For a second-order system,

$$2B = 1/\sqrt{(1 - \zeta^2)} \quad \text{and} \quad \sin\theta = \zeta$$

From this approximate equation it is possible to estimate the positions in which the dominant poles must be in order to meet a particular performance specification.

Consider each of the specifications from Section 4.6.1:

1. *Steady state error.* From Equation (4.33), the steady-state error is found to be

$$e_{ss}(t) = A - Ay_{ss} \tag{4.34}$$

2. *100 per cent rise time.* This will occur the first time $\cos(\omega_d t - \theta)$ becomes equal to zero, that is when

$$\omega_d t - \theta = \tfrac{1}{2}\pi \tag{4.35a}$$

Hence the rise time is

$$t_r = (\tfrac{1}{2}\pi + \theta)/\omega_d \tag{4.35b}$$

The rise time is a function of the damped frequency ω_d.

3. *Peak overshoot.* This is obtained by substituting the expression for the peak time into the response $y(t)$:

$$y(t_p) = Ay_{ss}(1 + e^{-\zeta\omega_n t_p}) \tag{4.36}$$

and is seen to be a function of the real part of the complex poles. Equation (4.36) gives the maximum or peak value of the response. The percentage maximum overshoot is given by

$$\left(\frac{y(t_p)}{Ay_{ss}} - 1\right) \times 100\% \tag{4.37}$$

4. *Peak time.* The peak time may be found by differentiating the respone $y(t)$ and setting the derivative dy/dt equal to zero. This will be found to occur when $\sin(\omega_d t)$ first becomes zero, hence the peak time is

$$t_p = \pi/\omega_d \tag{4.38}$$

which is again a function of the damped frequency.

5. *Subsidence ratio.* Assume that the subsidence ratio is $R:1$, such that the first peak, $y(t_{p1}) - y_{ss}$, is R times greater than the second peak, $y(t_{p2}) - y_{ss}$. It may be shown that for any two adjacent peaks the subsidence ratio of a second order system will be the same. From Equation (4.38), the first peak occurs when $t_{p1} = \pi/\omega_d$, and the second when $t_{p2} = 3\pi/\omega_d$. Hence

$$y(t_{p1}) = Ry(t_{p2}) \tag{4.39}$$

Substituting for $y(t)$ and taking natural logarithms yields

$$\zeta/\sqrt{(1 - \zeta^2)} = \log_e(R)/(2\pi) \tag{4.40}$$

which indicates that the subsidence ratio is a function of the damping ratio ζ alone.

6. *Setting time.* This is obtained by considering the decay of the response envelope. For a 2 per cent settling time,

$$0.02 = e^{-\zeta\omega_n t_s} \tag{4.41a}$$

or, taking natural logarithms,

$$t_s = 4/\zeta\omega_n \tag{4.41b}$$

For a 5 per cent settling time the expression becomes

$$t_s = 3/\zeta\omega_n \tag{4.42}$$

Settling time is a function of the real part of the dominant poles. The reciprocal of $\zeta\omega_n$ has units of time, and is referred to as the time constant of the equivalent second-order system. Equation (4.42) may also be used to define dominance. Non-dominant poles decay within their own response envelope, and consequently, if there were a real pole at $-\sigma$, the transient due to this pole would have the form $Ke^{-\sigma t}$. This term would decay to within 5 per cent of its initial value when $t = 3/|\sigma|$. If the rise time t_r is greater than or equal to this value, then the exponential term will have only a small effect on the performance measures. A similar argument holds for complex poles. Therefore, in general, a dominant pole analysis requires the non-dominant poles to be close to a zero, or to have a negative real part of magnitude greater than or equal to $3/t_r$.

7. *Number of oscillations to settling time.* Give the periodic time of the damped oscillations and the settling time of the system, then

$$\text{Number of oscillations} = \frac{\text{Settling time}}{\text{Periodic time}} \tag{4.43}$$

and is a function of the damping ratio.

Using dominant pole analysis, the above performance specifications have all been shown to be related to either the damped frequency, the damping ratio or the time constant of the dominant poles. In the *s*-plane, lines of constant damped frequency ω_d are lines parallel to the real axis of the *s*-plane. A line of constant damping ratio is a radial line emanating from the origin of the *s*-plane and is at an angle θ to the negative real axis; that is,

$$\zeta = \cos \theta \tag{4.44}$$

Lines of constant time constant, $\zeta\omega_n$-lines, are parallel to the imaginary axis. For completeness, note that lines of constant ω_n describe circles centred on the origin of the *s*-plane. All these lines are shown in Fig. 4.17.

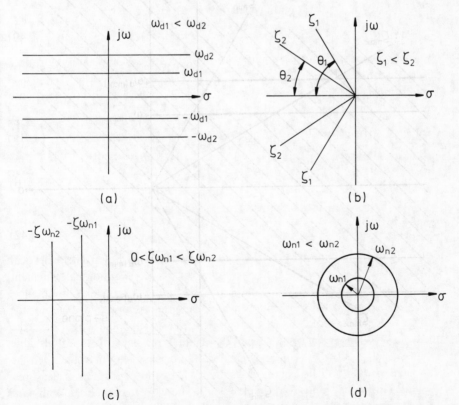

Figure 4.17 Performance lines in the *s*-plane: (a) constant ω_d, (b) constant ζ, (c) constant $\zeta\omega_n$ and (d) constant ω_n

In order to obtain an optimal step response it would be normal to impose limits on the value of ω_d; this defines bands on the rise time, the peak time and the damped frequency. A minimum and a maximum value of the damping ratio ζ specify the subsidence ratio and the number of oscillations to the settling time. A minimum time constant defines the peak overshoot and, probably more importantly, the settling time.

The design of a system using s-plane performance criteria typically reduces to the problem of ensuring that the system's dominant poles are not within the shaded area in Fig. 4.18. Again, in terms of the design problem it is desirable to keep the specifications to a minimum. Over-specification may create unnecessary difficulties, and seldom leads to any significant improvement in the transient response. Removal of the constraints on the maximum value of ζ, and/or on the minimum value of ω_d, would ease the design problem considerably (see Fig. 4.18).

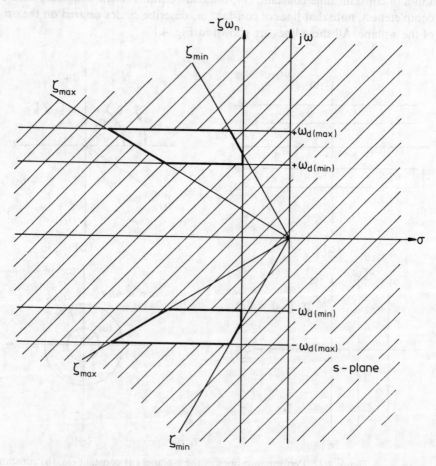

Figure 4.18 Dominant pole positions and s-plane performance criteria

Before leaving this section, it will be useful to consider the effect on the time response of shifting the dominant poles along the s-plane performance lines. Figure 4.19 illustrates this shift for two second-order systems. When the systems have the same $\zeta\omega_n$ value, the responses are contained within the same exponential envelopes, but the frequency of oscillation is greater for the system having the larger ω_d value (see Fig. 4.19(a)). The same ζ value indicates that the damping in the two systems is the same (see Fig. 4.19(b)), but the speed of response will be much faster for the system whose poles have the more negative real part. Figure 4.19(c) shows two systems with the same value of ω_d; the more negative poles produce the faster and more heavily damped response.

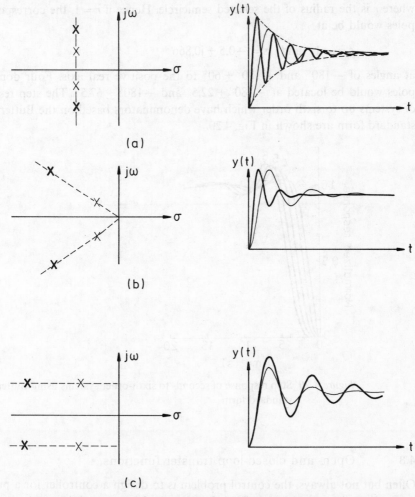

Figure 4.19 A system's pole positions and its step response for (a) the same $\zeta\omega_n$ value, (b) the same ζ value and (c) the same ω_d value

4.7   Mixed dominance

When there are more than two dominant poles in a system, it is said to have mixed dominance. For such a system the performance criteria based on the step response of a second-order system tend to break down, and alternative criteria must be sought. Probably the simplest of the available techniques is to locate (or attempt to locate) the poles equally spaced on a semicircle contained within the left half s-plane and centred on the origin. Such a distribution of poles is referred to as the Butterworth standard form. For example, if there were three dominant poles their contribution to the characteristic equation would be

$$D(s) = s^3 + 2rs^2 + 2r^2s + r^3 \tag{4.45}$$

where r is the radius of the selected semicircle. Hence if $r = 1$, the corresponding poles would be at

$$s = -1 \quad \text{and} \quad s = -0.5 \pm j0.866$$

at angles of $-180°$ and $-180° \pm 60°$ to the positive real axis. Four dominant poles would be located at $-180° \pm 22.5°$ and $-180° \pm 67.5°$. The step response of systems up to sixth order which have denominators based on the Butterworth standard form are shown in Fig. 4.20.

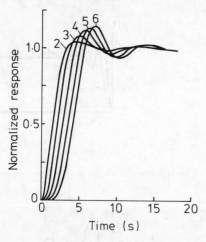

Figure 4.20 Step response of second- to sixth-order systems using Butterworth standard form

4.8   Open- and closed-loop transfer functions

Often but not always, the control problem is to design a controller for a piece of equipment whose structure is fixed. Since control is required, it may be assumed that the open-loop system does not meet the overall performance objectives. Some sort of control strategy must then be adopted which will modify the response of the system. Typically this will involve some form of feedback loop (see Section 3.5.2).

Consider an open-loop system described by the transfer function

$$G(s) = \frac{N(s)}{D(s)} \tag{4.46}$$

Assume that the system is to be controlled by means of a feedback loop and a simple proportional controller of gain K, as shown in Fig. 4.21. The closed-loop transfer function between the demand input $R(s)$ and the measured output $Y(s)$ is given by

$$\frac{Y(s)}{R(s)} = \frac{KN(s)}{D(s) + KN(s)} \tag{4.47}$$

Examination of Equations (4.46) and (4.47) indicates that the introduction of a feedback loop does not alter the position of the zeros. However, the closed-loop pole positions do change, becoming a function of the system's open-loop poles and zeros. This shows that feedback may be used to adjust a system's stability and dynamics. These concepts are best illustrated by means of an example.

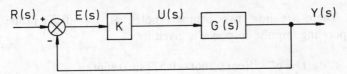

Figure 4.21 A simple closed-loop control system

Example 4.7

A system has an open-loop transfer function given by

$$G(s) = \frac{s}{s^2 + 1}$$

and is to be controlled by a simple proportional controller of the kind shown in Fig. 4.21. Determine the closed-loop pole positions, and hence the response of the system, for various values of gain K.

Solution

The closed-loop transfer function is given by

$$\frac{Y(s)}{R(s)} = \frac{Ks}{s^2 + Ks + 1}$$

Notice that both the open- and the closed-loop system have a zero at $s = 0$. The closed-loop poles are at

$$s = \tfrac{1}{2}K \pm \sqrt{(\tfrac{1}{4}K^2 - 1)}$$

For values of K between 0 and 2 the poles are complex, and for K greater than 2 the poles are real. At $K = 2$ there is a double pole at $s = -1$. For any negative value of gain the response is unstable.

Consider the following cases:

1. $K = -1$. There are two complex poles, at $s = +0.5 \pm j0.866$, and the time response for a unit impulse is given by

$$y(t) = e^{+0.5t}(\cos 0.866t - 0.577 \sin 0.866t)$$

The positive real part of the poles produces a positive exponential, and consequently the system is unstable.

2. $K = 0$. This essentially breaks the loop (see Fig. 4.21) since at $K = 0$ a change in $E(s)$ has no effect on $U(s)$. However, an impulse response may be obtained for the open-loop system and is given by

$$y(t) = \cos t$$

3. $K = 1$. There are now two complex poles, at $s = -0.5 \pm j0.866$, and the corresponding impulse response is given by

$$y(t) = e^{-0.5t}(\cos 0.866t - 0.577 \sin 0.866t)$$

The response is now stable, as indicated by the negative exponential.

4. $K = 2$. There is a double pole at $s = -1$ and the impulse response is given by

$$y(t) = e^{-t} - te^{-t}$$

5. $K = 4$. This gives poles at $s = -3.732$ and $s = -0.268$, and the closed-loop impulse response is

$$y(t) = -0.0774e^{-0.268t} + 1.0774e^{-3.732t}$$

Note that the effect of the dominant pole is reduced because of its proximity to the closed-loop zero.

Since a simple feedback loop affects only the closed-loop pole positions, it is useful to plot the pole trajectories, or loci, for all possible changes in the loop gain K. Figure 4.22 shows the closed-loop pole plot, or root loci plot, for the system considered in Example 4.7. The indicated poles and zeros are those of the open-loop system. The arrow on each locus indicates the direction of increasing loop gain. In general, only positive loop gains (for $K > 0$) need be considered. This topic is discussed in more detail in Chapter 6.

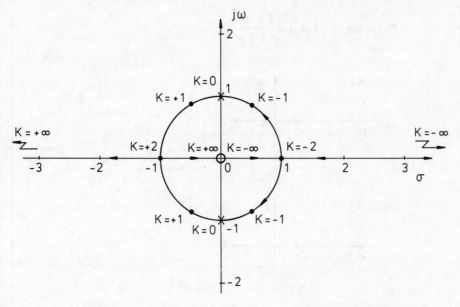

Figure 4.22 Root loci plot for the system considered in Example 4.7

Problems

4.1. A number of pole–zero plots are shown in Fig. 4.23 with the gain K indicated. As usual, K is the multiplier of the pole–zero function. For each plot:
(a) Find the transfer function $F(s)$.
(b) Graphically determine the inverse Laplace transform.
(c) Sketch the general shape of the transient response.

4.2. Evaluate the following function $F(s)$ at the s-plane locations $(0, 0)$, $(0, -4)$ and $(-2, 2)$:

$$F(s) = \frac{6(s+1)}{(s+2)(s+3)}$$

4.3. Use the graphical technique to evaluate the functions $F_1(s)$ and $F_2(s)$, given below, for any positive value of ω on the imaginary s-plane axis. Plot the resulting loci $F_1(j\omega)$ and $F_2(j\omega)$ on Argand diagrams. (Such Argand diagrams are called $F(s)$-planes, and such plots are called polar plots.)

Phase lead $$F_1(s) = \frac{1 + 10s}{1 + s}$$

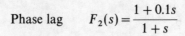

Phase lag $F_2(s) = \dfrac{1 + 0.1s}{1 + s}$

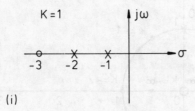

(i)

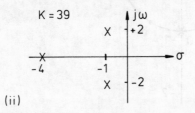

(ii)

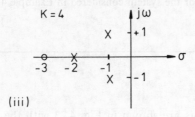

(iii)

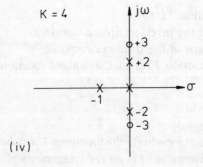

(iv)

Figure 4.23

4.4. From the following time domain performance specifications, establish s-plane performance criteria. In each case use these criteria to sketch the response to a unit step. Assume a second-order system with unity gain.

(a) 100 per cent rise time 0.5 s, peak time 0.7 s. What would be the peak overshoot and steady-state error for this system?
(b) Subsidence ratio 3:1, 5 per cent settling time 9 s. What is the peak overshoot for this system?
(c) Peak overshoot 30 per cent, 5 per cent settling time of 9 s. What is the subsidence ratio?

4.5.   Figure 4.22 at the end of this chapter shows the root loci plot of a system with the open-loop transfer function

$$G(s) = \frac{Ks}{s^2 + 1}$$

For any point s_0 on a locus, evaluate $G(s_0)$ and show that

$$|G(s_0)| = \left| \frac{1}{K} \right|$$

and

$$\arg G(s_0) = \begin{cases} -180° & \text{for } K > 0 \\ 0° & \text{for } K < 0 \end{cases}$$

4.6.   A system has an open-loop transfer function given by

$$G(s) = \frac{25}{s(s + 20)}$$

To control this system two alternative closed-loop designs have been proposed as shown in Fig. 4.24. For both designs determine:

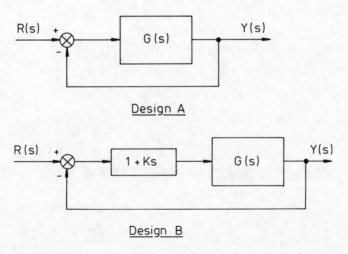

Design A

Design B

Figure 4.24

(a) the rise time,
(b) the peak time,
(c) the 5 per cent settling time and
(d) the percentage overshoot.

State which design has the best overall time domain specifications. (Note that the variable K in Design B should be set so that the closed-loop damping ratio is 0.5.)

CHAPTER 5

STABILITY

5.1   Introduction

From the previous chapters, it is evident that the possible modes of the transient response of a single input/single output system are determined by the nature of the roots of its characteristic equation. In this chapter more formal definitions of stability are given.

5.2   Stability defined

Stability is usually defined in terms of a system's impulse response. In Section 3.3 it was shown that any forcing input may be represented by a series of impulse responses, and therefore a definition based on impulse may be extended to any forcing input.

Many different definitions of stability are in common use, but for most engineering systems only four need be considered. These are the definitions of asymptotic stability, marginal stability, conditional stability and instability.

A linear system is said to be stable if its impulse response produces a bounded steady-state condition, and unstable otherwise. In this sence a sine wave is bounded, but a positive exponential is not. Stable systems are said to be asymptotically stable if the impulse response of the system tends to the original steady-state condition, and marginally stable otherwise. It is assumed that the original steady-state output of the system was some constant, non-dynamic value. Clearly, an engineering system should be asymptotically stable and not merely stable. At this point, it would be useful to have a look at Table 5.1 on p.114, which summarizes the stability of systems with various dominant pole positions.

Conditional stability is considered in detail in Chapters 6 and 7. It describes a condition in which the stability of a closed-loop system repeatedly switches between stable and unstable behaviour with increasing gain.

5.3   Open- and closed-loop systems, poles and zeros

Consider the closed-loop system depicted in Fig. 5.1. If the measurement path $B(s)$ is broken, the system is said to be open-loop. Under these conditions the

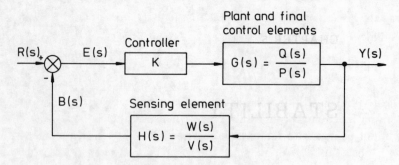

Figure 5.1 Closed-loop SISO system under P control

relationship between $R(s)$ and $B(s)$ is given by

$$B(s) = KG(s)H(s)R(s) = K\frac{Q(s)W(s)}{P(s)V(s)}R(s)$$ (5.1)

where $KG(s)H(s)$ (often abbreviated to $KGH(s)$) is the open-loop transfer function.

In Equation (5.1) the values of s which make the polynomial product $Q(s)W(s)$ zero are called the zeros of the open-loop transfer function, and the values of s which make the product $P(s)V(s)$ zero are called the poles of the open-loop transfer function. The measured time response to some forcing input is obtained by applying inverse Laplace transforms:

$$b(t) = \mathcal{L}^{-1}\left[K\frac{Q(s)W(s)}{P(s)V(s)}R(s)\right]$$ (5.2)

Note that the gain K could be taken outside the square brackets since it affects only the magnitude of the response, not the dynamics. Furthermore, from Chapter 2 it is clear that a partial-fraction expansion would yield terms whose denominators are of the form $s + p_i$, where $-p_i$ is one of the poles of the transfer function. Consequently the time domain response consists of exponential terms of the form $e^{-p_i t}$. Thus the poles define the characteristics, or essential modes, of the response, and accordingly the equation

$$P(s)V(s) = 0$$ (5.3)

is called the characteristic equation. The zeros of this characteristic equation are the poles of the open-loop transfer function. The residue (or gain) associated with each dynamic mode (or exponential) may be found from a Heaviside partial-function expansion. In this context the open-loop zeros will be seen to adjust the various gains associated with each exponential; that is, they affect the system's performance.

The closed-loop transfer function between the forcing input $R(s)$ and plant output $Y(s)$ is given by

$$\frac{Y(s)}{R(s)} = \frac{KG(s)}{1 + KGH(s)}$$ (5.4)

and the corresponding transfer function between the forcing input and measured output by

$$\frac{B(s)}{R(s)} = \frac{KGH(s)}{1 + KGH(s)} \qquad (5.5)$$

Clearly, the closed-loop characteristic function is always one plus the open-loop transfer function. This must always be true, since the characteristics of the system will not change with the observation position. The closed-loop zeros do change with observation position, since the response will change from one element to the next.

For the closed-loop transfer function given by Equations (5.4) and (5.5), the characteristic function is

$$1 + KGH(s) = 1 + K\frac{Q(s)W(s)}{P(s)V(s)} = 0 \qquad (5.6a)$$

or

$$F(s) = P(s)V(s) + KQ(s)W(s) = 0 \qquad (5.6b)$$

Hence the zeros of $F(s)$, which give the system's closed-loop poles, are functions of the open-loop poles and zeros and of the open-loop controller gain K.

5.4   System stability

If the open-loop system associated with the closed-loop system shown in Fig. 5.1 is asymptotically stable, then the roots of the polynomial $P(s)V(s) = 0$ must have negative real parts. Open-loop stability, although desirable, is not essential for closed-loop stability. For closed-loop stability the controller gain K must be set such that all the roots of the closed-loop characteristic equation

$$P(s)V(s) + KQ(s)W(s) = 0 \qquad (5.7)$$

have negative real roots. Clearly the closed-loop system could be stable, and the open-loop system unstable.

The correspondence between system stability and the nature of the characteristic roots may be interpreted graphically. The Laplace operator s is a complex variable, and the roots of a polynomial in s will have real and complex parts. Typically $s = \sigma + j\omega$, and therefore any value (or root) of s could be plotted on an Argand diagram, the so called s-plane plot (see Chapter 4). If all the poles of a system are contained in the left half s-plane, then it is asymptotically stable (see Fig. 5.2).

If any of the poles are in the right half s-plane, the system is unstable. Poles on the imaginary axis are on the boundary between stability and instability; only systems having dominant unrepeated poles on the imaginary axis are consistent with the definition of stability, and are referred to as being marginally stable.

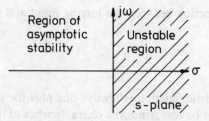

Figure 5.2 Region of asymptotic stability in the *s*-plane

Table 5.1 Impulse responses and *s*-plane pole plots

	Pole locations in s-plane	Impulse response	Comment
1			Asymptotically stable
2			Marginally stable
3			Unstable
4			Unstable
5			Asymptotically stable
6			Marginally stable
7			Unstable
8			Unstable

Table 5.1 gives a number of *s*-plane pole plots and the corresponding impulse responses. In this table multiple poles (two or more poles occupying the same location) are indicated by a number above the pole. For example, system No. 8

has two poles occupying the location $(0,0)$ in the s-plane. Also (see section 4.5), this table indicates the response of higher-order systems, where the poles indicated are the dominant poles.

5.5   Direct stability tests

If the root locations of a system's characteristic equation are known, then it is a simple matter to establish stability by visually checking that all the roots have negative real parts. Also, if the zero locations between the forcing input and the measured output are known, the response may be determined. Indeed, it is normal practice, when finding a system's time response using inverse Laplace transform techniques, to start by factoring the system's denominator polynomial and checking for asymptotic stability. Such a check is called a direct stability test. Program 3 of Chapter 11 may be used to find the roots of a system's characteristic equation, and can therefore be used for direct stability checks. However, root solving routines – particularly those designed for microcomputers – can be slow and may even fail to converge. For these reasons it is desirable to test for stability before attempting to determine root locations.

5.6   A necessary condition for asymptotic stability

In general a characteristic equation may be written in the form

$$F(s) = \sum_{i=0}^{n} a_i s^{n-i} = 0 \tag{5.8}$$

or

$$F(s) = a_0 s^n + a_1 s^{n-1} + \cdots + a_{n-1} s + a_n = 0 \tag{5.9}$$

If $F(s)$ has real coefficients, then its complex roots will occur in conjugate pairs. Let $F(s)$ have p real roots, $-\alpha_1, -\alpha_2, \ldots, -\alpha_p$, and q pairs of complex roots, $-\beta_1 \pm j\phi_1, -\beta_2 \pm j\phi_2, \ldots, -\beta_q \pm j\phi_q$, so that $n = 2q + p$. In factored form $F(s)$ may be written as

$$F(s) = a_0(s + \alpha_1) \cdots (s + \alpha_p)(s^2 + 2\beta_1 s + \beta_1^2 + \phi_1^2)$$
$$\cdots (s^2 + 2\beta_p s + \beta_q^2 + \phi_q^2) \tag{5.10}$$

All the αs and βs will be positive if the system is stable. By expansion, it is found that all the as must exist, and must all be of the same sign. This is a necessary condition for stability, but it may not be sufficient, for if one of the as is negative or zero the system is unstable (if $a_n = 0$ the system is at best marginally stable).

Example 5.1

Comment on the stability of the following system's characteristic equations:

(a) $F(s) = s^5 + 5s^4 + 3s^3 - 2s^2 + s + 8 = 0$

(b) $F(s) = s^5 + 5s^4 + 3s^3 + 2s^2 + 8 = 0$

(c) $F(s) = s^5 + 5s^4 + 3s^3 + 2s^2 + s = 0$

(d) $F(s) = -s^5 - 5s^4 - 3s^3 - 2s^2 - s - 8 = 0$

Solution

System (a) is unstable since the coefficient associated with s^2 is negative and therefore it does not satisfy the necessary condition for stability.

The s term is missing in system (b), so the system is unstable since it does not satisfy the necessary condition for stability.

In system (c) there is a root at $s = 0$, which means that the system is at best marginally stable. The remainder polynomial does satisfy the necessary condition for stability, and therefore Routh's criterion should be used to test for sufficiency (see Section 5.7). Routh's criterion would indicate that all the remaining roots are in the left half s-plane.

System (d) satisfies the necessary condition for stability. The system may or may not be stable, and a further test is required. Routh's criterion will show that the system is unstable.

5.7   The Routh stability criterion

The Routh stability criterion provides a quick and easy method of establishing a system's stability. It involves calculating the number (not the location) of characteristic roots within the unstable right half s-plane. The number of roots in the stable left half s-plane and the number of roots on the imaginary axis may also be found. Apart from the stability checks, usually carried out before determining root locations, the method may also be used to establish the limiting values for a variable parameter beyond which a system would become unstable. This section does not attempt to prove Routh's stability criterion, but just gives the basic results.

The characteristic equation of the system whose stability is to be tested must be expressed in the form

$$F(s) = \sum_{i=0}^{n} a_i s^{n-i} = 0 \tag{5.11}$$

so that Routh's criterion may be applied. For this polynomial the Routh array is

Row						
0	s^n	a_0	a_2	a_4	a_6	$\cdots$
1	s^{n-1}	a_1	a_3	a_5	a_7	$\cdots$
2	s^{n-2}	b_1	b_2	b_3	b_4	$\cdots$
3	s^{n-3}	c_1	c_2	c_3	c_4	$\cdots$
$\vdots$		$\vdots$	$\vdots$	$\vdots$	$\vdots$	
$n-1$	s^1	y_1	y_2			
n	s^0	z_1				

$$\tag{5.12}$$

where each element in the third row onwards is obtained from elements in the previous two rows in the following manner:

$$b_1 = \frac{a_1 a_2 - a_0 a_3}{a_1}, \qquad b_2 = \frac{a_1 a_4 - a_0 a_5}{a_1}, \dots \qquad (5.13a)$$

$$c_1 = \frac{b_1 a_3 - a_1 b_2}{b_1}, \qquad c_2 = \frac{b_1 a_5 - a_1 b_3}{b_1}, \dots \qquad (5.13b)$$

$$\vdots$$

The number of sign changes in the first column $(a_0, a_1, b_1, c_1, \dots, y_1, z_1)$ of the Routh array is equal to the number of roots of $F(s)$ with positive real parts.

The criterion and the array are named after Edward John Routh (1831–1907), in whose 1877 paper the array first appeared.

Example 5.2

Use the Routh array to test the stability of a system having the characteristic equation

$$F(s) = s^4 + s^3 + s^2 + 2s + 3 = 0$$

Solution
The system satisfies the necessary conditions for stability, so the Routh array (Equation (5.12)) may be used to test for sufficiency:

Row				
0	s^4	1	1	3
1	s^3	1	2	
2	s^3	−1	3	
3	s	5	0	
4	s^0	3		

The first column of the array has two sign changes, so the system is unstable and has two poles with positive real parts.

Program 3 in Chapter 11 may be used to show that the roots of $F(s)$ are at $-1.074 \pm j0.706$ and $+0.574 \pm j1.219$.

5.7.1   The Routh array: special cases

A zero appearing in the first element of any of the n rows indicates instability or, at best, marginal stability. The array could not be continued by the normal method of construction, since this would involve division by zero. If the root distribution is required an alternative technique must be adopted, and this gives rise to the so-called special cases.

(i)   *The first column term of any row vanishes, but some of the remaining terms in the row are not all zero*

There are several methods available for dealing with this particular case; some are better suited to hand calculation, and some to machine computation. The method given here is best used with hand calculation, and if applied with care will always yield the correct result.

In this method the first-column zero is replaced with an arbitrary small number, δ say, and the array is continued in the normal way. The limit as $\delta \to 0$ is then found, and the first column of the array is checked for sign changes. Again, the number of sign changes equals the number of roots with positive real parts.

Example 5.3

Test the stability of a system having the characteristic equation

$$F(s) = s^5 + 2s^4 + 2s^3 + 4s^2 + s + 1 = 0$$

Solution
For this polynomial the Routh array is

Row				
0	s^5	1	2	1
1	s^4	2	4	1
2	s^3	0	0.5	

The zero is replaced with δ and the array continued. After taking the limit as $\delta \to 0$, the final array, from row 2, becomes

2	s^3	δ	0.5
3	s^2	$-1/\delta$	1
4	s	0.5	0
5	s^0	1	

Since there are two sign changes in the first column, the system is unstable with two positive roots. The root-solving routine, Program 3 in Chapter 11, indicates that the roots of $F(s)$ are $-0.090 \pm j0.533$, $+0.069 \pm j1.274$ and -1.957.

(ii)   *All the coefficients of a row become zero*

This indicates the presence of a divisor polynomial $N(s)$ whose roots are all symmetrically located about the origin of the real and imaginary axes.

Assume that, in the general Routh array of Section 5.7, row 4 is found to be an all-zero row. The coefficients of the required divisor polynomial are obtained from the previous row, in this case row 3, to give

$$N(s) = \sum_{i=0}^{(n-3)/2} c_{i+1} s^{(n-3)-2i} = 0 \qquad (5.14)$$

Since the roots of $N(s)$ are symmetrically located around the real and imaginary axes, they will be of the form

$$s = \pm\sigma \quad \text{or} \quad s = \pm j\beta \quad \text{or} \quad s = -\alpha \pm j\beta \quad \text{and} \quad s = +\alpha \pm j\beta$$

Clearly $N(s)$ will always be of even order, and consequently the all-zero row will always be associated with an odd power of s.

In order to complete the array, $N(s)$ is differentiated with respect to s and the coefficients of $N(s)$ substituted into what was the all-zero row. Using these new coefficients, the array is completed in the normal way.

The array may now be interpreted as follows. As far as the all-zero row, the number of sign changes in the first column indicates the number of roots of the remainder polynomial with positive real parts. From what was the row before the all-zero row, each change in sign in the first column of the array indicates the number of roots of the divisor polynomial with positive real parts. However, as the roots are symmetrical, any sign changes associated with the divisor polynomial $N(s)$ will indicate the number of roots in the right half s-plane, and also the number in the left half s-plane. Roots of the divisor polynomial which are not accounted for in this way must lie on the imaginary axis.

Example 5.4

A closed-loop control system with unity negative feedback has an open-loop transfer function given by

$$KGH(s) = \frac{K}{s(s+1)(s^2+s+1)}$$

For what value of K will the system be marginally stable, and what is the corresponding frequency of oscillation?

Solution

For this system the closed-loop characteristic equation is

$$s^4 + 2s^3 + 2s^2 + s + K = 0$$

and the corresponding Routh array is

Row				
0	s^4	1	2	K
1	s^3	2	1	
2	s^2	$\frac{3}{2}$	K	
3	s	$(\frac{3}{2}-2K)/\frac{3}{2}$		

For $K = \frac{3}{4}$, row 3 becomes an all-zero row and the divisor polynomial of row 2 is

$$\tfrac{3}{2}s^2 + \tfrac{3}{4} = 0, \text{ or}$$

$$N(s) = 2s^2 + 1 = 0$$

By dividing $N(s)$ into $F(s)$ it is found that $N(s)$ is the required divisor polynomial, and that

$$F(s) = (2s^2 + 1)(\tfrac{1}{2}s^2 + s + \tfrac{3}{4})$$

In order to complete the array when $K = \frac{3}{4}$, $N(s)$ is differentiated with respect to s and the coefficients of $N'(s)$ are used to replace the zero coefficients of row 3. Now $N'(s) = 4$, and the completed array is

Row				
0	s^4	1	2	$\frac{3}{4}$
1	s^3	2	1	
2	s^2	$\frac{3}{2}$	$\frac{3}{4}$	
3	s	4		
4	s^0	$\frac{3}{4}$		

Since there are no sign changes up to and including row 2, there are no roots of the remainder polynomial in the right half s-plane. Since there are no sign changes from row 2, all roots of the divisor polynomial must lie on the imaginary axis.

Hence when $K = \frac{3}{4}$ the system will have a pair of complex conjugate roots on the imaginary axis, and will therefore be marginally stable. To find the location of these roots, and hence the frequency of oscillation s is set equal to $j\omega$ and the value of ω found from $N(j\omega)$:

$$N(j\omega) = -2\omega^2 + 1 = 0$$

hence

$$\omega = 1/\sqrt{2} \text{ rad per unit time}$$

which is the required frequency of oscillation.

5.8   Stability programs

The material in this chapter may be covered using two types of stability program; those based on direct stability testing through root solving, and those based on indirect testing based either on the Routh array or Hurwitz inequalities (not considered here). Program 3, Chapter 11 is a root solving routine which may be used for direct stability checks, and Program 4 generates a Routh array for indirect stability checks.

5.9   Routh stability and s-plane performance criteria

Often it is not sufficient to state merely that a system is stable, and it is desirable to give as well some degree, or measure, of stability. In this section the concepts of absolute and relative stability are introduced. Absolute stability indicates whether a system meets a minimum time constant specification; relative stability indicates whether it meets a minimum damping requirement. The other main s-plane performance criterion (see Section 4.7.2) is the maximum damped frequency, the ω_d lines, and it is shown below that the Routh array may be used to test whether a response has frequency components less than some specified maximum ω_d value.

5.9.1   Absolute stability

In Chapter 4 it was shown that the minimum time constant performance specification could be drawn as a line parallel to, but to the left of, the imaginary axis in the s-plane. If all the system's poles are to the left of this performance line, then the slowest rate of decay for any transient is established. The system's absolute stability may be defined as the distance δ of this performance line from the imaginary axis.

In order to use Routh's criterion to test a system's absolute stability, it is necessary to consider a shift in the imaginary axis from $s = 0$ to $s = -\delta$. This is achieved by finding $F(s-\delta)$, and testing the shifted polynomial $F(s-\delta)$ in the normal way. Any sign changes indicate the root distribution with respect to the new imaginary axis, which now passes through the point $-\delta$ in the s-plane (see Fig. 5.3).

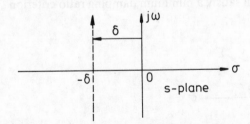

Figure 5.3 Translation of the imaginary axis

A shift in the imaginary axis may be found by using Horner's method (see Program 5 in Chapter 11) or by evaluating $F(s-\delta)$. Consider the function

$$F(s) = s^3 + 4s^2 + 3s + 1 = 0 \tag{5.15}$$

In order to shift the imaginary axis so that it passes through the $(-1, 0)$ point in the s-plane, $F(s-1)$ must be evaluated:

$$F(s-1) = (s-1)^3 + 4(s-1)^2 + 3(s-1) + 1 = 0 \tag{5.16}$$

Expanding and regrouping yields

$$F(s-1) = s^3 + s^2 - 2s + 1 = 0 \tag{5.17}$$

5.9.2   Relative stability

Relative stability indicates a system's minimum damping ratio. In the s-plane, this performance measure is represented by two symmetrical lines radiating from either side of the negative real axis (see Section 4.7.2). Provided all the system's roots are contained within these radiating lines (see Fig. 5.4), the effective damping ratio will be greater than some minimum value. The minimum damping ratio is specified by the cosine of the angle between the negative axis and either of the radiating lines.

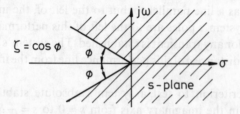

Figure 5.4 Relative stability and damping ratio

A system's relative stability may be tested by considering a rotation of the s-plane axes, as shown in Fig. 5.5. With this rotation a system having all its poles in the left half Σ, $j\Omega$-plane will satisfy a minimum damping ratio criterion $\zeta = \sin \theta$.

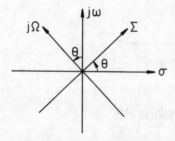

Figure 5.5 Rotation of s-plane axes

Rotation of the s-plane axes is achieved by making the substitution

$$s = e^{j\theta} \bar{s} \tag{5.18}$$

where the $\bar{s}$-plane is defined by the Σ, Ω axes. In general, a term s^n will be transformed according to:

$$s^n = e^{jn\theta}\bar{s}^n = [\cos(n\theta) + j\sin(n\theta)]\bar{s}^n \tag{5.19}$$

and therefore, after rotation of the axes, the characteristic function $F(s)$ becomes

$$F(e^{j\theta}\bar{s}) = H(\bar{s}) \tag{5.20}$$

The new characteristic polynomial $H(\bar{s})$ has complex coefficients, so the normal Routh technique cannot be used. It is therefore necessary to adapt the Routh technique for application to polynomials having complex coefficients. Only the results are given here, not the underlying theory from which they are derived.

Consider the transformation of a complex characteristic polynomial $H(\bar{s})$ such that

$$H(\bar{s}) = H(j\Omega) \tag{5.21}$$

With this transformation the characteristic polynomial becomes

$$H(j\Omega) = A_0\Omega^n + A_1\Omega^{n-1} + \cdots + A_n$$
$$+ j(B_0\Omega^n + B_1\Omega^{n-1} + \cdots + B_n) \tag{5.22}$$

and the first two rows of the Routh array may be written as

$$
\begin{array}{c|cccc}
\text{Row} & & & & \\
\hline
0 & A_0 & A_1 & \cdots & A_n \\
1 & B_0 & B_1 & \cdots & B_n
\end{array}
\tag{5.23}
$$

Provided B_0 is non-zero, the array may be completed in the normal way. Indeed, $B_0 = 0$ is the only special case that need be considered, since a first-column zero in any of the following rows would indicate that the roots of $F(s)$ are not contained within the effective damping ratio performance lines. Fortunately, this case is easily dealt with by considering the polynomial

$$G(\bar{s}) = jH(\bar{s}) \tag{5.24}$$

This new transformation switches the first two rows of the Routh array and multiplies the new row 0 by -1. As $H(\bar{s})$ and $G(\bar{s})$ have the same root locations, the method removes the first-column zero from row 1.

Once the array has been completed, the stability check may be carried out. It consists of taking the product of rows 1 and 2, of rows 3 and 4, of rows 5 and 6, and so on. If the system has all its roots in the negative half of the $\bar{s}$-plane, then all these products will be non-zero and will have the same sign.

Example 5.5

Test whether the characteristic equation

$$F(s) = s^3 + 4s^2 + 6s + 4 = 0$$

will satisfy:

(a) An effective damping ratio criterion of 0.5.

(b) An effective damping ratio criterion of 0.866.

Solution

Performing the standard Routh test on $F(s)$ indicates that all the roots are contained in the left half s-plane, and therefore the system's relative stability may be checked.

(a) An effective damping ratio of 0.5 requires an s-plane rotation of axes through $\sin^{-1}(0.5)$, or $30°$. Letting $s = e^{j30°}\bar{s}$,

$$H(\bar{s}) = F(e^{j30°}\bar{s}) = (\cos 90° + j\sin 90°)\bar{s}^3 + 4(\cos 60° + j\sin 60°)\bar{s}^2$$

$$+ 6(\cos 30° + j\sin 30°)\bar{s} + 4 = 0$$

Hence

$$H(\bar{s}) = j\bar{s}^3 + (2 + j3.4641)\bar{s}^2 + (5.1962 + j3)\bar{s} + 4 = 0$$

and

$$H(j\Omega) = \Omega^3 - 2\Omega^2 - 3\Omega + 4 + j(-3.4641\Omega^2 + 5.1962\Omega) = 0$$

The Routh array for this system is obtained by finding $G(j\Omega)$, which gives the array

Row				
0		0	3.464	-5.196 0
1	+ve	1	-2	-3 4
2		3.464	-5.196	0
3	+ve	-0.5	-3	4
4		-25.98	27.71	
5	+ve	-3.533	4	
6		-1.701		

Since the products of rows 1 and 2, 3 and 4, and 5 and 6 are all positive, the system has an effective minimum damping ratio of at least 0.5.

(b) For an effective damping ratio of 0.866, the s-plane axes are rotated through $60°$, whereupon

$$H(j\Omega) = 2\Omega^2 - 5.196\Omega + 4 + j(\Omega^3 - 3.464\Omega^2 + 3\Omega) = 0$$

The corresponding Routh array for this system is

Row

		0	2	−5.196	4
0		0	2	−5.196	4
1	+ve {	1	−3.464	3	0
2		2	−5.196	4	
3	+ve {	−0.866	1		
4		−2.190	4		
5	−ve {	−0.582			
6		4			

This system does not have the required effective damping ratio of 0.866, since the first-row products contain a sign change.

As the roots of the characteristic polynomial $F(s)$ are at $s = -2$ and $s = -1 \pm j$, it would be instructive to produce a Routh array for an effective damping ratio of 0.707; that is, to test the system's relative stability under a rotation of axes through 45°.

5.9.3   Maximum ω_d test

A characteristic polynomial may be tested for maximum ω_d values by making a number of transformations of axes, and then testing the resulting complex polynomial using the techniques described in Section 5.9.2. However, this will result in the damped frequency of all the characteristic roots being tested, including those of the non-dominant poles, and for this reason the method is of more academic than practical interest.

The first transformation translates the real axis up the imaginary axis, so that it passes through the point $j\omega_d$ on the imaginary axis. This is accomplished by letting

$$s = S + j\omega_d \tag{5.25}$$

so that the characteristic polynomial becomes

$$P(S) = F(S + j\omega_d) \tag{5.26}$$

The axes may now be rotated through 90° by making the substitution

$$S = e^{j\theta}\bar{s} \tag{5.27}$$

where $\theta = 90°$. It may be desirable to make θ less than 90°, and thus permit the damped frequency component of the non-dominant poles to be greater than ω_d.

After rotation of axes the complex characteristic polynomial is

$$H(\bar{s}) = P(e^{j\theta}\bar{s}) \tag{5.28}$$

and the techniques described in Section 5.9.2 for testing the complex polynomial $H(\bar{s})$ may be used.

5.10   The Routh array and PID controller design

Many industrial controllers for single input/single output systems consist of three elements: proportional (P), integral (I) and derivative (D) action. The transfer function of a controller which includes all three terms, the so-called three-term controller, is given by

$$G_c(s) = K[1 + T_d s + 1/(T_i s)] \qquad\qquad (5.29)$$

where

K = gain of the proportional channel
T_d = derivative action time
T_i = integral action (or reset)

There are a number of controllers based on the three-term PID controller. The most common are the P or PI type which together account for the majority of industrial control elements. These controllers are derived from the three-term controller $G_c(s)$ by making adjustments to T_d and T_i:

$T_d = 0$ and $T_i = \infty$ gives a P controller

$T_d = 0$ and T_i finite   gives a PI controller

Commercially available controllers, normally pneumatic or electronic, may be non-interacting (in which case an adjustment of any one parameter does not affect the others) or interacting (as above), depending on the principles of action. It is also important to note that derivative action is never implemented exactly in practice. The PID controllers are considered further in Chapter 8.

The next section examines the empirical Ziegler and Nichols methods for establishing settings for PID controllers.

5.10.1   Ziegler–Nichols rules for controller tuning

These results were designed to be used when a process or model is available and amenable to a few simple experiments. They are named after J. G. Ziegler and Nathaniel Burgess Nichols, who developed them in the 1940s.

(i)   First approach

The process to be controlled is shown in Fig. 5.6. Assume that it has the property that, under purely proportional control, it is asymptotically stable in the range $0 \leqslant K < K_c$, and goes unstable in an oscillatory manner for $K > K_c$. For this type of system the following experimental procedure is devised:

1. Turn up the gain K until the onset of continuous oscillations. At this critical gain K_c the closed-loop system is marginally stable – on the boundary between

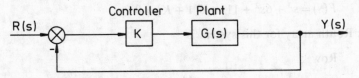

Figure 5.6 System under pure P control

stable and unstable behaviour – so any gain adjustments must be carried out with extreme care.

2. Note the value K_c and the period of oscillations T.
3. The recommended settings are given by:

P control	$K = 0.5\ K_c$	(5.30)
PI control	$K = 0.45\ K_c$	(5.31a)
	$T_i = 0.83\ T$	(5.31b)
PID control	$K = 0.6\ K_c$	(5.32a)
	$T_i = 0.5\ T$	(5.32b)
	$T_d = 0.125\ T$	(5.32c)

If a Laplace transform model of the plant is available, Routh's array may be used to establish the critical gain K_c and the corresponding period of oscillation T. The procedure is:

1. Find the system's closed-loop characteristic equation under pure proportional control.
2. Form the Routh array and establish the gain K_c that produces an all-zero row. If the system goes unstable in an oscillatory manner, the all-zero row will be the row associated with s^1, the divisor polynomial will be second order and there will be no roots of the remainder polynomial with positive real parts. Note that the system should remain stable for all positive values of K below the critical value.
3. Use the divisor polynomial to find the period of oscillation T, and apply the recommended settings given above.

Example 5.6

Use the Ziegler–Nichols rules to find P, PI and PID controller settings for a plant having the open-loop transfer function

$$G(s) = \frac{6}{(s+1)(s+2)(s+3)}$$

Solution
Under proportional control, the closed-loop characteristic equation is

$$F(s) = s^3 + 6s^2 + 11s + 6(1 + K) = 0$$

and the Routh array for this system is

Row			
0	s^3	1	11
1	s^2	6	$6(1 + K)$
2	s	$11 - (1 + K)$	

For row 2 to be an all-zero row, K must equal 10 (that is, $K_c = 10$). The divisor polynomial is obtained from row 1 and is

$$s^2 + 11 = 0$$

Letting $s = j\omega$ and solving for ω gives the frequency of oscillation as

$$\omega = \sqrt{11}$$

and

$$T = 2\pi/\omega = 1.895$$

The recommended settings are therefore as follows:

For P control $K = 5$; this gives the closed-loop transfer function as

$$\frac{Y(s)}{R(s)} = \frac{30}{s^3 + 6s^2 + 11s + 35}$$

For PI control $K = 4.5$ and $T_i = 1.57$, to give the closed-loop transfer function as

$$\frac{Y(s)}{R(s)} = \frac{42.5s + 27}{1.572s^4 + 9.434s^3 + 17.3s^2 + 51.89s + 27}$$

For PID control $K = 6$, $T_i = 0.947$ and $T_d = 0.237$, which gives the closed-loop transfer function as

$$\frac{Y(s)}{R(s)} = \frac{8.525s^2 + 34.1s + 36}{0.947s^4 + 5.683s^3 + 18.5s^2 + 39.78s + 36}$$

Figure 5.7 shows the output response $y(t)$ produced by a forcing step input $r(t)$ for this system under P, PI and PID control. Note the relatively fast response of the system under proportional control, and that the final steady-state output $y(t)$ is offset from the final demand input of 1. In order to eliminate this steady-state error, a PI controller may be used. However, as shown, the introduction of integral action reduces the absolute stability of the system. The inclusion of derivative action has a stabilizing effect on the plant, but derivative action could not be used if the controller were being fed with measurement noise. (In this context noise refers to spurious signals.)

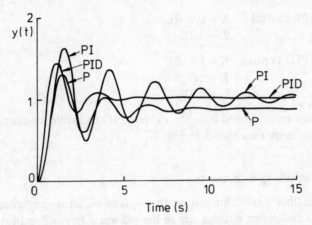

Figure 5.7 Time response under P, PI and PID control

(ii)   Second approach

This tuning method is included for the sake of completeness; it is not based on any stability test. The method is particularly suitable for open-loop systems having a measured step response containing appreciable time lag. This response is often referred to as the plant's signature. For a process plant, which will usually have large inertias, this is normally non-oscillatory with a fairly well-defined point of inflection.

The test procedure is normally carried out as follows:

1. Generate a step response from the open-loop plant. Typically, the response obtained will be similar to that shown in Fig. 5.8.

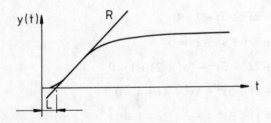

Figure 5.8 Open-loop step response plot

2. Measure the slope of the response R and the lag time L (see Fig. 5.8). Note that R should pass through the response's point of inflection, and should therefore have the maximum possible slope.
3. The recommended settings are:

$$P \text{ control} \qquad K = 1/RL \qquad\qquad (5.33)$$

PI control $K = 0.9/RL$ (5.34a)
               $T_i = 3.3L$ (5.34b)

PID control $K = 1.2/RL$ (5.35a)
              $T_i = 2L$ (5.35b)
              $T_d = 0.5L$ (5.35c)

The types of response obtained from the various PID controller combinations will be similar to the responses shown in Fig. 5.7.

5.11   Concluding comments

The stability methods used for testing the s-plane performance criteria, or for producing PID controller settings, are of limited use. They will indicate whether a particular design meets some given performance specification, or provides initial settings for a PID controller. However, if the design does not meet the specifications, or if the initial controller settings are inadequate, they provide no indication of what modifications are required. In the following chapters, design techniques are presented which not only indicate stability, but also provide information on the controller structure and adjustments needed to meet given performance specifications.

Problems

5.1. Indicate the root distribution and hence the stability of the following characteristic equations:

(a) $s^3 + 2s^2 + 3s + 1 = 0$

(b) $s^4 + 2s^3 + s^2 + 2s = 0$

(c) $s^5 + 2s^4 + 3s^2 + 6s^2 + 2s + 1 = 0$

(d) $s^5 + s^4 + 5s^3 + 5s^2 + 4s + 4 = 0$

(e) $s^6 + s^5 + 3s^4 + 3s^3 + 2s + 1 = 0$

(f) $s^{10} + s^9 + 2s^8 + 2s^7 + s^6 + 2s^5 + 6s^4 + 7s^3 + 10s^2 + 6s + 4 = 0$

5.2. A system's characteristic equation may be written in the form

$$F(s) = a_0 s^n + a_1 s^{n-1} + a_2 s^{n-2} + \cdots + a_{n-1} s + a_n = 0$$

If all the coefficients a_i are positive and exist, show, using Routh's criterion, that:

(a) A second-order system is always stable.

(b) A third-order system is stable if $a_1 a_2 - a_0 a_3 > 0$.

What is the stability condition for a system of fourth order?

5.3. The characteristic equation of a certain system is

$$F(s) = s^4 + 7s^3 + 18s^2 + 22s + 12 = 0$$

Use Routh's criterion to establish whether this system meets the following two s-plane performance criteria:

(a) a decay rate of at least $e^{-0.5t}$,

(b) a damping ratio between 0.5 and 0.866.

5.4. The block diagram shown in Fig. 5.9 consists of an inner loop with feedback gain K_T and an outer loop with forward path gain K_B. Find the gain K_T which makes the inner loop critically damped.

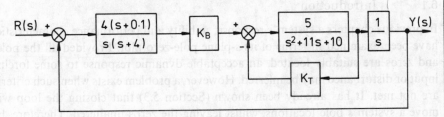

Figure 5.9

Using this value of K_T, determine:

(a) The range of gain K_B for which the closed-loop system is stable.

(b) The gain K_B which results in a marginally stable sinusoidal response. What is the frequency of this response?

5.5. A process control system with unity negative feedback has plant dynamics given by

$$G_p(s) = \frac{4.0}{(s^2 + 8s + 80)(s + 1)}$$

and incorporates a three-term controller with transfer function

$$G_c(s) = 20\left(1 + \frac{1}{T_i s} + T_d s\right)$$

Establish the values of T_i and T_d that will ensure the system's closed-loop stability.

5.6. Use the Ziegler–Nichols rules to design a three-term controller for a plant model having an open-loop transfer function given by

$$G(s) = \frac{1}{s(s + 1)(s + 2)(s + 3)}$$

Show that the resulting closed-loop system is stable.

CHAPTER 6

ROOT LOCI

6.1   Introduction

In previous chapters a system's response, stability and performance characteristics have been examined in terms of its s-plane pole-zero plot. Provided all the poles and zeros are suitably located, an acceptable dynamic response to some forcing input or disturbance may be inferred. However, a problem exists when such criteria are not met. It has already been shown (Section 5.3) that closing the loop will move a system's pole locations, whilst leaving the zeros unaltered. Therefore, by closing the loop and introducing dynamic elements and gains it is possible to create new or modified designs which may meet the required specifications. In such cases the root loci technique provides a design method, based on the open-loop s-plane pole–zero plot, which will give the closed-loop pole positions for all possible changes in a single variable, normally the loop gain. Since the open- and closed-loop zeros between any two points in the loop are identical, the root loci method effectively provides all possible closed-loop s-plane plots for variations of a single variable.

6.2   Preliminaries

In control system design, a knowledge of the transient response characteristics associated with variations in one parameter is often required. For example, the open-loop gain has been shown to have a crucial effect on stability. Root loci provide a very convenient means of analysing systems where this critical parameter occurs linearly in the characteristic equation – that is, when this equation is of the form

$$F(s) = D(s) + KN(s) = 0 \tag{6.1}$$

In Equation (6.1) K could be any positive real parameter, and $D(s)$ and $N(s)$ the corresponding real polynomials in s. By dividing through by $D(s)$, this equation may be written as

$$1 + KM(s) = 0, \qquad M(s) = \frac{N(s)}{D(s)} \tag{6.2}$$

Consider the closed-loop system depicted in Fig. 6.1. With the measurement path $B(s)$ broken, the system is open loop, and the relationship between $R(s)$ and $B(s)$ is given by

$$B(s) = KG(s)H(s)R(s) = K\frac{Q(s)W(s)}{P(s)V(s)}R(s) \tag{6.3}$$

where $KG(s)H(s)$ (or $KGH(s)$) is the open-loop transfer function and

$$P(s)V(s) = 0 \tag{6.4}$$

is the open-loop characteristic equation.

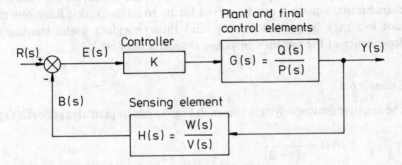

Figure 6.1 Closed-loop SISO system under proportional control

In Section 5.3 it was shown that the closed-loop transfer function between the forcing input $R(s)$ and the plant output $Y(s)$ is given by

$$\frac{Y(s)}{R(s)} = \frac{KG(s)}{1 + KGH(s)} \tag{6.5}$$

and that between the forcing input and measured output by

$$\frac{B(s)}{R(s)} = \frac{KGH(s)}{1 + KGH(s)} \tag{6.6}$$

For both Equations (6.5) and (6.6) the closed-loop characteristic function is the same, and is one plus the open-loop transfer function. This must be true since the characteristics of the system do not change with the observation position. The number of closed-loop zeros may change with observation position; that is, the numerators of $G(s)$ and $GH(s)$ may not be of the same order.

For the closed-loop transfer functions (6.5) and (6.6), the corresponding characteristic equations are

$$1 + KGH(s) = 1 + K \frac{Q(s)W(s)}{P(s)V(s)} = 0, \text{ or} \tag{6.7}$$

$$F(s) = P(s)V(s) + KQ(s)W(s) = 0 \tag{6.8}$$

The zeros of $F(s)$, which give the system's closed-loop poles, are functions of the open-loop poles and zeros and of the open-loop controller gain K. From a comparison of Equations (6.7) and (6.8) with Equations (6.1) and (6.2), it will be seen that K is the gain of the proportional controller and that

$$M(s) = \frac{Q(s)}{P(s)} \frac{W(s)}{V(s)} = GH(s) \tag{6.9}$$

In this system, $M(s)$, which describes the plant and transducer dynamics, is fixed, and only the controller gain K may be adjusted. As K varies, the characteristic roots move about the s-plane and their trajectories constitute the root loci. If the characteristic equation can be solved for its roots as explicit functions of K, the root loci may be plotted directly, and thus provide a useful summary of the dependence of the system's response characteristics on K.

Example 6.1

The negative feedback system shown in Fig. 6.1 has its plant dynamics $G(s)$ given by

$$G(s) = \frac{1}{s(s+2)}$$

and has unity negative feedback ($H(s) = 1$). Plot the root loci diagram for this system.

Solution
The closed-loop characteristic equation for this system is

$$s^2 + 2s + K = 0$$

Solving directly for s gives

$$s = -1 \pm \sqrt{(1 - K)}$$

These two values of s may be plotted in the s-plane for all positive values of K, as in Fig. 6.2. There are a number of points to note about this plot:

1. Open-loop poles are indicated by small crosses; open-loop zeros, if present, would be indicated by small enclosing circles. This is the accepted convention for representing open-loop poles and zeros on a root loci plot.
2. Since the coefficients of the characteristic equation are real, the root loci plot is symmetrical with respect to the real axis – that is, all complex poles occur in complex conjugate pairs.

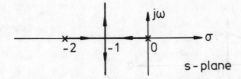

Figure 6.2 Root loci plot for a system with the open-loop transfer function
$$KGH(s) = K/[s(s + 2)]$$

3. When $K = 0$ the closed-loop poles will coincide with the open-loop poles. With $K = 0$ in Equation (6.8) the closed-loop characteristic equation is the same as the open-loop characteristic equation, Equation (6.4).

4. When $K = \infty$ the closed-loop poles will be either infinite or coincident with the open-loop zeros. Dividing both sides of Equation (6.8) by K and letting $K = \infty$ gives the closed-loop characteristic equation as

$$Q(s)W(s) = 0$$

Comparing this with Equation (6.3) shows that some of the closed-loop poles will coincide with the open-loop zeros.

5. In Fig. 6.2 the arrows denote increasing values of K. At $K = 0$ the closed-loop poles are at $s = 0$ and $s = -2$. As K increases, the pole at $s = 0$ moves along the real axis until $K = 1$, at which point the pole is at $s = -1$. Simultaneously, the pole at $s = -2$ moves along the real axis to meet the other pole at $s = -1$. Points at which two or more poles coalesce are called singular points, and if a singular point occurs on the real axis it is also known as a break point. For $K > 1$, the closed-loop roots become complex and diverge, to become infinite as $K \to \infty$.

Example 6.2

Plot the root loci for the system shown in Fig. 6.1 if

$$GH(s) = \frac{2s}{s^2 + 1}$$

and determine the gain K which will give the closed-loop system its maximum margin of absolute stability.

Solution

The characteristic equation is

$$s^2 + 2Ks + 1 = 0$$

Solving directly for s gives

$$s = -K \pm \sqrt{(K^2 - 1)}$$

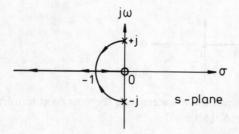

Figure 6.3 Root loci plot for a system with the open-loop transfer function
$$KGH(s) = K2s/(s^2 + 1)$$

from which the root loci plot shown in Fig. 6.3 is obtained. From this figure, the maximum margin of absolute stability (that is, when the slowest exponential decay has its most negative value) occurs when $K = 1$, and the system has two real poles at $s = -1$. For $K < 1$ the poles are complex and their real parts will always be greater than -1. For $K > 1$, one pole tends to the origin and the other tends to minus infinity, and consequently the margin of absolute stability is again reduced.

It is interesting that, with this system, better transient responses are likely to be obtained when $K \neq 1$. For $K < 1$, the margin of relative stability can be improved to achieve acceptable damping characteristics. Another point to note is that if the open-loop zero occurs between the forcing input and the measured output, the closed-loop pole–zero map will have a zero at the origin. For large values of K this zero will give approximate closed-loop pole–zero cancellation. For a step input this would result initially in a very fast measured response which would quickly approach its final steady-state value, but would take a considerable amount of time to reach that value. However, if the open-loop zero occurs in some other part of the system, the measured closed-loop response will become progressively more sluggish as K is increased.

6.3   Rules for constructing root loci

Although it has been assumed that the roots of the closed-loop characteristic equation may be calculated directly, it is possible to determine the salient features of root loci by applying simple rules. These rules obviate the necessity of solving the characteristic equation, and thus make the method applicable to quite complicated systems.

These rules, often known as Evans' rules (after Walter R. Evans, whose original paper on the subject of root loci appeared in 1948), are stated below for the closed-loop single input/single output system shown in Fig. 6.1 and described in Section 6.2. Only those rules required for determining changes in root positions resulting from positive changes in K are given. It should be noted that there is a complementary set of rules for determining root positions resulting from negative changes in K. Also, note that the open-loop function $M(s)$ is assumed to have m zeros and n poles.

(i)   *Number of root loci*

      The number of root loci is equal to the order of the closed-loop characteristic equation.

For engineering systems $M(s)$ will be rational, and the closed-loop characteristic function $F(s)$ must therefore be of the same order as the denominator of $M(s)$; that is, of order n. Since $F(s)$ is an nth-order polynomial it will have n roots, each with its own locus.

(ii)   *Symmetry of loci*

      The root loci of a characteristic equation having real coefficients are symmetrical with respect to the real axis.

This follows from the fact that the complex roots of a real characteristic equation occur in conjugate pairs.

(iii)   *Poles of $M(s)$*

      The poles of $M(s)$ lie on the root loci and correspond to $K = 0$.

This follows directly from Equations (6.8) and (6.9).

(iv)   *Zeros of $M(s)$*

      The zeros of $M(s)$ lie on the root loci and correspond to $K = \infty$.

Again, this follows directly from Equations (6.8) and (6.9). If there are r more poles than zeros, then r of the loci will become infinite as $K \to \infty$. These loci are dealt with in rules (v) and (vi).

(v)   *Asymptotes of root loci*

      If $M(s)$ has r more poles than zeros, the root loci are asymptotic to r straight lines making angles

$$\frac{(2j+1)\pi}{r}, \quad j = 0, 1, 2, \ldots, r - 1 \tag{6.10}$$

with the real axis. The root loci approach symptotes when $K \to \infty$.

(vi)　　*Point of intersection of asymptotes*

　　　　Asymptotes intersect on the real axis at a point with abscissa

$$\sigma_0 = \frac{1}{r}\left(\sum_{j=1}^{n} p_j - \sum_{j=1}^{m} z_j \right) \tag{6.11}$$

where the p_j and z_j are respectively the poles and zeros of $M(s)$.

(vii)　　*Root loci on the real axis*

　　　　If $M(s)$ has one or more real poles or zeros, then the segment of the real axis having an odd number of real poles and zeros to its right will be occupied by a root locus.

This rule is best demonstrated by means of the example shown in Fig. 6.4, in which $M(s)$ has six poles and three zeros on the real axis.

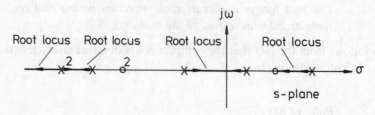

Figure 6.4 Root loci on the real axis

(viii)　　*Singular points*

　　　　Singular points indicate the existence of multiple characteristic roots, and occur at those values of s which satisfy $dK/ds = 0$.

This rule is probably best demonstrated by an example. Consider a system having the closed-loop characteristic equation

$$s^2 + 2Ks + 1 = 0$$

as in Example 6.2. The rule states that there will be singular points when $dK/ds = 0$, and since

$$K = \frac{-(s^2 + 1)}{2s}$$

differentiation of K with respect to s indicates that

$$s^2 - 1 = 0$$

Therefore there are two singular points, at $s = +1$ and $s = -1$. Substituting both values of s back into the characteristic equation indicates that when $s = +1$,

$K = -1$, and when $s = -1$, $K = +1$. Since only the rules for positive values of K are being considered, the singular point at $s = +1$ will not appear on the root loci plot (see Fig. 6.3).

Typically, not all the singular points found using this rule will be for positive values of K; some will be associated with zero or negative values. When producing the root loci plot it is usually obvious which of the singular points are required. In the few cases where it is not clear, the associated value of s may be substituted back into the closed-loop characteristic equation, and the sign of K determined.

(ix)   *Intersection of root loci with the imaginary axis*

   The intersections of root loci with the imaginary axis can be determined by calculating the values of K which result in the existence of imaginary characteristic roots.

These values of K, together with the corresponding imaginary roots, can be found from Routh's array using the method described in the previous chapter.

(x)   *Slopes of root loci at complex poles and zeros of $M(s)$*

   The slope of a root locus at a complex pole or zero of $M(s)$ can be found at a point in the neighbourhood of the pole or zero.

This technique can be illustrated by considering the complex pole p_1 shown in Fig. 6.5, where Ω is the unknown slope of the locus at p_1. The arguments of the complex numbers, represented by vectors drawn from the other poles p_2, p_3 and p_4 and the zero z_1 to a point on the root locus near p_1, obviously differ very little from the angles ϕ_2, ϕ_3, ϕ_4 and θ_1.

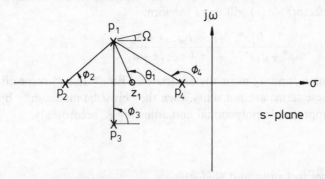

Figure 6.5 Determination of the slope of a root locus at a complex pole

Now, it has been shown (see Problem 4.5) that for some point s on the root loci

$$\sum_{j=1}^{m} \arg(s - z_j) - \sum_{j=1}^{n} \arg(s - p_j) = (2i + 1)\pi \qquad (6.12)$$

where $\arg(s - z_j)$ and $\arg(s - p_j)$ are the angles of the vectors from the zeros and poles of $M(s)$ to the point s, and i is some real integer. Equation (6.12) states that the sum of the arguments of all the vectors drawn from s to all the zeros, minus the sum of the arguments of all the vectors drawn from s to all the poles, is given by an odd integer multiple of π for all points on the root loci.

Therefore it follows that

$$\theta_1 - (\phi_2 + \phi_3 + \phi_4 + \Omega) = (2i + 1)\pi \tag{6.13}$$

where i is an appropriate integer. This equation can be solved for Ω since the angles ϕ_2, ϕ_3, ϕ_4 and θ_1 are easily measured.

(xi)   *Calculation of K on the root loci*

The absolute magnitude of the value of K corresponding to any point s_0 on a root locus can be found by measuring the lengths of the vectors drawn to s_0 from the poles and zeros of M(s), and then evaluating

$$|K| = \prod_{j=1}^{n} |s_0 - p_j| \bigg/ \prod_{j=1}^{m} |s_0 - z_j| \tag{6.14}$$

This rule states that

$$K = \frac{\text{Product of lengths of vectors from open-loop poles to } s_0}{\text{Product of lengths of vectors from open-loop zeros to } s_0}$$

If the open-loop system has no zeros, then

$$K = \text{Product of lengths of vectors from open-loop poles to } s_0$$

In applying this rule, it is assumed that, on multiplying out all the numerator and denominator factors, $M(s)$ will be of the form

$$M(s) = \frac{s^m + b_1 s^{m-1} + \cdots b_{m-1} s + b_m}{s^n + a_1 s^{n-1} + \cdots + a_{n-1} s + a_n} \tag{6.15}$$

The coefficients b_0 and a_0 normally associated with s^m and s^n, respectively, must be unity. If these terms are not unity, then they must be made unity by dividing through the appropriate polynomial and adjusting K accordingly.

6.3.1   Root loci rules and algorithms

Rules (x) *and* (xi) are of particular interest in that they are often used as the basis of an algorithm for producing root loci plots. Consider the closed-loop characteristic function given by Equations (6.7) and (6.9), which may be written as

$$1 + KM(s) = 0 \tag{6.16a}$$

or

$$KM(s) = -1 \tag{6.16b}$$

Taking the modulus and argument of each side of Equation (6.16b) – that is, using the Euler (polar) representation – the modulus is given by

$$K|M(s)| = 1 \tag{6.17}$$

and the argument by

$$\arg(KM(s)) = \arg(M(s)) = 180° \tag{6.18}$$

Equation (6.18) is called the angle criterion, of which rule (x) is a special case. Any point in the s-plane may be tested, using the angle criterion, to see whether it lies on a root locus. If this criterion is satisfied, then from Equation (6.17) the modulus criterion

$$|K| = 1/|M(s)| \tag{6.19}$$

may be used to determine the value of K. Again, this is simply a restatement of rule (xi).

Example 6.3

A system has the open-loop transfer function

$$\frac{K(s^2 + 1.5s + 1.5625)}{(s - 0.75)(s + 0.25)(s + 1.25)(s + 2.25)}, \quad K > 0$$

Use Evans' rules to plot the root loci diagram for this system, given that, for positive values of K, there are only four singular points, at $s = +0.26$, $s = -1.76$ and $s = -0.75 \pm j1.74$.

Determine the values of gain for a stable closed-loop system.

Solution

By inspection, the denominator and numerator polynomials are of the correct form. The coefficient associated with s^2 in the numerator is unity, and the coefficient associated with s^4 in the denominator is also unity.

Evans' rules as given in Section 6.3 are applied to yield the following results:

(i) Number of roots = 4.
(ii) The loci are symmetrical about real axis.
(iii) When $K = 0$, loci start from poles at $s = 0.75$, $s = -0.25$, $s = -1.25$ and $s = -2.25$.
(iv) When $K = \infty$, loci terminate at infinity and the zeros $s = -0.75 \pm j1$.
(v) Asymptotes: $M(s)$ has two more poles than zeros, so $r = 2$, and two loci approach asymptotes with angular slopes given by

$$\left(\frac{2j+1}{r}\right)\pi, \quad j = 0 \text{ or } 1$$

that is, at $\pi/2$ and $3\pi/2$ (90° and 270°).

(vi)   Asymptotes intersect on the real axis at

$$\sigma_0 = \frac{1}{r}\left(\sum p - \sum z\right)$$

$$= \frac{(+0.75 - 0.25 - 1.25 - 2.25) - (-0.75 + j - 0.75 - j)}{2}$$

$$= -0.75$$

(vii)   Loci are on the real axis between $s = 0.75$ and $s = -0.25$, and also between $s = -1.25$ and $s = 2.25$.

(viii)   The singular points are given. (It would be a useful exercise for the reader to show that these points are singular points, and to find the associated closed loop gain K.)

(ix)   To determine the points of intersection of the loci with the imaginary axis, using Routh's array, the closed-loop characteristic equation must be found. It is

$$K(s^2 + 1.5s + 1.5626)$$

$$+ (s^4 + 3s^3 + 0.875s^2 - 2.063s - 0.527) = 0$$

or

$$s^4 + 3s^3 + (0.875 + K)s^2$$

$$+ (1.5K - 2.063)s + (1.563K - 0.527) = 0$$

The Routh array for this equation is then

Row				
0	s^4	1	$(0.875 + K)$	$(1.563K - 0.527)$
1	s^3	3	$(1.5K - 2.063)$	
2	s^2	$\dfrac{3(0.875 + K) - (1.5K - 2.063)}{3}$	$(1.563K - 0.527)$	

To be on the imaginary axis, row 3 must be an all-zero row – that is, the first and only element must be zero:

$$\frac{3(0.875 + K) - (1.5K - 2.063)}{3}(1.5K - 2.063)$$

$$- 3(1563K - 0.527) = 0$$

which may be rearranged to give

$$0.75K^2 - 3.377K - 1.643 = 0$$

from which

$$K = 2.251 \pm 2.694$$

Since only positive values of K are required, then $K = 4.945$.

Note that this analysis also indicates that the closed-loop system is asymptotically stable for all values of K greater than 4.945.

Letting $s = j\omega$ and equating the resulting imaginary part of the characteristic equation to zero gives

$$-3\omega^3 + (1.5K - 2.063)\omega = 0$$

Therefore

$$\omega = \sqrt{\left(\frac{1.5 \times 4.945 - 2.063}{3}\right)} = 1.34$$

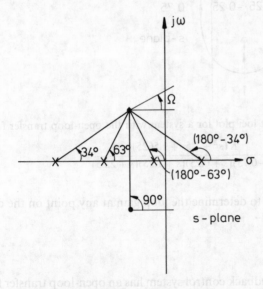

Figure 6.6 Application of angle criterion

(x)   This rule will be used to find the slope of the locus at the complex zero, $s = -0.75 + j$. Let Ω be the unknown slope of the locus at the zero $s = -0.75 + j$ (see Figure 6.6). From the angle criterion (Equation (6.12)),

$$(90° + \Omega) - [(180° - 34°) + 34° + (180° - 63°) + 63°]$$

$$= (2i + 1)180°$$

where i is some real integer. If i is set equal to -1, then $\Omega = +90°$, which is the required answer. With $i = 0$, $\Omega = 450°$ and, since this is a $360°$ rotation plus a $90°$ rotation, the results are identical. Indeed, any value of the integer i will give the correct result once any superfluous $360°$ rotations have been removed.

From the above rules the root loci plot may now be produced, and is shown in Fig. 6.7.

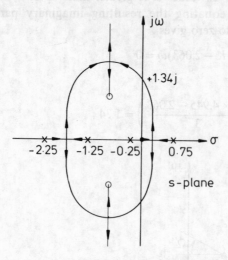

Figure 6.7 Root loci plot for a system with the open-loop transfer function:

$$\frac{K(s^2 + 1.5s + 1.5625)}{(s - 0.75)(s + 0.25)(s + 1.25)(s + 2.25)}$$

Rule (xi) could be used to determine the loop gain at any point on the diagram.

Example 6.4

A closed-loop, negative feedback control system has an open-loop transfer function given by

$$KGH(s) = \frac{K(s + 1)}{s^2(s + 9)}$$

Plot the root loci diagram, and hence determine the gain K which will give the closed-loop system its maximum value of absolute stability.

Why would this particular control system have good tracking characteristics?

Solution

The closed-loop characteristic equation is

$$1 + KGH(s) = 0$$

or

$$s^3 + 9s^2 + Ks + K = 0$$

Evans' rules yield the following results:

(i)   The characteristic equation is of degree three, and hence has three roots.

(ii)   The characteristic equation has real coefficients, so the loci must be symmetrical with respect to the real axis.

(iii)   When $K = 0$, there are loci at points $s = 0$ (a double root and therefore also a singular point) and $s = -9$.

(iv)   When $K = \infty$, there are loci at $s = -1$ and infinity.

(v)   The asymptotes have angular slopes

$$\tfrac{1}{2}(2j + 1)\pi, \quad j = 0 \text{ or } 1$$

that is, at $\pi/2$ and $3\pi/2$.

(vi)   The asymptotes intersect the real axis at

$$\sigma_0 = \frac{0 - 9 + 1}{2} = -4$$

(vii)   Loci are on the real axis between -1 and -9.

(viii) The axis crossing points are determined using the Routh array:

Row			
0	s^3	1	K
1	s^2	9	K
2	s	$8K$	
3	s^0	K	

Since there are no sign changes in the first column of the array, it may be concluded that the loci do not cross the real axis for positive K.

(ix)   Break points occur when $dK/ds = 0$. The closed-loop characteristic equation may be solved for K to give

$$K = \frac{-(s^3 + 9s^2)}{s + 1}$$

Differentiating K with respect to s and setting $dK/ds = 0$ yields

$$(s^3 + 9s^2) - (s + 1)(3s^2 + 18s) = 0$$

On expansion the above equation becomes

$$-2s^3 - 12s^2 - 18s = 0$$

which may be solved directly to give the singular points $s = -3$, -3 and 0.

Substitution of any of these values of s back into the closed-loop characteristic equation indicates that all are associated with positive values of K. Note also that the analysis has indicated that the double open-loop pole at $s = 0$ is a singular point.

(x)   This rule is not required.

The root loci plot for this system may now be drawn, and is shown in Fig. 6.8. In this figure, the two loci breaking from the double pole at $s = 0$ coalesce on the real axis at $s = -3$, together with the real pole emanating from $s = -9$. Two of the three loci break from this point, and become infinite, with asymptotes passing through the $s = -4$ point on the real axis. The third locus breaks from the $s = -3$ point, and moves with increasing K along the real axis until, at $K = \infty$, it reaches the zero at $s = -1$. From the root loci plot it is clear that the maximum value of absolute stability would be at $s = -3$; for at any other point on the plot at least one-closed-loop pole would have a more positive real part.

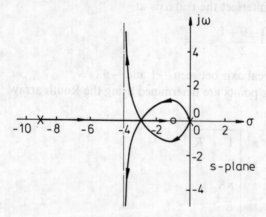

Figure 6.8 Root loci plot for a system with the open-loop transfer function:

$$KGH(s) = K(s+1)/[s^2(s+9)]$$

The loop gain which gives the maximum value of absolute stability is found from rule (xi) to be

$$K = \frac{|-3||-3||6|}{|2|} = 27$$

The tracking properties of this system are most easily evaluated by using the final value theorem to find the closed-loop transfer function between the forcing input and the error signal. Applying the final value theorem for various forcing inputs shows that the system will track both a step and a ramp input, with zero steady-state error. More will be said on the relationship between an open-loop transfer function and a system's closed-loop tracking capabilities in Section 7.5.1.

6.4   Root loci and PID controllers

In Chapter 5 the empirical Ziegler and Nichols methods were used to establish settings for PID controllers. However, as shown by Fig. 5.7, for some applications the resulting time responses tend to be under-damped, and the methods provide no indication of how the performance may be modified. Here it is shown how the root loci method can be used to select controller settings, and how the introduction of control elements within a closed-loop system affects performance.

Consider again Example 5.6, in which the plant transfer function is given by

$$G(s) = \frac{6}{(s+1)(s+2)(s+3)} \tag{6.20}$$

and the PID controller's transfer function is given by

$$G_c(s) = K\left(1 + T_d s + \frac{1}{T_i s}\right) \tag{6.21}$$

Under proportional control, $T_d = 0$ and T_i is infinite. For the closed-loop system the characteristic equation is

$$1 + \frac{6K}{s^3 + 6s^2 + 11s + 6} = 0 \tag{6.22}$$

The root loci plot is shown in Fig. 6.9(a). In this plot K_c, the critical gain, and K_{ZN}, the Ziegler–Nichols gain, are indicated by their corresponding dominant closed-loop pole positions. Clearly, the response dynamics would be improved by reducing the controller gain K to 0.6. However, this increases the offset caused by a demand step change in the reference input (see Fig. 6.9(b)).

Under proportional plus integral (PI) control, the closed-loop characteristic equation is

$$1 + \frac{6K(1 + T_i s)}{T_i s(s^3 + 6s^2 + 11s + 6)} = 0 \tag{6.23}$$

Since the root loci method can only cope with adjustments to one variable at a time, either K or T_i must be fixed. If T_i is fixed at its recommended Ziegler–Nichols value of 1.57, then

$$1 + \frac{K(3.82 + 6s)}{s(s^3 + 6s^2 + 11s + 6)} = 0 \tag{6.24}$$

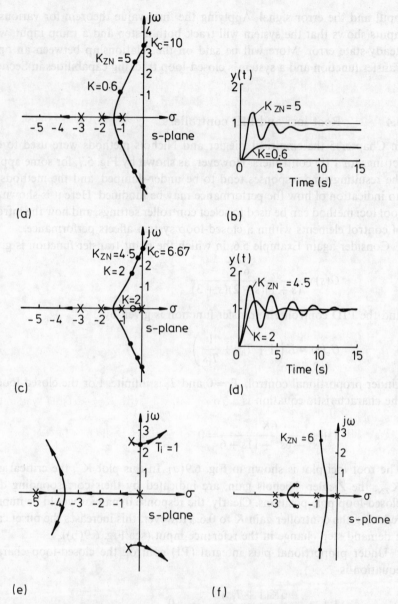

Figure 6.9 (a) Root loci for P control and (b) step responses for $K_{ZN} = 5$ and $K = 0.6$. (c) Root loci for PI control with $T_i = 1.57$ and (d) step responses for $K_{ZN} = 4.5$ and $K = 2$. (e) Root loci for PI control with $K = 4.5$. (f) Root loci for PID control with $T_i = 0.947$ and $T_d = 0.237$

and the corresponding root loci plot is as shown in Fig. 6.9(c). Note that the PI controller introduces a pole at the origin of the *s*-plane, and a zero at $s = -0.64$. The controller pole eliminates the offset for a demand step change in input. Provided the controller gain is high enough, the zero will give approximate closed-loop pole–zero cancellation and make the complex poles dominant. Again, the Ziegler–Nichols gain K_{ZN} and the gain K_c which makes the closed-loop system marginally stable are indicated on the root loci plot.

From this plot, it would appear that the response could be improved by reducing the gain K to 2, as shown. Further reductions in gain will move the real pole away from the zero, and the complex poles closer to the zero, and hence reduce the effective dominance of the complex poles. The step responses for the system with both the Ziegler–Nichols settings and the reduced gain setting of $K = 2$ are shown in Fig. 6.9(d).

To see the effect of adjustments to T_i, the characteristic equation must be modified so that it is of the form

$$1 + K_1 M(s) = 0 \tag{6.25}$$

Again, the controller gain K must be fixed and, as previously, its Ziegler–Nichols value is used. Hence, from Equation (6.23) and with K set at 4.5 it is easily shown that

$$1 + \frac{27/T_i}{s(s^3 + 6s^2 + 11s + 33)} = 0 \tag{6.26}$$

If K_1 is now set equal to $27/T_i$, the above closed-loop characteristic equation is of the required form. The root loci plot for $0 \le K_1 \le \infty$ is shown in Fig. 6.9(e), in which arrows on the root loci indicate decreasing values of T_i. When $T_i = 1.57$, its Ziegler–Nichols value, the loci on this plot coincide with the points on the loci of Fig. 6.9(c), for which $K_{ZN} = 4.5$. Also, if $T_i < 1$ (with $K = 4.5$), the system becomes unstable.

With full PID control there are three terms which could be adjusted, and therefore two must be fixed. For convenience T_i and T_d are set at their Ziegler–Nichols values of 0.947 and 0.237, respectively, and the root loci plot for variations in K is drawn. This is shown in Fig. 6.9(f), from which the closed-loop system is seen to be stable for all values of loop gain K. With the proposed Ziegler–Nichols value, $K_{ZN} = 6$, there are four closed-loop poles, in the positions shown. Further, since the open-loop and closed-loop zeros between the forcing input and measured output are the same, there is approximate pole–zero cancellation and the dominant poles are at $s = -0.85 \pm j2.67$. By comparing these closed-loop pole positions with those obtained from the P and PI controllers (Fig. 6.9(a) and 6.9(c)), this control system is seen to have the better absolute stability and damping ratio, whilst the damped frequencies of all three systems are similar. A plot of the Ziegler–Nichols responses to a step input for P, PI and PID control is shown in Fig. 5.7.

6.5   Root loci and systems with time delays

A phenomenon often encountered in process equipment is that of transportation lag, or time delay. For example, if a fluid is flowing with a velocity v in a pipe of length L, the time taken for any individual element of fluid to flow from the entrance to the exit of the pipe is L/v. This time, usually given the symbol τ, represents a delay between an action occurring and its observed response. The real shift theorem may be used to show that the open-loop Laplace transform model of such a system is

$$KGH(s)e^{-s\tau} \qquad (6.27)$$

and the corresponding closed-loop characteristic equation is

$$1 + KGH(s)e^{-s\tau} = 0 \qquad (6.28)$$

One of the limitations of the root loci method is that a time delay cannot be handled conveniently using the standard Evans' rules. For this reason it is advisable to use one of the frequency domain techniques described in Chapter 7. However, the root loci technique gives such insight into the system's closed-loop stability and performance that on occasions it may be worth the inconvenience of using it.

Two ways of dealing with time delays are considered here. The first, which is frequently used but is often not very accurate, replaces the time delay with a first-order Padé approximation. Thus the time delay $e^{-s\tau}$ is replaced by

$$e^{-s\tau} = \frac{1 - \frac{1}{2}\tau s}{1 + \frac{1}{2}\tau s} \qquad (6.29)$$

and the closed-loop characteristic equation becomes

$$1 + KGH(s)\frac{1 - \frac{1}{2}\tau s}{1 + \frac{1}{2}\tau s} = 0 \qquad (6.30)$$

Equation (6.30) may now be dealt with using Evans' rules in the normal way. If necessary, $e^{-s\tau}$ may be approximated by a power series expansion, although for convergence τ should be small.

The second and more accurate method is to base the root loci plot on a modified form of the algorithm given in Section 6.3.1. As before, the modulus and arguments of the characteristic equation (Equation (6.28)) are found, and the modulus criterion for positive K is then

$$K = \frac{1}{|GH(s)|e^{-\tau\sigma}} \qquad (6.31)$$

and the angle criterion for positive K becomes

$$GH(s) = (2i + 1)\pi + \omega\tau \qquad (6.32)$$

where the point being tested is given by the rectangular coordinates $\sigma + j\omega$, and i is a positive integer.

Note that the angle criterion now depends on ω, which varies along the vertical axis in the s-plane. Hence, for a given σ-line (a line parallel to the j-axis) there will be an infinite number of points which satisfy the angle criterion. This is in fact correct since the characteristic equation (Equation (6.28)) is transcendental and can be shown to have an infinite number of roots. Consequently, Evans' rules become invalid when a time delay is introduced, and a modified set of rules is required. Root loci for simple low-order systems can be produced by using the angle criterion to track any locus emanating from an open-loop pole. In general, this plot would be dominant and all other loci could be ignored.

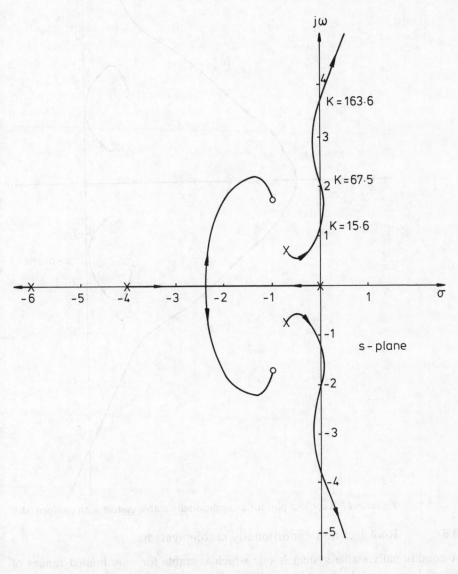

Figure 6.10 Root loci plot for a conditionally stable system

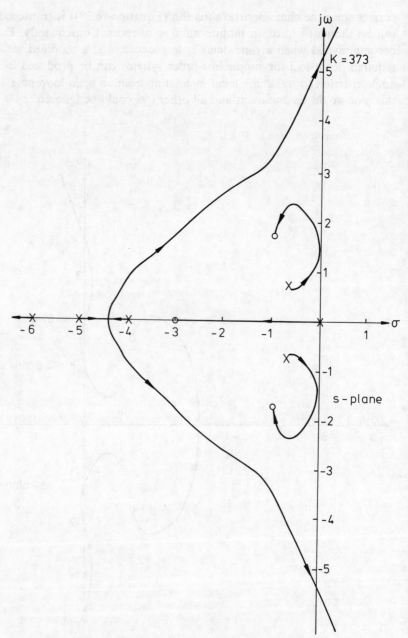

Figure 6.11 Root loci plot for a conditionally stable system with compensator

6.6   Root loci and conditionally stable systems

A conditionally stable system is one which is stable for only limited ranges of closed-loop gain. In practice, such conditions should be avoided since any disturbance producing a variation in the loop gain could cause the system to become unstable.

Figure 6.10 shows the root loci plot of a conditionally stable closed-loop system with an open-loop transfer function given by

$$GH(s) = \frac{K(s^2 + 2s + 4)}{s(s+4)(s+6)(s^2 + 1.4s + 1)} \tag{6.33}$$

The closed-loop system is stable for $0 < K < 15.6$ and also for $67.5 < K < 163.6$; for all other values of loop gain it is unstable. Problems could arise if the designer stipulated a gain between 67.5 and 163.6, as this level of gain would produce oscillatory transient response characteristics. Normally a reduction in loop gain would dampen the response, but in this case a reduction in gain would cause the system to go unstable. The problem can usually be overcome by using a suitable compensating network. For example, a forward path compensator having the transfer function

$$G_c(s) = \frac{s+3}{s+5} \tag{6.34}$$

would produce the root loci plot of Fig. 6.11 which is stable for all values of gain less than 373.

6.7   Root loci and non-minimum phase systems

If all the poles and zeros of a system lie in the left half s-plane, then the system is termed minimum phase. If a system has one or more poles or zeros in the right half s-plane, then the system is termed non-minimum phase (see Section 7.2.1). An unstable system is therefore a non-minimum phase system, since at least one of its

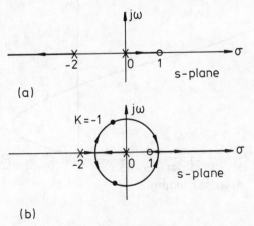

(a)

(b)

Figure 6.12 Root loci plots for the open-loop system $GH(s)$ for (a) $0 < K < \infty$ and (b) $-\infty < K < 0$

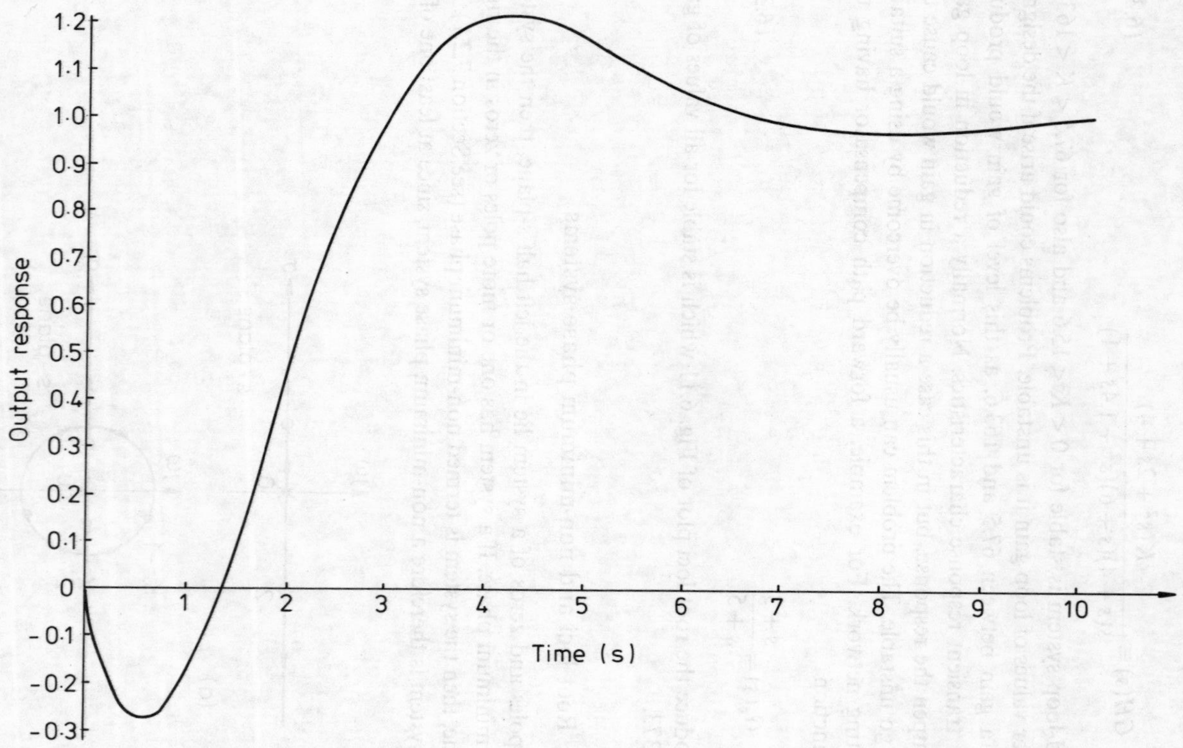

Figure 6.13 Unit step response of a system with a non-minimum phase zero

poles is in the right half s-plane. A zero in the right half s-plane does not affect stability, but it may produce some rather unexpected results. A non-minimum phase zero will produce a tendency for the system's step response to behave, during the initial phase of the transient, in the opposite sense to that expected. This is one of the problems associated with the Padé approximation, which uses a non-minimum phase zero to approximate a time delay (see Section 6.5).

In this section the effect of non-minimum phase elements on the root loci plot is examined by means of two simple examples. First, consider the following open-loop system $GH(s)$ with a non-minimum phase zero:

$$GH(s) = \frac{K(s-1)}{s(s+2)} \qquad (6.35)$$

The root loci plots for this system are shown in Fig. 6.12. For all positive values of K the system is unstable, and the locus from the open-loop pole at the origin of the s-plane moves to the zero at -1 as K increases (see Fig. 6.12(a)). However, if positive feedback is used – that is, if the root loci plot for decreasing values of K is produced (see Fig. 6.12(b)) – then the system is stable for $0 > K > -2$. From this root loci plot, a gain of $K = -1$ indicates that the two closed-loop poles are at $-0.5 \pm j0.866$. What may not have been expected is the resulting closed-loop system's unit step response plot, shown in Fig. 6.13.

Now consider the following open-loop system $GH(s)$ with a non-minimum phase pole:

$$GH(s) = \frac{K(s+1)}{s(s-3)} \qquad (6.36)$$

The root loci plot for this system is shown in Fig. 6.14. The closed-loop system is seen to be stable for all values of K greater than 3, and unstable otherwise. When $K = 10$ the closed-loop poles are at -2 and -5, and the system's time response

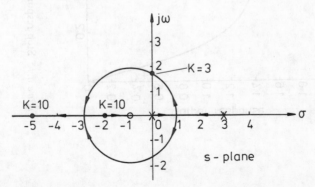

Figure 6.14 Root loci plot for the open-loop system $GH(s) = K(s+1)/[s(s-3)]$

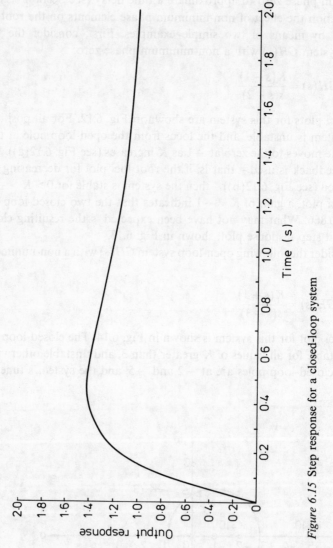

Figure 6.15 Step response for a closed-loop system

to a unit step input is given by

$$y(t) = 1 + 1.67e^{-2t} - 2.67e^{-5t} \tag{6.37}$$

The unit step response plot of this system exhibits an overshoot (see Fig. 6.15), even though the dynamics Equation (37) are not oscillatory.

6.8   Comments on the root loci design procedure

The root loci design procedure is clearly an iterative one, and the more design variables there are, the greater the required number of root loci plots. For three or more variables the method becomes unwieldy as a design technique, although it is still useful as a test of the system's sensitivity to changes in variables and loop integrity. (Loop integrity is concerned with the stability and response of a system in the face of some failure: when a loop breaks, how will the system respond?)

Since the root loci method gives the closed-loop pole positions, it provides the designer with considerable insight into a system's stability, performance and response characteristics. With experience it is possible to make a quick sketch of the general shape of a root loci diagram that will show what type of controller or compensator network is required to meet a particular design criterion.

The method's main limitations are its inability to deal with more than one variable at a time, and the difficulty normally experienced when dealing with time delays.

Problems

6.1. Produce the root loci plots for closed-loop control systems with unity negative feedback and having the open-loop transfer functions given below. In each case the roots of the characteristic equation may be solved directly, and positive values of K assumed.

(a) $\dfrac{K}{s(s+2)}$;   (b) $\dfrac{K}{s^2 + 4s + 3}$;   (c) $\dfrac{K(s+1)}{s^2 + 2s + 1}$

6.2. A feedback system has the open-loop transfer function

$$GH(s) = \frac{K(s+3)}{s^2 - 2s + 2}$$

Use Evans' rules to plot the root loci diagram for the closed-loop system. Find the gain K that will give the closed-loop system a damping factor of 0.707. With this value of gain, what would be the system's 5 per cent settling time?

6.3. For the following closed-loop characteristic equations, plot the root loci diagrams for positive values of K:

(a) $\qquad s^3 + (2+K)s^2 + (4-2K)s + K = 0$

(b) $\qquad s^3 + (5+K)s^2 + (6+K)s + 2K = 0$

(c) $\qquad s^4 + Ks^3 + (7K-5)s^2 + 12Ks + 4 = 0$

6.4. A control system with unity negative feedback has an open-loop transfer function given by

$$GH(s) = \frac{K}{s(s+3)(s+6)}$$

(a) Show that the points $s_1 = -0.55 + j3.0$ and $s_2 = -1.1 + j1.4$ are, to a sufficient degree of accuracy, on a root locus. In each case give the value of K.

(b) Determine the gain K that would give the dominant closed-loop poles a real value of -0.8.

(c) With the gain calculated in (b), what is the closed-loop transfer function?

(d) Use the graphical method to find the response of this closed-loop system to a unit step input.

6.5. The open-loop transfer function of a closed-loop control system with unity negative feedback is

$$G(s) = \frac{K}{s(s+3)(s^2+6s+64)}$$

Plot the root loci diagram for this system, and hence determine the closed-loop gain that gives an effective damping factor of 0.707.

6.6. The open-loop transfer function of a closed-loop control system with unity negative feedback is given by

$$G(s) = \frac{90s+9}{800s^2(5s+1)}$$

Construct the root loci diagram for this system and hence determine the closed-loop poles.

CHAPTER 7

FREQUENCY RESPONSE
DESIGN METHODS

7.1   Introduction

Design methods based on a system's steady-state frequency response characteristics
have many attractions. High-order systems, complex controller dynamics and time
delays are all easily handled. The main disadvantage is that for systems higher
than second order there is no direct relationship between the transient response
in the time domain and the various frequency response plots. However, experience
shows that, provided certain frequency conditions are met, good transient response
characteristics may be expected. These conditions are usually known as frequency
performance criteria.

In this chapter, the relationship between the Laplace and frequency domain is
explored. Conditions for closed-loop instability are established and used to gain
insight into various frequency performance criteria. The Nyquist, Bode and Nichols
techniques are also examined.

7.2   The steady-state frequency response

A linear control system subjected to a continuous sinusoidal forcing input will
produce a transient and a steady-state response. The transient response is produced
by the dynamics of the system, and will decay to zero with time provided the
system is stable. When the transient has decayed, the output resulting from the
forcing input remains, and is referred to as the steady-state response. It is a study
of this steady-state sinusoidal response over a range of frequencies that provides
the basis for the stability and design techniques presented in this chapter.

7.2.1   Transfer functions and the steady-state frequency response

Consider the open-loop system shown in block diagram form in Fig. 7.1. This
system is assumed to be governed by ordinary linear differential equations which

Figure 7.1 A forced open-loop system

are Laplace-transformable, so that

$$Y(s) = \frac{Q(s)}{P(s)} R(s) + \frac{I(s)}{P(s)} \qquad (7.1)$$

where $I(s)$ is a function of the initial conditions on $y(t)$ and $G(s) = Q(s)/P(s)$ is the system's open-loop transfer function.

The output $y(t)$ has a transient and a steady-state response. In mathematical terms, this is equivalent to saying that the solution of an ordinary linear differential equation is given by the complementary function plus the particular integral. The complementary function gives the transient response, and the particular integral the steady-state response.

Before examining the steady-state frequency response, it is useful first to consider a system's step response. The forcing input producing the step response is given by

$$r(t) = \begin{cases} 0 & \text{for } t < 0 \\ A & \text{for } t \geqslant 0 \end{cases} \qquad (7.2)$$

If the system is asymptotically stable, all its poles and zeros are contained within the left half s-plane. The forcing input $r(t)$ introduces a further pole at the origin of this s-plane plot. By using this plot the output $y(t)$ may be obtained graphically. If, though, only the steady-state response is required, this may be obtained from the residue associated with the pole at $s = 0$, and found to be

$$y(\infty) = A \frac{Q(s)}{P(s)} \bigg|_{s=0} = AG(s)|_{s=0} \qquad (7.3)$$

It was shown in Chapter 4 that all the poles in the left half s-plane produce decaying exponential terms in the time domain, and that their contribution to $y(t)$ therefore diminishes with increasing time. Equation (7.3) indicates that the steady-state response to a forcing step input is simply the system's gain, $G(s)|_{s=0}$, times the amplitude A on the step. Since only the residue associated with the pole at the origin of the s-plane is required, the steady-state response is easily calculated. Also, note that this residue is dependent on the pole–zero locations.

A similar but more detailed analysis of a system's steady-state response to a harmonic forcing input is now developed. The input term is of the form

$$r(t) = a \sin \omega t \qquad (7.4)$$

which, on Laplace transformation, becomes

$$R(s) = \frac{\omega a}{s^2 + \omega^2} \qquad (7.5)$$

Substituting for $R(s)$ in Equation (7.1) gives

$$Y(s) = \frac{Q(s)}{P(s)} \frac{\omega a}{s^2 + \omega^2} + \frac{I(s)}{P(s)} \qquad (7.6)$$

Since the system is assumed to be stable, the effect of the initial conditions diminishes with time and

$$\mathscr{L}^{-1}\left[\frac{I(s)}{P(s)}\right] \to 0 \quad \text{as } t \to \infty$$

Therefore

$$y(t) = \mathscr{L}^{-1}\left[\frac{Q(s)}{P(s)} \frac{\omega a}{s^2 + \omega^2}\right] \quad \text{as } t \to \infty \qquad (7.7)$$

In order to solve for the steady-state response of Equation (7.7), it is first necessary to make a partial-fraction expansion of the terms requiring inverse transformation, namely the terms within the square brackets:

$$\frac{Q(s)}{P(s)} \frac{\omega a}{s^2 + \omega^2} = \frac{A}{s - j\omega} + \frac{B}{s + j\omega} + (\text{all terms arising from } P(s)) \qquad (7.8)$$

As the system is stable, all the terms arising from the system's characteristic function $P(s)$ must be functions which disappear with time as $t \to \infty$. Hence, the steady-state response may be found by solving for A and B. Using the Heaviside formula,

$$A = \left(\frac{Q(s)}{P(s)} \frac{\omega a(s - j\omega)}{(s - j\omega)(s + j\omega)}\right)\Bigg|_{s = j\omega} \qquad (7.9a)$$

which reduces to

$$A = \frac{a}{2j} \frac{Q(j\omega)}{P(j\omega)} = \frac{a}{2j} G(j\omega) \qquad (7.9b)$$

Similarly,

$$B = -\frac{a}{2j} G(-j\omega) \qquad (7.10)$$

The terms $G(j\omega)$ and $G(-j\omega)$ are complex, and their real and imaginary parts are functions of ω. Hence, by the normal rules of complex algebra, these terms may be written in polar form as

$$G(j\omega) = M(\omega)e^{j\phi(\omega)} \quad \text{or} \quad Me^{j\phi} \qquad (7.11a)$$

and

$$G(-j\omega) = M(\omega)e^{-j\phi(\omega)} \quad \text{or} \quad Me^{-j\phi} \tag{7.11b}$$

Thus, from Equations (7.9) and (7.10),

$$A = \frac{a}{2j}Me^{j\phi} \tag{7.12a}$$

$$B = \frac{a}{2j}Me^{-j\phi} \tag{7.12b}$$

Substituting the expansion of Equation (7.8) back into Equation (7.7) yields the steady-state response as

$$y(t)_{ss} = \mathcal{L}^{-1}\left[\frac{A}{s-j\omega} + \frac{B}{s+j\omega}\right] \tag{7.13}$$

which, on inverse Laplace transformation, becomes

$$y(t)_{ss} = Ae^{+j\omega t} + Be^{-j\omega t} \tag{7.14}$$

Substituting Equations (7.12) for A and B into Equation (7.14) gives

$$y(t)_{ss} = \frac{a}{2j}Me^{+j\phi}e^{j\omega t} - \frac{a}{2j}Me^{-j\phi}e^{-j\omega t}$$

or

$$y(t)_{ss} = \frac{a}{2j}M(e^{j(\phi+\omega t)} - e^{-j(\phi+\omega t)})$$

Since $\sin x = (1/2j)(e^{jx} - e^{-jx})$, then

$$y(t)_{ss} = aM\sin(\omega t + \phi) \tag{7.15}$$

On comparing the steady-state output (Equation (7.15)) with the input (Equation (7.4)), it is evident that the input signal has been multiplied by the factor M, known as the magnitude ratio or magnification factor. As M (or, more correctly, $M(\omega)$) is a function of frequency, for any given frequency it may be determined directly from the open-loop transfer function $G(s)$. Also, the steady-state input and output signals have the same frequency of oscillation, but the output is shifted by an amount ϕ. This phase shift $\phi(\omega)$, like the magnification factor, is frequency-dependent and may be determined directly from the open-loop transfer function $G(s)$.

The above analysis is simply a calculation of the residues associated with a pair of complex conjugate poles on the imaginary s-plane axis (see Section 4.2). Provided the system is linear and stable, the steady-state response is characterized by $M(\omega)$ and $\phi(\omega)$, which in turn depend on the transfer function $G(s)$. From Equations (7.9) and (7.11),

$$G(s)|_{s=j\omega} = G(j\omega) \tag{7.16a}$$

and

$$G(j\omega) = \frac{Q(j\omega)}{P(j\omega)} \tag{7.16b}$$

One way to find $G(j\omega)$ is to replace every value of s in $G(s)$ by $j\omega$, and then to reduce $G(j\omega)$ to its simplest form using the normal rules of complex algebra. For example, if

$$G(s) = \frac{s+4}{s^2+5s+6} \tag{7.17}$$

then

$$G(j\omega) = \frac{j\omega+4}{(j\omega)^2+5j\omega+6} \tag{7.18}$$

which may be simplified to give

$$G(j\omega) = \frac{\omega^2+24}{\omega^4+13\omega^2+36} + j\frac{\omega(\omega^2-14)}{\omega^4+13\omega^2+36} \tag{7.19}$$

Equation (7.19) uses rectangular coordinates – that is, it has the form

$$G(j\omega) = U(\omega) + jV(\omega) \tag{7.20}$$

To convert this equation to polar form, let

$$G(j\omega) = M(\omega)e^{j\phi(\omega)} \tag{7.21a}$$

where

$$M(\omega) = \sqrt{(U(\omega)^2 + V(\omega)^2)} \tag{7.21b}$$

and

$$\phi(\omega) = \tan^{-1}[V(\omega)/U(\omega)] \tag{7.21c}$$

This procedure can become quite complex, particularly if the transfer function is of high order. There is a simpler way to find $M(\omega)$ and $\phi(\omega)$, and this is best demonstrated by considering the system's s-plane pole–zero map.

For the transfer function

$$G(s) = \frac{s+4}{s^2+5s^2+6} \tag{7.22}$$

the pole–zero map is shown in Fig. 7.2. By using the graphical technique described in Chapter 4, the function $G(j\omega_1)$ at the point $j\omega_1$ is evaluated as

$$G(j\omega_1) = M(\omega_1)e^{j\phi(\omega_1)} \tag{7.23}$$

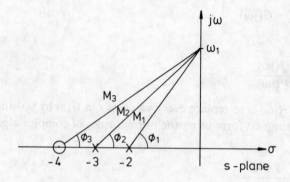

Figure 7.2 Pole–zero map for the system $G(s) = (s+4)/(s^2 + 5s + 6)$

where

$$M(\omega_1) = \frac{M_3}{M_1 M_2}$$

which becomes

$$M(\omega_1) = \frac{\sqrt{(\omega_1^2 + 4^2)}}{\sqrt{(\omega_1^2 + 2^2)}\sqrt{(\omega_1^2 + 3^2)}} \tag{7.24a}$$

and

$$\phi(\omega_1) = \phi_3 - (\phi_1 + \phi_2)$$

which becomes

$$\phi(\omega_1) = \tan^{-1}(\omega_1/4) - \tan^{-1}(\omega_1/2) - \tan^{-1}(\omega_1/3) \tag{7.24b}$$

Note that when ω_1 is set equal to zero, $M(0)$ gives the system's gain and $\phi(0)$ is zero. This is equivalent to the evaluation of the constant associated with a unit step input – that is, with the calculation of the residue-associated with a pole at the origin of the s-plane (see Equation (7.3)).

In general, then, a transfer function in which s is set equal to $j\omega$ may be represented by an expression of the form

$$G(j\omega) = K \frac{\prod\limits_{i=1}^{m} (a_i + jc_i)}{\prod\limits_{i=1}^{n} (g_i + jh_i)} \tag{7.25}$$

where m and n are the number of terms in the numerator and denominator, respectively. For this system

$$M(\omega) = K \frac{\prod\limits_{i=1}^{m} \sqrt{(a_i^2 + c_i^2)}}{\prod\limits_{i=1}^{n} \sqrt{(g_i^2 + h_i^2)}} \tag{7.26}$$

and

$$\phi(\omega) = \sum_{i=1}^{m} \tan^{-1}(c_i/a_i) - \sum_{i=1}^{n} \tan^{-1}(h_i/g_i) \qquad (7.27)$$

Before leaving this section, note that in general a forcing input excites many harmonic forcing signals, producing a response which, depending on the dynamics of the plant, is either attenuated or amplified. Furthermore, these signals will be phase-shifted relative to each other, with the effect that the recombined signals will tend to weaken or strengthen each other.

Example 7.1

A temperature recording system with transfer functions as shown in Fig. 7.3 is used to measure a time-varying fluid temperature in an experiment. For steady-state operation a Fourier analysis of the recorder output trace yields the expression

$$Q_0 = 1.5 + \sin 10t + 0.2 \sin 60t$$

Determine the corresponding expression for the true input temperature.

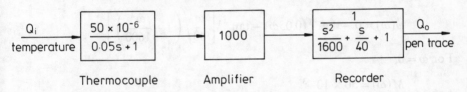

Figure 7.3 A temperature recording system

Solution

The overall transfer function for the temperature recording system is

$$G(s) = \frac{50 \times 10^{-3}}{(0.05s + 1)\left(\dfrac{s^2}{1600} + \dfrac{s}{40} + 1\right)} = \frac{Q_o(s)}{Q_i(s)}$$

Since the steady-state output was given by

$$Q_{o,ss} = 1.5 + \sin 10t + 0.2 \sin 60t$$

the corresponding input must have been of the form

$$Q_i = A + B \sin(10t - \phi(10)) + C \sin(60t - \phi(60))$$

In general, for an input of the form

$$q_i(t) = a \sin \omega t$$

the corresponding steady-state output will be

$$(q_o(t))_{ss} = aM(\omega)\sin(\omega t + \phi(\omega))$$

To find $M(\omega)$ and $\phi(\omega)$ from $G(s)$, set $s = j\omega$; then

$$G(j\omega) = \frac{50 \times 10^{-3}}{(0.05j\omega + 1)\left(-\dfrac{\omega^2}{1600} + \dfrac{j\omega}{40} + 1\right)}$$

$$= \frac{50 \times 10^{-3}}{(1 + 0.05j\omega)\left(1 - \dfrac{\omega^2}{1600} + \dfrac{j\omega}{40}\right)}$$

Hence, from Equation (7.26),

$$M(\omega) = \frac{50 \times 10^{-3}}{\sqrt{[1 + (0.05\omega)^2]}\sqrt{\left[\left(1 - \dfrac{\omega^2}{1600}\right)^2 + \dfrac{\omega^2}{1600}\right]}}$$

and from Equation (7.27)

$$\phi(\omega) = 0 - \tan^{-1}(0.05\omega) - \tan^{-1}\left[\frac{\omega}{40}\bigg/\left(1 - \frac{\omega^2}{1600}\right)\right]$$

For $\omega = 0$,

$$M(\omega) = 50 \times 10^{-3}$$

which is the system's steady-state gain. Hence, to obtain a constant output signal of 1.5 the input signal must be

$$1.5/50 \times 10^{-3} = 30$$

For $\omega = 10$,

$$M(10) = \frac{50 \times 10^{-3}}{\sqrt{(1.25)}\sqrt{(0.9414)}} = 46.1 \times 10^{-3}$$

and

$$\phi(10) = -0.464 - 0.261 = -0.725 \text{ rad}$$

For $\omega = 60$,

$$M(60) = \frac{50 \times 10^{-3}}{\sqrt{(10)}\sqrt{(3.8125)}} = 8.10 \times 10^{-3}$$

and

$$\phi(60) = -1.249 + 0.876 = -0.373 \text{ rad}$$

Consequently, the true input temperature was

$$Q_i = 30 + 21.7\sin(10t + 0.725) + 24.7\sin(60t + 0.373)$$

7.2.2   Pole–zero plots and the steady-state frequency response

It has been shown in Section 7.2.1 that $G(j\omega)$ may be found for any value of ω by making direct measurements from each pole and zero in the s-plane to the point ω on the imaginary axis. This relationship between the s-plane and the frequency response is worth further examination. From Equations (7.25)–(7.27) it may be seen that the magnitude of $G(j\omega)$ is the product of the magnitude of each zero divided by the product of the magnitude of each pole. Similarly, the phase of $G(j\omega)$ is the sum of the phases of each zero less the sum of the phases from each pole. Consequently, the contribution from each pole and zero in a transfer function may be considered separately.

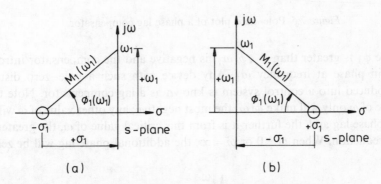

Figure 7.4 (a) Minimum and (b) non-minimum phase zeros

Figure 7.4 shows the difference between a minimum and a non-minimum phase zero. In both cases, the magnitude contribution $M_1(\omega_1)$ to the frequency response $G(j\omega_1)$ is the length of the vector from the zero to the point ω_1 on the imaginary axis. The phase contribution is found from Equation (7.27). For the minimum phase zero located at $s = -\sigma_1$, the phase contribution is

$$\tan^{-1}(\omega_1/\sigma_1) = \phi_1(\omega_1) \tag{7.28}$$

and for the non-minimum phase zero at $s = +\sigma_1$ the phase contribution is

$$\tan^{-1}(\omega_1/-\sigma_1) = -\tan^{-1}(\omega_1/\sigma_1) = \pi - \phi_1(\omega_1) \text{ or } -\phi_1(\omega_1) \quad (7.29)$$

In Fig. 7.4(b) the phase angle of the non-minimum phase zero is measured relative to the negative σ-axis, and since the vector is rotated clockwise the phase angle is negative.

With poles, the magnitudes and phase angles are measured in exactly the same way, but the contribution to $G(j\omega)$ (Equations (7.26) and (7.27)) will be the reciprocal of the magnitude, and the negative of the measured phase angle.

In control work the phase of a system often needs to be modified around a given frequency. This may be accomplished by means of a pole–zero pair, as shown in Figs 7.5 and 7.6. In Fig. 7.5, the net phase contribution to the system at frequency ω_1 is

$$\phi(\omega_1) = \phi_2 - \phi_1 \tag{7.30}$$

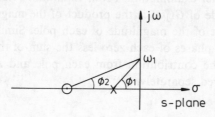

Figure 7.5 Pole–zero plot of a phase lag compensator

Since ϕ_1 is greater than $\phi_2, \phi(\omega_1)$ is negative and the compensator introduces a lag in phase at frequency ω_1. Any device with such a pole–zero distribution introduced into a control system is known as a lag compensator. Note that one value of ω only will make $\phi(\omega)$ the most negative; any other value of ω will reduce the phase lag and, the further it is from this critical value of ω, the greater will be the reduction. When $\omega = 0$ or $\omega = \infty$, the additional phase lag will be zero.

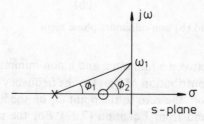

Figure 7.6 Pole–zero plot of a phase lead compensator

The phase lead compensator shown in Fig. 7.6 also has a net phase contribution given by Equation (7.30). Here, however, ϕ_2 is greater than ϕ_1, and there will be a positive shift in phase; a phase lead. Again, there will be one value of ω for which the phase lead is a maximum.

A transfer function $G(s)$ may be plotted in the s-plane in terms of its pole–zero map. Once plotted, $G(s)$ may be evaluated at any point in the s-plane. In particular, when $s = j\omega$, $G(j\omega)$ gives the steady-state frequency response at frequency ω. Since $G(j\omega)$ is complex, the locus of the $G(j\omega)$ points, for increasing values of ω, may be plotted on an Argand diagram. This Argand diagram is called the $G(s)$-plane, since any point in the s-plane – and not just points on the $j\omega$-axis – may be mapped onto a corresponding point on the diagram. Familiarity with the mapping of points from the s-plane onto the $G(s)$-plane using a graphical (or some other) method is a prerequisite to the understanding of the Nyquist stability criterion.

7.3   Conditions for closed-loop marginal stability

Marginal stability in a closed-loop system occurs when the controller and other control elements in the loop are adjusted so as to produce a self-sustained steady-state cycling in the response. Such a condition requires the closed-loop system to have two complex conjugate poles on the imaginary axis of the s-plane, and all its other poles to have negative real parts. By quantifying this condition in terms of a system's frequency response, it is shown to provide a basis for the definition of performance criteria.

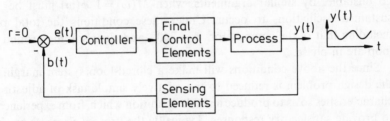

Figure 7.7 A marginally stable closed-loop system

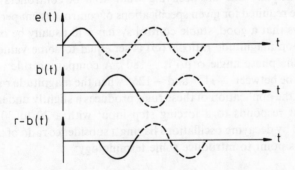

Figure 7.8 Time response of a marginally stable system

Consider the marginally stable closed-loop system shown in Fig. 7.7, which is performing self-sustained oscillations. A time profile of the system's response is shown in Fig. 7.8, in which the first half cycle of a sine wave is shown as a solid line in the $e(t)$ plot. This signal passes through the various elements in the loop and emerges at $b(t)$ with the same amplitude but with a phase lag of 180°, as shown. The signal then enters the comparator, which in this case inverts the signal

$b(t)$. Wave inversion is equivalent to a phase shift of $-180°$, since

$$-b = -A \sin \omega t = A \sin(\omega t - 180°) \tag{7.31}$$

The output of the comparator, $r - b(t)$, is in phase with the error signal $e(t)$ and provides the second half of the sine wave. This process repeats itself continuously, and the system performs self-sustained oscillations.

Clearly, two conditions must be met if sustained oscillations are to occur: at the signal frequency ω, $M(\omega)$ must be unity and $\phi(\omega)$ must be $-180°$ (that is, the phase lag must be $180°$). If $M(\omega)$ were less than unity, 0.5 say, then the feedback signal $b(t)$ would be half the amplitude of the input signal $e(t)$, phase-shifted by $-180°$. This signal would be inverted to form the new error signal to $e(t)$. As the process continued, each successive peak in the error signal would be a quarter the height of its predecessor, and the system would have quarter-amplitude damping – that is, the subsidence ratio would be 4:1 (see Section 4.6.1). If $M(\omega)$ were greater than unity, the amplitude of successive peaks would increase and the system would go unstable. By similar arguments, when $M(\omega) = 1$ $\phi(\omega)$ must be $-180°$ for sustained oscillations to occur. Under these conditions the total phase shift, including that of the comparator, is $-360°$, and signals going around the closed loop are in phase.

Since the above conditions will make a closed-loop system marginally stable, the design problem is reduced to the relatively simple task of adjusting the loop characteristics so as to produce a stable condition which, from experience, is known to provide satisfactory responses. Typically, the process elements are known but are not easily accessible and cannot be altered to any great extent. Consequently, the problem is principally one of specifying what kind of controllers and final control elements are required for given specifications of control system performance.

Experience shows that a good, stable control system will usually be obtained if the overall open-loop magnitude ratio $M(\omega)$ is set equal to some value between 0.4 and 0.5 when the phase angle $\phi(\omega)$ is $-180°$. A companion rule is that the phase angle should be between $-115°$ and $-125°$ when the magnitude ratio $M(\omega)$ is unity. In general, the application of these rules produces a slightly under-damped control system that responds to a forcing step input with a 20 to 30 per cent overshoot, followed by decaying oscillations having a subsidence ratio of about 3:1.

It is useful at this point to introduce some terminology:

Phase crossover is the frequency at which the overall open-loop phase angle $\phi(\omega)$ first reaches the critical value of $-180°$.

Gain crossover is the frequency at which the overall open-loop magnitude ratio $M(\omega)$ first reaches the value of unity.

Phase margin is the number of degrees by which the phase angle is numerically smaller than the critical angle of $-180°$ at gain crossover.

Gain margin is the factor by which the magnitude ratio must be multiplied at phase crossover to make it unity.

The empirically derived design rules may now be restated in terms of the more commonly used measures of gain and phase margin, as:

Gain margin between 2 and 2.5, phase margin between 45° and 65°.

In using these rules the following points should be noted:

1. The above design rules do not give accuracy of control. It is possible to produce a satisfactory transient response, but still have an unacceptable steady-state error.
2. The rules say nothing about desirable speeds of response.

Example 7.2

The open-loop frequency response characteristics of a simple single-loop control system, including the controller but excluding the comparator, are found to be as follows:

Frequency, cycles/min	0.01	0.06	0.10	0.06	1.0
Magnitude ratio	4.82	1.00	0.47	0.08	0.02
Phase angle	−10°	−122°	−180°	−272°	−316°

What are the gain margin and phase margin for this system?

Solution

From the data in the table, the gain margin (which is the safety factor in the magnitude ratio when the phase angle is −180°) is 1.00/0.47, or 2.13, and the phase margin (which is the margin of safety in the phase angle at gain crossover, overall magnitude equal to unity) is 180° − 122°, or 58°.

7.4   Frequency response plots

A frequency response consists of three variables – frequency, magnitude and phase – which may be represented in several ways. The most common are described below.

7.4.1   Logarithmic plots of Bode diagrams

These are two plots in rectangular coordinates in which the magnitude is expressed in decibels (dB), and the phase angle in degrees, both plotted as functions of the logarithm of frequency in rad/unit time.

Bode diagrams are normally plotted on semi-log graph paper. The frequency response magnitude ratio is expressed in decibels,

$$M(\omega)_{dB} = 20 \log_{10} M(\omega) \qquad (7.32)$$

where $M(\omega)_{dB}$ is the log modulus in decibels and $M(\omega)$ is the magnitude ratio.

One of the main advantages of using a logarithmic scale for the magnitude ratio is the ease with which the dynamic elements in a control loop can be manipulated. At a given frequency, the magnitude ratio of an open-loop transfer function is obtained by multiplying together the individual magnitude ratios of the elements, and the phase angle is obtained by summing the individual angles. By using Bode diagrams, both these tasks are accomplished by graphical summation.

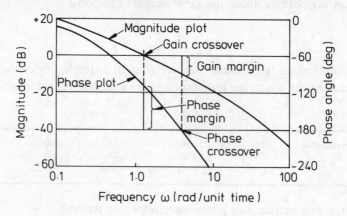

Figure 7.9 A typical open-loop Bode diagram

Typical Bode plots are shown in Fig. 7.9. For convenience, the two types of diagram are shown together. As usual, the magnitude ratio falls off and the phase angle becomes increasingly lagging with increasing frequency.

7.4.2   Polar and inverse polar plots (Nyquist and Inverse Nyquist diagrams)

Nyquist diagrams, named after Harry Nyquist (1889–1976), are mappings of the Nyquist contour from the s-plane onto the $G(s)$-plane (see Section 7.7.2). Since the Nyquist contour includes the imaginary s-plane axis, and mappings along this axis produce the polar plot, then polar plots form part of the Nyquist diagram. Furthermore, after testing for stability only the polar plot portion of the diagram is used for design work. Similarly, the inverse Nyquist diagram contains the inverse polar plot. Again, after checking for stability only that portion of the diagram containing the inverse polar plot is used.

The polar plot is a plot of the gain (magnitude) and phase locus for all frequencies. The inverse polar plot is similar and shows the reciprocal function, the locus of the inverse of the gain and negative phase for all frequencies. Both diagrams may be plotted on polar or normal linear graph paper.

Nyquist diagrams are particularly useful as they can be used to determine the closed-loop stability of a system directly from the open-loop polar plot. It should be noted that, except in specific cases, satisfactory gain and phase margin do not guarantee closed-loop stability. This means that with Bode diagrams (see Section 7.6), for example, it may be necessary to test the stability of a particular design using the Nyquist or the Routh stability criterion.

Inverse Nyquist diagrams are particularly useful when the system has a minor feedback loop, or if feedback compensation (the deliberate introduction of dynamic elements into the feedback path in order to meet a particular design specification) is required. Familiarity with inverse Nyquist diagrams is a prerequisite for a frequency domain study of multivariable systems (systems with several inputs and outputs).

A typical polar plot is shown in Fig. 7.10, and a typical inverse polar plot in Fig. 7.11, for a system having an open-loop transfer function $G(s)$. In both the Nyquist and inverse Nyquist diagrams (or polar and inverse polar plots) a unit circle is the locus of all points at which the respective magnitudes will be unity, and the negative real axis is the locus of all points at which the phase lag is 180°. Arrows on the $G(j\omega)$ and $[G(j\omega)]^{-1}$ loci indicate the directions of increasing frequency.

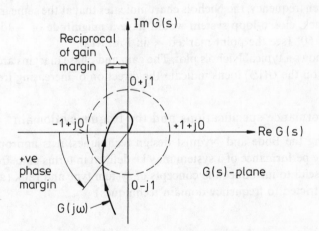

Figure 7.10 A typical open-loop polar plot

7.4.3   Rectangular coordinate plot of magnitude and phase, or Nichols plot

In this plot the y-axis represents the magnitude expressed in decibels, and the x-axis the phase angle in degrees. Linear graph paper could be used to produce a Nichols plot, although it would be more common to use a Nichols chart (both named after Nathaniel Burgess Nichols, b. 1914).

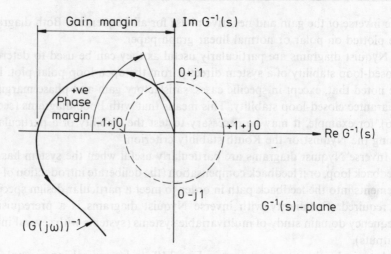

Figure 7.11 A typical open-loop inverse polar plot

The Nichols chart consists of specially prepared linear graph paper (see Fig. 7.12). By plotting the open-loop magnitude and phase locus on this chart, the system's closed-loop frequency response characteristics may be determined. For example, if an open-loop system has a magnitude ratio of -8dB and a phase shift of $-80°$ at a given frequency, the Nichols chart indicates that at the same frequency the unity feedback, closed-loop system would have a magnitude of -9dB and a phase shift of $-60°$ (see the point marked × in Fig. 7.12).

Figure 7.13 shows a typical Nichols plot. The gain and phase margins are shown, and the arrows on the $G(\text{j}\omega)$ locus indicate the direction of increasing frequency.

7.5   Performance specifications and the frequency domain

Before examining the Bode and Nyquist design techniques, it is appropriate to consider how the performance of a system may be defined in terms of its frequency response. It is useful to introduce the concepts of system type and rank (although they are not restricted to frequency domain techniques).

7.5.1   System type and rank

The open-loop transfer function of a system between its forcing input and measured output may be written as

$$KGH(s) = \frac{K \prod_{i=1}^{m} (s + z_i)}{s^r \prod_{i=1}^{n-r} (s + p_i)} \tag{7.33}$$

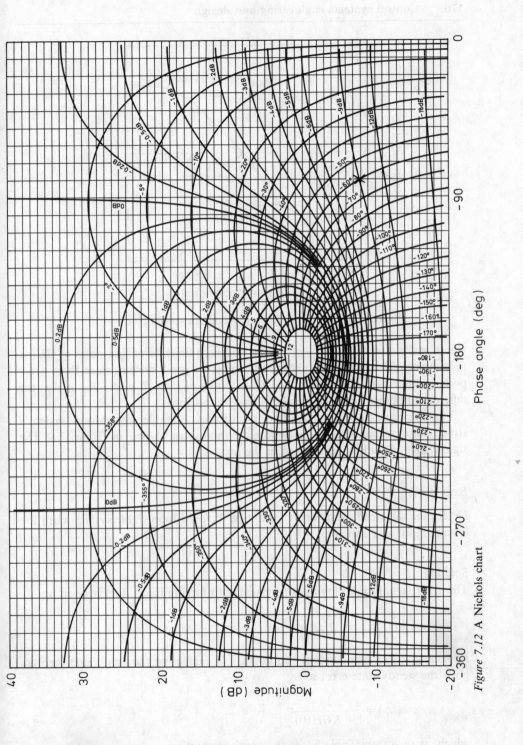

Figure 7.12 A Nichols chart

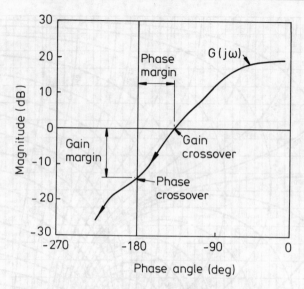

Figure 7.13 A typical open-loop magnitude–phase plot

where $m \leqslant n$, and $-z_i$ and $-p_i$ are the non-zero finite zeros and poles of $GH(s)$, respectively. This transfer function represents a type r system, since there are r poles at the origin of the s-plane. The system is of rank R, where $R = m - n$, the difference between the total number of poles and zeros.

The importance of type is that it indicates a stable closed-loop system's steady-state response to various forcing inputs, as shown in Table 7.1. The significance of rank is highlighted in the text.

Example 7.3

Determine the steady-state error for a type 0 closed-loop stable system when the forcing input is a unit step.

Solution

The error signal for a closed-loop system with forcing input $R(s)$ is given by

$$E(s) = \frac{R(s)}{1 + KGH(s)}$$

Replacing $R(s)$ by its Laplace transform, $1/s$, and using the final value theorem gives the steady-state error as

$$e(\infty) = \frac{1}{1 + KGH(0)}$$

which, as seen from Table 7.1, is the required solution.

Table 7.1 Steady-state error for system types 0, 1 and 2

System type	Input		
	Unit step $\left(\dfrac{1}{s}\right)$	Unit ramp $\left(\dfrac{1}{s^2}\right)$	Unit parabola $\left(\dfrac{1}{s^2}\right)$
Type 0	$\dfrac{1}{1+KGH(0)}$	∞	∞
Type 1	0	$\dfrac{1}{KGH(0)}$	∞
Type 2	0	0	$\dfrac{1}{KGH(0)}$

7.5.2   Frequency domain specifications

The most commonly used open-loop frequency domain specifications are the gain margin and phase margin, introduced in Section 7.3, and described below. For open-loop stable systems of types 0, 1 and 2, a gain margin which is greater than unity combined with a phase margin greater than 0° will guarantee closed-loop stability (see Section 7.7.3). With unstable open-loop systems, or systems having any other type number, alternative methods for checking closed-loop stability must be employed.

(i)   Gain Margin

For a stable closed-loop system the gain margin gives the amount by which the loop gain may be increased before the system becomes unstable. From Chapter 6 it is clear that this change in gain will move a pair of complex conjugate closed-loop poles onto the imaginary axis. However, pole positions, or the distance moved in the s-plane by any of the system poles, are not indicated.

(ii)   Phase Margin

For second-order systems, phase margin and damping ratio can be shown to be related. A simple rule of thumb states that the numerical value of the phase margin in degrees divided by 100 gives the closed-loop damping ratio. With systems higher than second order this rule gives quite a good indication of the damping ratio, but it should be used in conjunction with the other frequency domain specifications.

The remaining frequency domain performance specifications relate to the frequency response of the closed-loop system.

(iii)   *Maximum Magnitude Ratio*

This is usually referred to as 'M peak' (M_p), and gives an indication of a system's relative stability. For a second-order system, it may be shown that

$$\zeta^2 = \frac{1}{2} - \frac{1}{2}\sqrt{\left(1 - \frac{1}{M_p^2}\right)}, \quad M_p \geqslant 1 \tag{7.34}$$

Normally, large M_p corresponds to a large peak overshoot in the step response. For most design problems an optimum value of M_p would be somewhere between 1.1 and 1.5, which for a second-order system would give a damping ratio between 0.54 and 0.36, respectively.

(iv)   *Frequency at M_p*

For a second-order system ω_p, the resonant frequency is given by

$$\omega_p = \omega_n\sqrt{(1 - 2\zeta^2)} \tag{7.35}$$

and, like bandwidth, indicates the system's speed of response.

(v)   *Bandwidth*

Bandwidth is normally defined as the frequency at which the magnitude ratio drops to 0.707 ($1/\sqrt{2}$) of its zero-frequency level. On the decibel scale, this is equivalent to a fall of 3 dB from the zero frequency gain. Bandwidth gives a measure of the transient response properties. A large bandwidth corresponds to a faster rise time, since higher-frequency signals are passed to the output; if the bandwidth is small, only signals of relatively low frequencies are passed, and the time response will generally be slow and sluggish. The bandwidth of a system is also an indicator of its noise filtering characteristics: an unnecessarily wide bandwidth would produce a system with poor noise rejection characteristics.

(vi)   *Cutoff Rate*

This is the rate of decrease in the magnitude ratio outside the system's bandwidth. A high-frequency cutoff rate which is high would indicate a system with good signal-to-noise ratio. However, high cutoff characteristics may be accompanied by a large M_p, which in turn corresponds to a system with low relative stability. A typical cutoff rate specification is 6 dB/octave (that is, doubling the frequency should halve the magnitude ratio).

The closed-loop performance criteria (iii)–(vi) are illustrated on the frequency response plot shown in Fig. 7.14.

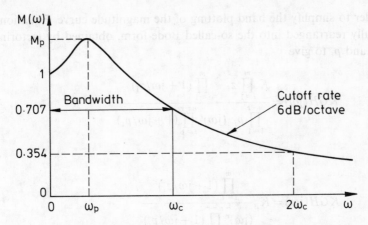

Figure 7.14 Closed-loop frequency response performance specifications

7.6   Bode plots

One of the main attractions of Bode diagrams (named after Hendrik K. Bode, b. 1905) is the ease with which they may be produced and modified by using pencil-and-paper methods. A system's transfer function, together with any necessary compensating elements, is approximated by a number of straight-line functions which may be summed graphically to produce plots accurate enough for most design studies. Typically, once the steady-state accuracy, gain and phase margin criteria are satisfied, open-loop data is transferred to a Nichols chart, from which the closed-loop response is found and checked for performance. Nowadays, simple computer programs like those given in Chapter 11 can produce accurate open-loop Bode plots in seconds, and a two-line algorithm will then convert the open-loop data into its closed-loop form. A computer thus removes the tedium of producing the various plots, leaving the engineer free to concentrate on the design problem.

This section considers the traditional method of producing Bode plots based on asymptotic (straight-line) approximations to the various elements within the loop. The prime reason for using this approach is that it gives insight into the selection and design of compensation elements.

7.6.1   Magnitude and phase equations

A system's open-loop frequency response may be obtained by setting $s = j\omega$ in the system's loop transfer function; thus, from Equation (7.33),

$$KGH(j\omega) = \frac{K_i \prod_{i=1}^{m} (j\omega + z_i)}{(j\omega)^r \prod_{i=1}^{n-r} (j\omega + p_i)} \tag{7.36}$$

In order to simplify the hand plotting of the magnitude curve, Equation (7.36) is normally rearranged into the so-called Bode form, obtained by factoring out all the z_i and p_i to give

$$KGH(j\omega) = \frac{K \prod\limits_{i=1}^{m} z_i \;\; \prod\limits_{i=1}^{m} (1+j\omega/z_i)}{\prod\limits_{i=1}^{n-r} p_i \; (j\omega)^r \prod\limits_{i=1}^{n-r} (1+j\omega/p_i)} \qquad (7.37a)$$

or

$$KGH(j\omega) = K_B \frac{\prod\limits_{i=1}^{m} (1+j\omega/z_i)}{(j\omega)^r \prod\limits_{i=1}^{n-r} (1+j\omega/p_i)} \qquad (7.37b)$$

In this form, the coefficient K_B is called the Bode gain.

The magnitude of $KGH(j\omega)$ for any frequency ω is the product of the Bode gain, the magnitude of each factor in the numerator and the reciprocal of the magnitude of each factor in the denominator. However, if the magnitude of $KGH(j\omega)$ is expressed in decibels, then

$$20\log|KGH(j\omega)| = 20\log|K_B| + \sum_{i=1}^{m} 20\log|1+j\omega/z_i|$$

$$- 20\log|(j\omega)^r| - \sum_{i=1}^{n-r} 20\log|1+j\omega/p_i| \qquad (7.38)$$

(Note that in logarithmic form all products become summations.) The phase angle of $KGH(j\omega)$ at any frequency is found in the normal way as

$$\arg KGH(j\omega) = \sum_{i=1}^{m} \arg(1+j\omega/z_i) - \arg(j\omega)^r$$

$$- \sum_{i=}^{n-r} \arg(1+j\omega/p_i) \qquad (7.39a)$$

or

$$\arg KGH(j\omega) = \sum_{i=1}^{m} \tan^{-1}(\omega/z_i) - (90r)^\circ - \sum_{i=1}^{n-r} \tan^{-1}(\omega/p_i) \qquad (7.39b)$$

Both the magnitude and phase plots are obtained by summing the contribution provided by each term in the open-loop transfer function. If all the coefficients of the unfactored transfer function are real, then each complex pole or zero in the factored transfer function will have a complex conjugate. This means that there are only four possible terms that need be considered: the gain term K_B, poles (or zeros) at the origin, real poles (or zeros) not at the origin and complex conjugate poles (or zeros).

7.6.2   Bode plots for a gain term

A constant K_B provides a magnitude contribution of $20 \log|K_B|$, and a phase angle of $0°$ if K_B is positive or $-180°$ if K_B is negative. If $K_B = 1$, then $20 \log|K_B|$ is zero; if $K_B = 2$, then the dB magnitude is 6; and if $K_B = 0.5$, the dB magnitude is -6. In all three cases the phase angle contribution is $0°$. For $K_B = -2$ or -0.5, the dB magnitude would be 6 or -6, respectively, but the phase contribution would be $-180°$. A Bode plot for a positive gain term is shown in Fig. 7.15.

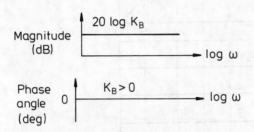

Figure 7.15 Bode plots for a gain term

7.6.3   Bode plots for poles or zeros at the origin of the s-plane

A system with poles or zeros at the origin of the s-plane will have an expression in its transfer function of the form

$$\frac{1}{s^r} \tag{7.40}$$

and the magnitude and phase contributions will be, respectively,

$$20 \log \frac{1}{(j\omega)^r} = -20r \log \omega \text{ dB} \tag{7.41}$$

$$\arg \frac{1}{(j\omega)^r} = -90r° \tag{7.42}$$

where r is a positive integer equal to the number of poles at the origin of the s-plane (that is, the system's type number is r). Also, r is a negative integer for zeros at the origin.

The equation for the Bode magnitude plot, Equation (7.41), describes a straight line of slope $-20r$ dB/decade passing through the 0 dB point when $\omega = 1$. (A decade indicates a tenfold increase in frequency.) The Bode phase plot, obtained from Equation (7.42), indicates that the phase angle is independent of frequency, and has a value which is dependent on r. Bode plots for one, two or three poles at the origin are shown in Fig. 7.16.

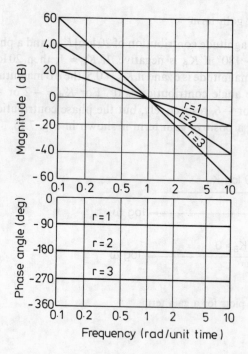

Figure 7.16 Bode plots for poles at the origin of the *s*-plane

Equations (7.41) and (7.42) are also valid for zeros at the origin of the *s*-plane. The Bode plots for a system with one, two or three zeros at the origin of the *s*-plane would be the reflections of the Bode plots in Fig. 7.16 about the 0 dB and 0° lines.

7.6.4   Bode plots for real poles and zeros

A real pole or zero not at the origin of the *s*-plane contributes to the system's transfer function a term of the form

$$\frac{1}{(1 + a/\tau)^r} \tag{7.43}$$

In Equation (7.43), *r* is any positive or negative integer. Its introduction serves two purposes; if *r* is positive the contribution is from system poles, and if it is negative the contribution is from system zeros. Also, if the magnitude of *r* is greater than unity, there are *r* poles or zeros occupying the same *s*-plane location, and the pole or zero is said to be of order *r*.

The dB magnitude and phase angle given by Equation (7.38) and (7.39) are now

$$-10r \log(1 + (\omega/\tau)^2) \tag{7.44}$$

and

$$-r\tan^{-1}(\omega/\tau) \tag{7.45}$$

Asymptotic approximations for the magnitude plot are obtained by considering the normalized frequency ω/τ. When ω/τ is very small, the dB magnitude (Equation (7.44)) approximates to

$$-10r\log 1 = 0\,\mathrm{dB} \tag{7.46a}$$

and for large values of ω/τ to

$$-20r\log(\omega/\tau)\,\mathrm{dB} \tag{7.46b}$$

The corresponding phase angle for small ω/τ, from Equation (7.45), is approximately

$$-r\tan^{-1}(0) = 0° \tag{7.47a}$$

and when ω/τ is very large, the corresponding phase angle tends to

$$-90°r \tag{7.47b}$$

From Equations (7.46), it is apparent that the Bode magnitude plot asymptotically approaches a horizontal straight line at 0 dB as $\omega/\tau \rightarrow 0$, and $-20r\log(\omega/\tau)\,\mathrm{dB}$

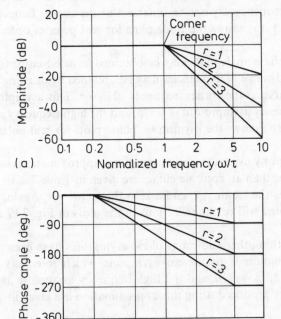

Figure 7.17 (a) dB magnitude and (b) phase angle Bode plots for real poles at $s = -1/\tau$

Table 7.2 Asymptotic errors for a real pole (or zero) of order r, and corner frequency τ, and mid-range phase asymptote between $\tau/5$ and 5τ

	$\tau/10$	$\tau/5$	$\tau/2$	τ	2τ	5τ	10τ
Magnitude error (dB)	$-0.043r$	$-0.17r$	$-0.96r$	$-3r$	$-0.96r$	$-0.17r$	$-0.043r$
Phase angle error	$-5.7r°$	$-11.3r°$	$-0.8r°$	$0°$	$+0.8r°$	$+11.3r°$	$+5.7r°$

Table 7.3 Asymptotic phase angle error for a real pole (or zero) of order r and corner frequency τ and mid-range asymptote between $\tau/10$ and 10τ

	$\tau/10$	$\tau/5$	$\tau/2$	τ	2τ	5τ	10τ
Phase angle error	$-5.7r°$	$-2.3r°$	$-4.9r°$	$0°$	$4.9r°$	$+2.3r°$	$+5.7r°$

as $\omega/\tau \to \infty$. When plotted on a logarithmic frequency scale, the high-frequency asymptote is a straight line with a slope of $-20r$ dB/decade, or $-6r$ dB/octave. The low- and high-frequency asymptotes intersect at the corner frequency $\omega = \tau$ rad/unit time; Fig. 7.17(a) shows the Bode plots for real poles of order one, two and three.

Asymptotic Bode phase angle plots may be obtained from Equations (7.47). An asymptote for the mid-range of frequencies may be obtained by drawing a tangent to the exact phase curve at the corner frequency of $\omega = \tau$. This asymptote would intersect the low-frequency asymptote at $\omega = \tau/5$, and the high-frequency asymptote at $\omega = 5\tau$. Figure 7.1(b) shows the asymptotic phase plots for real poles of order one, two and three.

The errors produced by using the given asymptotic approximations for the real poles and zeros, rather than an accurate curve, are given in Table 7.2. In this table a pole of order r gives the factor $+r$, and a zero the factor $-r$. Again, the Bode plots for zeros of order r will be the reflections of the plots in Fig. 7.17 about the 0 dB and 0° lines.

It should be noted that other mid-range phase asymptotes have been proposed. A common approximation is to use an asymptote which intersects the low-frequency asymptote at $\omega = \tau/10$, and the high-frequency asymptote at $\omega = 10\tau$. The phase angle errors produced using this approximation are given in Table 7.3.

7.6.5   Bode plots for complex conjugate poles and zeros

Complex conjugate poles or zeros are normally multiplied together and expressed in the standard form

$$\frac{1}{\left(1+\dfrac{2\zeta s}{\omega_n}+\dfrac{s^2}{\omega_n^2}\right)^r}, \quad 0 \leqslant \zeta \leqslant 1 \tag{7.48}$$

where ω_n is the natural frequency, ζ the damping ratio and r an integer indicating the number of complex conjugate roots at a given location. Again, if r is positive the roots are poles, and if r is negative the roots are zeros.

The frequency response function corresponding to Equation (7.48) is given by

$$\frac{1}{[1 + j2\zeta(\omega/\omega_n) - (\omega/\omega_n)^2]^r} \tag{7.49}$$

which has a dB magnitude ratio of

$$-10r \log\left[\left(1 - \frac{\omega^2}{\omega_n^2}\right)^2 + 4\zeta^2 \frac{\omega^2}{\omega_n^2}\right] \tag{7.50}$$

and a phase angle of

$$-r \tan^{-1} \frac{2\zeta(\omega/\omega_n)}{1 - (\omega/\omega_n)^2} \tag{7.51}$$

An asymptotic magnitude plot is obtained by considering the frequency ratio ω/ω_n. When ω/ω_n is very small, Equation (7.50) gives the low-frequency asymptote as 0 dB. The high-frequency asymptote, obtained by letting ω/ω_n become very large, is

$$-40r \log(\omega/\omega_n) \tag{7.52}$$

which is a line of slope $-40r$ dB/decade emanating from the corner frequency ω_n.

Use of the asymptotic magnitude approximation will result in an error in this plot around the corner frequency, the size of which depends on the damping ratio, ζ. These errors are given in Table 7.4. For low values of ζ the magnitude ratio tends to a peak value near the corner frequency. It can be shown that the exact curve peaks when

$$\omega = \omega_n\sqrt{(1 - 2\zeta^2)} \tag{7.53}$$

and that at this frequency the peak value of the dB magnitude ratio is

$$20 \log \frac{1}{2\zeta\sqrt{(1 - \zeta^2)}} \tag{7.54}$$

Since frequency must be real, Equation (7.53) indicates that a specific maximum value occurs only if $\zeta < 1/\sqrt{2}$ – that is, less than 0.707. If the damping ratio is greater than 0.707, the maximum value of the magnitude curve is along the 0 dB line. Figure 7.18 shows the Bode plots and asymptotic Bode plots for four second-order systems.

The asymptotic phase plot is obtained from Equation (7.51). At low frequencies (when the frequency ratio ω/ω_n is small), the phase angle is approximately 0°; at

Table 7.4 Asymptotic magnitude errors for complex conjugate poles (or zeros) of order r and corner frequency ω_n

ζ	$\omega_n/10$	$\omega_n/5$	$\omega_n/2$	ω_n	$2\omega_n$	$5\omega_n$	$10\omega_n$
1	$-0.086r$	$-0.34r$	$-1.92r$	$-6r$	$-1.92r$	$+0.34r$	$-0.086r$
0.707	0	$-0.007r$	$-0.263r$	$-3r$	$-0.263r$	$-0.007r$	0
0.5	$+0.043r$	$+0.17r$	$+0.902r$	0	$+0.902r$	$+0.17r$	$+0.043r$
0.3	$+0.071r$	$+0.287r$	$+1.85r$	$+4.44r$	$+1.85r$	$+0.28r$	$+0.071r$
0.2	$+0.08r$	$+0.325r$	$+2.2r$	$+7.96r$	$+2.42r$	$+0.347r$	$+0.086r$

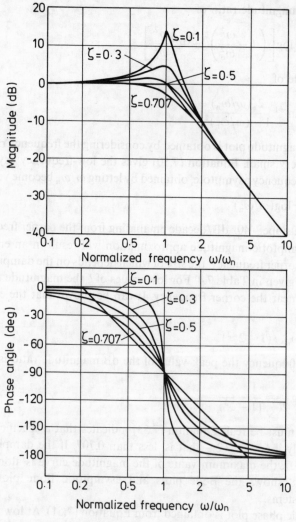

Figure 7.18 Bode plots for four second-order systems

Table 7.5 Asymptotic phase angle errors for complex conjugate poles (or zeros) of order r and corner frequency ω_n, and mid-range asymptote between $\omega_n/5$ and $5\omega_n$

ζ	$\omega_n/10$	$\omega_n/5$	$\omega_n/2$	ω_n	$2\omega_n$	$5\omega_n$	$10\omega_n$
1	$-11.4r°$	$-22.6r°$	$-1.6r°$	$0°$	$+1.6r°$	$+22.6r°$	$+11.4r°$
0.707	$-8.1r°$	$-16.4r°$	$-7.9r°$	$0°$	$+7.9r°$	$+16.4r°$	$+8.1r°$
0.5	$-5.8r°$	$-11.8r°$	$-17.5r°$	$0°$	$+17.5r°$	$+11.8r°$	$+5.8r°$
0.3	$-3.5r°$	$-7.1r°$	$-29.4r°$	$0°$	$+29.4r°$	$+7.1r°$	$+3.5r°$
0.2	$-2.3r°$	$-4.8r°$	$-36.3r°$	$0°$	$+36.3r°$	$+4.8r°$	$+2.3r°$
0.1	$-1.2r°$	$-2.4r°$	$-43.6r°$	$0°$	$+43.6r°$	$+2.4r°$	$+1.2r°$

Table 7.6 Asymptotic phase angle errors for complex conjugate poles (or zeros) of order r and corner frequency ω_n, and mid-range asymptote between $\omega_n/10$ and $10\omega_n$

ζ	$\omega_n/10$	$\omega_n/5$	$\omega_n/2$	ω_n	$2\omega_n$	$5\omega_n$	$10\omega_n$
1	$-11.4r°$	$-4.6r°$	$-9.8r°$	$0°$	$+9.8r°$	$+4.6r°$	$+11.4r°$
0.707	$-8.1r°$	$-10.7r°$	$-19.6r°$	$0°$	$+19.6r°$	$+10.7r°$	$+8.1r°$
0.5	$-5.8r°$	$-15.3r°$	$-29.2r°$	$0°$	$+29.2r°$	$+15.3r°$	$+5.8r°$
0.3	$-3.5r°$	$-20.0r°$	$-41.1r°$	$0°$	$+41.1r°$	$+20.0r°$	$+3.5r°$
0.2	$-2.3r°$	$-22.3r°$	$-48.0r°$	$0°$	$+48.0r°$	$+22.3r°$	$+2.3r°$
0.1	$-1.2r°$	$-25.9r°$	$-55.3r°$	$0°$	$+55.3r°$	$+25.9r°$	$+1.2r°$

high frequencies the phase angle tends to $-180r°$. For the frequencies around the corner frequency ω_n it is normal to draw an asymptote consistent with those drawn for real poles and zeros. Consequently the mid-range frequency asymptote crosses the low-frequency asymptote when $\omega = \omega_n/5$, and crosses the high-frequency asymptote when $\omega = 5\omega_n$ (see Fig. 7.18). Again, the discrepancy between the exact and asymptotic phase plot is a function of the damping ratio, as indicated in Table 7.5 and Fig. 7.18. For completeness, Table 7.6 gives the errors in the asymptotic phase plot when the mid-range asymptote crosses the low-frequency asymptote at $\omega_n/10$, and crosses the high-frequency asymptote at $10\omega_n$.

7.6.6   Bode plots for a time delay

It has already been indicated that time delays occur in the transportation of mass or energy along a particular path, and that a time delay is the interval of time, after the application of an input, during which no response is observable. Most physical systems have some time delay which, if it is significant, should be accounted for in the design procedure. The most convenient method of doing this is to use frequency methods.

A time delay has a Laplace transform of the form

$$e^{-\tau s} \tag{7.55}$$

The corresponding frequency response function is

$$e^{-j\tau\omega} \tag{7.56}$$

which has a magnitude of unity for all frequencies, and a phase angle in degrees of

$$-\tau\omega\frac{360}{2\pi} \tag{7.57}$$

A time delay does not affect the magnitude, since its contribution is 0 dB, but it does affect the phase plot by adding the negative angle, given by Equation (7.57), to the phase curve.

Example 7.4

Draw the asymptotic Bode plots for a system having the open-loop transfer function

$$GH(j\omega) = \frac{320(1 + j\omega)}{j\omega(j\omega + 2)[(j\omega)^2 + 4j\omega + 16]}$$

Solution

The open-loop frequency response of $GH(s)$ is obtained by setting $s = j\omega$:

$$GH(j\omega) = \frac{320(1 + j\omega)}{j\omega(j\omega + 2)[(j\omega)^2 + 4j\omega + 16]}$$

In Bode form, this is

$$GH(j\omega) = \frac{10(1 + j\omega)}{j\omega\left(1 + \dfrac{j\omega}{2}\right)\left[1 + \dfrac{j\omega}{4} - \left(\dfrac{\omega}{4}\right)^2\right]}$$

From Equation (7.38), the dB magnitude is obtained as

$$20\log|GH(j\omega)| = 20\log 10 + 20\log|1 + j\omega| - 20\log|j\omega|$$

$$- 20\log\left|1 + \frac{j\omega}{2}\right| - 20\log\left|1 + \frac{j\omega}{4} - \left(\frac{\omega}{4}\right)^2\right|$$

and from Equation (7.39a) the phase angle is

$$\arg GH(j\omega) = \arg(1 + j\omega) - 90° - \arg\left(1 + \frac{j\omega}{2}\right)$$

$$- \arg\left[1 + \frac{j\omega}{4} - \left(\frac{\omega}{4}\right)^2\right]$$

As $\omega \to \infty$, the phase angle tends to $-270°$. This could be deduced without resorting to the phase angle equation, by noting that the difference between the number of poles and zeros is 3; in other words, the system's rank is 3. For minimum phase systems (systems having no poles or zeros in the right half s-plane) with a positive loop gain, the ultimate phase shift is given by

$$\arg GH(\mathrm{j}\infty) = -90R°$$

where R is the system's rank.

An asymptotic plot for each of the terms in the magnitude and phase expressions may now be drawn (see Fig. 7.19). The asymptotic plots of each element may be summed graphically to produce the asymptotic Bode plot shown in Fig. 7.20. For reference, the actual Bode plot is also shown in this figure.

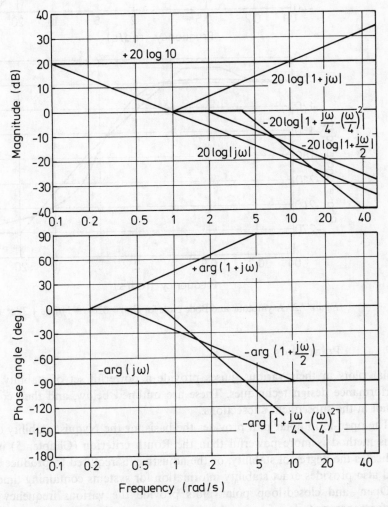

Figure 7.19 Asymptotic Bode element plots

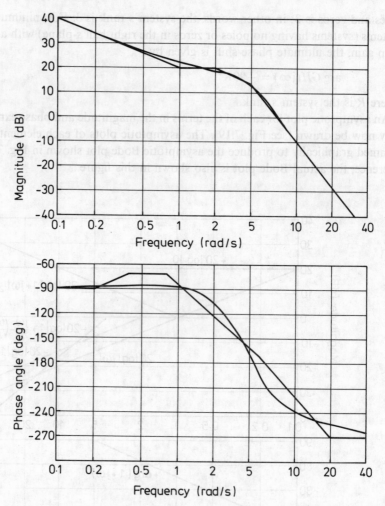

Figure 7.20 Asymptotic and Bode plots for $G(s) = 320(s+1)/[s(s+2)(s^2+4s+16)]$

7.7   Polar plots

Polar plots in their various guises provide a powerful set of stability and the performance design techniques. These are outlined below, and then covered in detail in the subsequent subsections.

The open-loop polar plot provides the basis for the Nyquist stability criterion. This method is more powerful than the Routh criterion (Chapter 5) in that it indicates the degree of stability, or the adjustments required to produce stability, and also provides exact stability information for systems containing time delays.

Open- and closed-loop polar plots provide the various frequency domain performance criteria of gain margin, phase margin, peak magnification and bandwidth, together with steady-state error information. The closed-loop frequency response may be obtained from the open-loop polar plot without resorting to the use of a Nichols chart.

Inverse polar plots are particularly useful in the design of systems having minor loops. If the inverse notation is used, the effect of the minor loop on the system normally resolves itself into a problem in vector addition. Inverse polar plots are also useful in the study of multivariable systems (systems having several interacting inputs and outputs), but this is outside the scope of this book.

If polar plots are used in experimental work, it is often possible to complete the design study without having to establish a transfer function model. A common method of doing this is to use an instrument known as a transfer function analyser. This instrument injects a sinusoidal signal of known frequency into the system and resolves the steady-state output response into in-phase and quadrature components. The data are then plotted on polar graph paper to give the system's polar plot, on which all the design work and stability tests may then be carried out.

Despite their many advantages, polar plots have in the past rarely been used exclusively. The reason for this is that, without computing facilities, polar plots can be difficult to generate in sufficient detail; consequently, they have tended to be used to supplement other methods. Their main disadvantage is that they do not yield frequency values directly.

7.7.1   The Nyquist stability criterion

The Nyquist stability criterion is a graphical method based on the polar form of the steady-state frequency response. It is assumed that the system may be represented in transfer function form, and that the open-loop transfer function has at least as many poles as zeros.

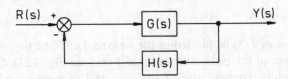

Figure 7.21 A closed-loop system

Consider the closed-loop system shown in Fig. 7.21. The closed-loop transfer function between the output $Y(s)$ and the input $R(s)$ is given by

$$\frac{Y(s)}{R(s)} = \frac{G(s)}{1 + GH(s)} \tag{7.58}$$

and the characteristic equation is

$$F(s) = 1 + GH(s) \tag{7.59}$$

Let

$$G(s) = K'\frac{Q_1(s)}{P_1(s)} \quad \text{and} \quad H(s) = \frac{W_1(s)}{V_1(s)} \tag{7.60}$$

Substituting for $G(s)$ and $H(s)$ into $F(s)$ produces

$$F(s) = \frac{P_1(s)V_1(s) + K'Q_1(s)W_1(s)}{P_1(s)V_1(s)} \tag{7.61}$$

A comparison of Equation (7.61) with the open-loop transfer function

$$GH(s) = K'\frac{Q_1(s)}{P_1(s)}\frac{W_1(s)}{V_1(s)} \tag{7.62}$$

reveals the following properties:

1. The poles of $F(s)$ are the poles of the open-loop transfer function (Equations (7.61) and (7.62)).
2. The zeros of $F(s)$ are functions of the open-loop poles and zeros and of any fixed gain K'.
3. Since the zeros of $F(s)$ are also the system's closed-loop poles, for closed-loop stability all the zeros of $F(s)$ must lie in the left half s-plane.

Nyquist's stability criterion establishes the number of zeros of $F(s)$ that are located in the right half s-plane. Clearly, for closed-loop stability this number should be zero.

Now, the closed-loop characteristic equation, Equation (7.59), may be written as

$$F(s) = \frac{K(s+a_1)(s+a_2)\cdots(s+a_m)}{(s+b_1)(s+b_2)\cdots(s+b_n)} \tag{7.63}$$

where $n \geqslant m$ and all the as and bs are constant. The polar form of Equation (7.63) is

$$F(s) = |F(s)|e^{j \arg F(s)} \tag{7.64}$$

Let the poles and zeros of $F(s)$ be plotted in the s-plane. Let Γ_s be by any closed contour enclosing some of the poles and zeros of $F(s)$ (see Fig. 7.22). Consider the zero at $-a_2$, which is representative of any zero, real or complex, which lies

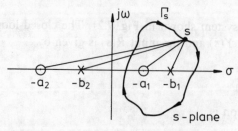

Figure 7.22 A closed s-plane contour Γ_s

outside the contour Γ_s. To determine the contribution of this zero to $F(s)$ at any point s on the contour Γ_s, draw a vector from the zero to the required point on Γ_s. The modulus contribution to $F(s)$ is given by the length of the vector, and the argument contribution is the angle the vector makes with the positive σ-axis. As usual, the normal mathematical convention is used in which a positive angle indicates that the vector is rotated anticlockwise from the positive real axis. Equation (7.63) indicates that the modulus contribution is $|(s + a_2)|$ and the argument contribution is $+\arg(s + a_2)$, evaluated at the point s on Γ_s. If the vector now traverses once around the contour, the net contribution to the argument of $F(s)$ must be zero; the vector oscillates about its starting position and its net displacement is zero degrees. Similarly, any pole which is outside the contour, for example $-b_2$, makes a net contribution to the argument of $F(s)$ of zero degrees every time the vector traverses once around Γ_s.

The only other possibilities that need to be considered are zeros or poles of $F(s)$ inside the contour. The zero at $-a_1$ is representative of such zeros and, as shown in Fig. 7.22, a_1 must be negative since it lies in the right half s-plane. However, the argument's contribution to $F(s)$ is independent of the sign of a_1, provided a_1 is within the contour. As the vector from a_1 traverses once around the contour in the clockwise direction, the net contribution to the argument of $F(s)$ is $-360°$. For each circuit of the contour the vector is displaced from its starting point by $-360°$.

A vector from any pole within the contour, say $-b_1$, will be displaced by $-360°$ for each circuit of the contour. However, since for a pole

$$\arg\left|\frac{1}{s + b_1}\right| = -\arg(s + b_1) \tag{7.65}$$

the net contribution to the argument of $F(s)$ is $+360°$.

A more formal statement of the above reasoning would conclude that the net angular rotation of the $F(s)$ locus about the origin of the $F(s)$-plane is some multiple of $\pm 360°$. Note that it may be shown that the mapping of the closed contour Γ_s onto the $F(s)$-plane produces a closed contour. Consequently,

$$\arg F(s) = 2\pi N = 2\pi(P^* - Z^*) \tag{7.66}$$

where

N = net number of anticlockwise encirclements of the $F(s)$ locus about the origin of the $F(s)$ plane

Z^* = number of zeros of $F(s)$ enclosed by Γ_s

P^* = number of poles of $F(s)$ enclosed by Γ_s

and Γ_s is taken in the clockwise direction.

Generalizing this idea, let Γ_s be a contour enclosing the whole of the right half s-plane, but avoiding poles on the $j\omega$ axis. This is the Nyquist contour – see Fig. 7.23. For obvious reasons it is often called the D-contour. The arm of the D

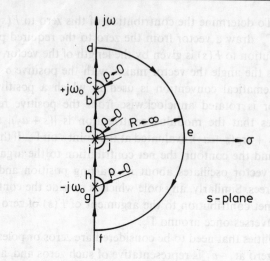

Figure 7.23 The general Nyquist contour, or D-contour

travels along the imaginary axis, and poles on the imaginary axis are avoided by introducing a semicircular arc of infinitesimally small radius ρ, as shown. The large semicircle of the D has a radius R which is infinitely large, and therefore must enclose all the right half s-plane poles and zeros of $F(s)$.

The Nyquist stability criterion may now be stated as

$$Z = P - N \tag{7.67}$$

where

> Z = number of zeros of $F(s)$ in the right half s-plane. Since $F(s)$ is the closed-loop characteristic equation, Z gives the number of closed-loop poles in the right half s-plane. For closed-loop stability Z must be zero
>
> P = number of poles of $F(s)$ in the right half s-plane, and is equal to the number of poles of $GH(s)$ in the right half s-plane. P may be obtained directly or from Routh's array. Note that for an open-loop stable system $P = 0$
>
> N = net number of encirclements by the $F(s)$ locus about the origin of the $F(s)$-plane. Anticlockwise encirclements are positive, and clockwise encirclements negative

In applying the Nyquist stability criterion the object is to find Z from a knowledge of P and N. P, the number of open-loop right half s-plane, poles, may be found directly from the open-loop transfer function. As N is the net number of encirclements of the $F(s)$ locus about the origin of the $F(s)$-plane, it may be found by mapping $F(s)$ as s traverses around the Nyquist D-contour. However, it is not

Table 7.7 Mathematical equations for the Nyquist contour in Fig. 7.23

Path	Equation	Range of validity
ab	$s = j\omega$	$0 < \omega < \omega_0$
bc	$s = \lim\limits_{\rho \to 0} (j\omega_0 + \rho e^{j\theta})$	$-90° \leqslant \theta \leqslant 90°$
cd	$s = j\omega$	$\omega_0 < \omega < \infty$
def	$s = \lim\limits_{R \to \infty} R e^{j\theta}$	$+90° \geqslant \theta \geqslant -90°$
fg	$s = j\omega$	$-\infty < \omega < -\omega_0$
gh	$s = \lim\limits_{\rho \to 0} (-j\omega_0 + \rho e^{j\theta})$	$-90° \leqslant \theta \leqslant +90°$
hi	$s = j\omega$	$-\omega_0 < \omega < 0$
ija	$s = \lim\limits_{\rho \to 0} \rho e^{j\theta}$	$-90° \leqslant \theta \leqslant +90°$

necessary to plot the $F(s)$ locus since the origin of the $F(s)$ plane is

$$F(s) = 0 + j0 \tag{7.68}$$

and, from Equation (7.59), it is evident that an equivalent point exists in the $G(s)H(s)$ plane, namely

$$GH(s) = -1 + j0 \tag{7.69}$$

The normal procedure is to plot the open-loop transfer function $GH(s)$ as s traverses the Nyquist contour; N is then determined from the net number of encirclements of the $GH(s)$ locus around the critical point $(-1 + j0)$ in the $GH(s)$-plane.

Closed-loop stability is therefore determined from the system's open-loop transfer function. If the closed-loop system is stable, the open-loop gain and phase margin measures indicate the degree of stability and the adjustments required to improve stability. Since a system's frequency response deals with time delays exactly, the method may be used to test the system for stability.

7.7.2   The Nyquist contour

Figure 7.23 shows a general Nyquist contour which encloses the whole of the right half s-plane. The contour has been deformed to avoid the real open-loop pole at the origin and the pair of complex conjugate poles on the imaginary axis. In plotting the open-loop transfer function locus $GH(s)$, appropriate values of s must be used as s travels round the Nyquist contour. For example, consider the path ab shown in Fig. 7.23. At any point on this path $s = j\omega$, and therefore $GH(j\omega)$ must be evaluated. This portion of the Nyquist plot is therefore the polar plot of the open-loop frequency response for all frequency values between points a and b.

The various paths along the Nyquist contour and their mathematical equations and range of validity are summarized in Table 7.7.

Example 7.5

A negative feedback closed-loop system has an open-loop transfer function given by

$$GH(s) = \frac{K}{(s+3)^3(s+1)}$$

Use the Nyquist stability criterion to determine the range of gains K for which the closed-loop system is asymptotically stable.

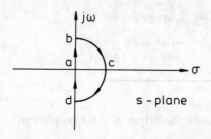

Figure 7.24 Nyquist contour for Example 7.5

Solution

Since the open-loop transfer function has no poles on the imaginary axis, the Nyquist contour takes the form shown in Fig. 7.24. Now, $GH(s)$ has none of its poles enclosed by this contour, so P (Equation (7.67)) equals zero. For stability Z must be zero, and therefore N must be zero. To determine N, the $GH(s)$ locus must be plotted as s travels round the Nyquist contour, and this is carried out as follows.

For path ab

Along the path ab, $s = j\omega$ for $0 \leqslant \omega \leqslant +\infty$, so

$$GH(j\omega) = \frac{K}{(j\omega + 3)^3(j\omega + 1)}$$

At this point it is useful to fix K at some value. A suitable choice is $K = 27$, which makes the static gain of the system unity. Consequently,

$$|GH(j\omega)| = \frac{27}{(\omega^2 + 3^2)^{3/2}(\omega^2 + 1^2)^{1/2}}$$

and

$$\arg GH(j\omega) = 0 - 3\tan^{-1}(\omega/3) - \tan^{-1}(\omega)$$

Note that K affects only $|GH(j\omega)|$, and not $\arg GH(j\omega)$.

Table 7.8 Frequency response points for path ab

ω	$GH(j\omega)$	$\arg GH(j\omega)$
0	1	$0°$
0.5	0.859	$-55°$
1	0.604	$-100°$
1.5	0.397	$-136°$
2	0.257	$-165°$
3	0.112	$-207°$
∞	0	$-360°$

A table of frequency response points, Table 7.8, may now be calculated from which to draw the polar plot.

For path bcd

This section of the Nyquist contour (see Table 7.7 and Fig. 7.23) is described by

$$s = \lim_{R \to \infty} Re^{j\theta}, \quad +90° \geqslant \theta \geqslant -90°$$

For large values of R

$$\lim_{R \to \infty} GH(Re^{j\theta}) = \lim_{R \to \infty} \frac{1}{R^4 e^{j4\theta}}$$

$$= 0e^{-j4\theta}$$

The infinite semicircular arc in the s-plane maps onto a point of zero radius in the $GH(s)$-plane. The $GH(s)$ locus approaches the origin of the $GH(s)$-plane at $-360°$, makes two complete anticlockwise revolutions and leaves the origin at $+360°$. That is, the inversion of $1/e^{-j4\theta}$ makes the expression $e^{-j4\theta}$ valid over the range $+90° \leqslant \theta \leqslant -90°$.

For path da

Over this path $s = -j\omega$, and its evaluation produces the mirror image of the $GH(j\omega)$ locus along the real axis.

The complete $GH(s)$ locus may now be plotted. This is shown in Fig. 7.25 where, for convenience, polar graph paper has been used. At $-180°$ the $GH(s)$ locus has a magnitude of 0.195, so it does not enclose the $(-1+j0)$ point, and therefore $N = 0$. From Equation (7.67),

$$Z = P - N = 0$$

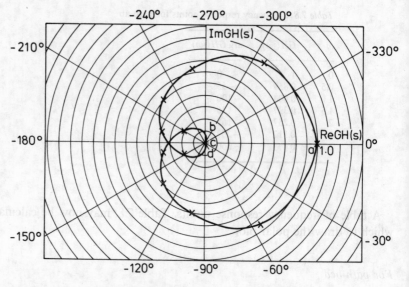

Figure 7.25 Nyquist plot for $GH(s) = 27/[(s+3)^3(s+1)]$

and the system is stable. If the system is to become marginally stable, then the gain K must be increased until the $GH(s)$ locus passes through the critical point $(-1+j0)$:

$$\frac{K}{27} \times 0.195 = 1$$

or $K = 139$. The system is asymptotically stable in the closed loop for $0 \leqslant K < 139$.

Example 7.6

Use the Nyquist stability criterion to test the closed-loop stability of the unstable open-loop system

$$G(s) = \frac{1}{s(s-1)}$$

Solution

The open-loop system has one pole on the imaginary s-plane axis at $s = 0$, which means that the Nyquist contour must be deformed as shown in Fig. 7.26. There is one pole at $s = 1$, which is inside this contour, and therefore $P = 1$ (see Equation (7.67)). The Nyquist contour is mapped onto the $G(s)$ plane as follows.

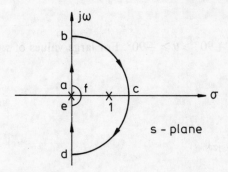

Figure 7.26 Nyquist contour for Example 7.6

For path ab

Along this path $s = j\omega$ for $0 < \omega \leqslant +\infty$. Therefore

$$G(j\omega) = \frac{1}{j\omega(j\omega - 1)}$$

from which

$$|G(j\omega)| = \frac{1}{\omega\sqrt{(\omega^2 + 1)}}$$

and

$$\arg G(j\omega) = -(\tan^{-1}(\infty) + \tan^{-1}(-\omega))$$

$$= -\left(\frac{\pi}{2} + \pi - \tan^{-1}(\omega)\right)$$

$$= -\frac{3\pi}{2} + \tan^{-1}(\omega)$$

$$= +\frac{\pi}{2} + \tan^{-1}(\omega)$$

Note that $\tan^{-1}(-\omega)$ has been rewritten as $\pi - \tan^{-1}(\omega)$; this is the normal procedure.

A scale plot of the $G(j\omega)$ locus is inappropriate since $|G(j\omega)| \to \infty$ as $\omega \to 0$. Also, the locus is confined to one quadrant of the $G(s)$-plane, so points do not need to be plotted accurately in order to determine stability.

For path bcd

On this path $s = \lim\limits_{R \to \infty} Re^{j\theta}$ for $+90° \geqslant \theta \geqslant -90°$. For large values of s

$$G(s) = \frac{1}{s^2 - s} \approx \frac{1}{s^2}$$

so

$$\lim\limits_{R \to \infty} G(Re^{j\theta}) = \frac{1}{\infty} e^{-j2\theta}$$

The modulus of this expression is zero, and the argument is valid for $-90° \leqslant \theta \leqslant +90°$.

For path de

This path is the mirror image of the path ab with respect to the real $G(s)$-plane axis.

For path efa

Here

$$s = \lim\limits_{\rho \to 0} \rho e^{j\theta}, \quad -90° \leqslant \theta \leqslant +90°$$

or

$$(-s) = \lim\limits_{\rho \to 0} \rho e^{-j\theta}, \quad -90° \leqslant \theta \leqslant +90°$$

For small values of s

$$G(s) \approx \frac{1}{-s}$$

so

$$\lim\limits_{\rho \to 0} G(\rho e^{-j\theta}) = \lim\limits_{\rho \to 0} \left(\frac{1}{\rho} e^{+j\theta} \right)$$

The modulus of this expression is infinite, and the argument is valid for $-90° \geqslant \theta \geqslant +90°$.

The complete $G(s)$ locus may now be plotted. This is shown in Fig. 7.27. In this sketch there is one clockwise encirclement of the $(-1, +j0)$ point, and hence $N = -1$. From the Nyquist stability formula

$$Z = P - N = 1 - (-1) = 2$$

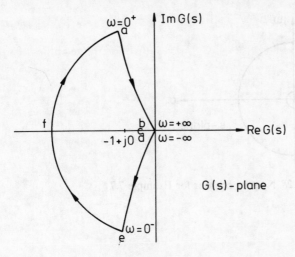

Figure 7.27 Nyquist plot for $G(s) = 1/[s(s-1)]$

The closed-loop system is clearly unstable since there are two closed-loop poles with positive real parts.

Example 7.7

Plot the Nyquist diagram for the system having the open-loop transfer function

$$G(s) = \frac{2(s+0.1)(s+0.6)(s^2+s+1)}{s^3(s-0.2)(s+1)}$$

and hence determine the stability of the closed-loop system.

Solution

The Nyquist contour must avoid the multiple-open loop poles at the origin of the s-plane, and therefore takes the form shown in Fig. 7.28. There is one open-loop pole within the Nyquist contour at $s = 0.2$, and therefore $P = 1$.

For path ab

Here $s = j\omega$ for $0 < \omega \leqslant +\infty$, so

$$G(j\omega) = \frac{2(j\omega + 0.1)(j\omega + 0.6)[j\omega + (1 - \omega^2)]}{(j\omega)^3(j\omega - 0.2)(j\omega + 1)}$$

from which

$$|G(j\omega)| = \frac{2\sqrt{(\omega^2 + 0.1^2)}\sqrt{(\omega^2 + 0.6^2)}\sqrt{[\omega^2 + (1 - \omega^2)^2]}}{\omega^3\sqrt{(\omega^2 + 0.2^2)}\sqrt{(\omega^2 + 1)}}$$

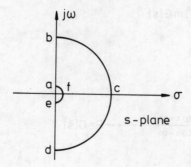

Figure 7.28 Nyquist contour for Example 7.7

and

$$\arg G(j\omega) = \tan^{-1}\left(\frac{\omega}{0.1}\right) + \tan^{-1}\left(\frac{\omega}{0.6}\right) + \tan^{-1}\left(\frac{\omega}{1-\omega^2}\right)$$

$$- 3\tan^{-1}\left(\frac{\omega}{0}\right) - \tan^{-1}\left(\frac{\omega}{-0.2}\right) - \tan^{-1}(\omega)$$

Remember that the term $1/s^n$ gives an argument contribution of $-n\tan^{-1}(\omega/0)$, or $-n \times (90°)$.

Now, for $\omega < 1$ the term $\omega/(1-\omega^2)$ is positive, so

$$\arg G(j\omega) = \tan^{-1}(10\omega) + \tan^{-1}(1.67\omega) + \tan^{-1}\left(\frac{\omega}{1-\omega^2}\right)$$

$$- \frac{3\pi}{2} - [\pi - \tan^{-1}(5\omega)] - \tan^{-1}(\omega)$$

For $\omega > 1$ the term $\omega/(1-\omega^2)$ is negative, and

$$\arg G(j\omega) = \tan^{-1}(10\omega) + \tan^{-1}(1.67\omega) + \left[\pi - \tan^{-1}\left(\frac{\omega}{\omega^2-1}\right)\right]$$

$$- \frac{3\pi}{2} - [\pi - \tan^{-1}(5\omega)] - \tan^{-1}(\omega)$$

To assist in sketching this locus a minimal table of points is drawn up (see Table 7.9).

For path bdc

The required transformation expression is

$$s = \lim_{R\to\infty} Re^{j\theta}, \quad +90° \geqslant \theta \geqslant -90°$$

Table 7.9 Points for $G(j\omega)$ locus

ω	$M(\omega)$	$\phi(\omega)$
0	∞	$-\pi/2$
0.1	761.8	$-8.915°$
0.2	120.2	$+37.33°$
0.5	9.54	$+105.6°$
1.2	2.2	$+199.2°$
∞	0	—

For large values of s, $G(s) \to 1/s$, so

$$\lim_{R \to \infty} G(Re^{j\theta}) = 0e^{-j\theta}$$

For path de

This is the mirror image with respect to the real $G(s)$-axis of the $G(j\omega)$ locus.

For path efa

For small values of s, $G(s)$ may be approximated by

$$G(s) \approx 1/(-s^3)$$

For $-s$ the Nyquist path in the s-plane is described by the expression

$$(-s) = \rho e^{-j\theta}, \quad -90° \leqslant \theta \leqslant +90°$$

Therefore

$$\lim_{\rho \to 0} G(\rho e^{-j\theta}) = \infty e^{j3\theta}$$

and the argument of this expression is valid for $-90° \geqslant \theta \geqslant +90°$.

The Nyquist plot of this system is shown in Fig. 7.29. There are two anticlockwise encirclements of the $(-1 + j0)$ point, the two inner circles of the Nyquist plot, and one clockwise encirclement of infinite radius from e to a. The net number of anticlockwise encirclements is therefore plus one (two anticlockwise minus one clockwise) so $N = 1$. From the Nyquist stability formula,

$$Z = P - N = 1 - 1 = 0$$

Hence it may be concluded that the closed loop system is stable.

In all the Nyquist stability examples it would be a useful exercise to check the closed-loop stability of the various systems using the Routh criterion.

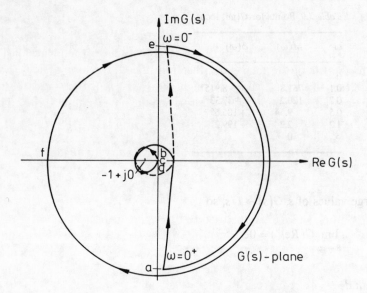

Figure 7.29 Nyquist plot for $G(s) = \dfrac{2(s + 0.1)(s + 0.6)(s^2 + s + 1)}{s^3(s - 0.2)(s + 1)}$

7.7.3   The left-hand rule

In the previous section the Nyquist stability criterion has been established. By using this criterion, it is possible to show that for open-loop stable systems of type 0, 1 or 2, only the $GH(\mathrm{j}\omega)$ locus need be plotted in order to determine closed-loop stability. Systems with complex open-loop poles on the imaginary axis are excluded.

Parts of three possible $GH(\mathrm{j}\omega)$ loci are shown in Fig. 7.30. In this figure the direction of increasing ω is indicated by means of arrows on the loci. It is also assumed that each locus is obtained from an open-loop stable system of type 0, 1 or 2.

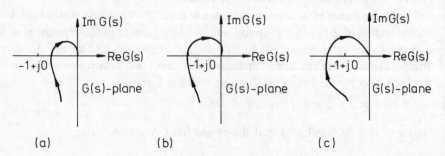

Figure 7.30 Polar plots of open-loop stable systems of type 0, 1 or 2: (a) stable, (b) marginally stable and (c) unstable

The left-hand rule states that if the $(-1 + j0)$ point lies to the left of the $GH(j\omega)$ locus, then the system is closed-loop stable. In Fig. 7.30(a), an observer looking along the $GH(j\omega)$ locus in the direction of increasing ω would place the $(-1 + j0)$ point to the left of the locus, and therefore the closed-loop system is stable. In Fig. 7.30(b), the locus passes through the $(-1 + j0)$ point and the system is marginally stable; the loop gain is unity when the phase shift is $-180°$. The system in Fig. 7.30(c) is closed-loop unstable since the $(-1 + j0)$ point lies to the right of the $GH(j\omega)$ locus.

Proof of the left-hand rule is left as an exercise for the reader.

7.7.4   Open-loop polar plots and performance criteria

Open-loop polar plots may be used to determine gain margin, phase margin and steady-state error. They may also be used to determine closed loop frequency data which, if plotted in polar form, may be used to determine peak magnification and bandwidth. How to determine the closed-loop frequency response is considered in Section 7.7.5.

Most systems are open-loop stable and of type 0, 1 or 2, and a satisfactory gain and phase margin will therefore ensure their closed-loop stability. These two criteria are often thought of as stability measures, but in fact they can be used only to test whether a particular class of system is stable or not. A graphical method of finding gain and phase margins from polar plots has been given in Section 7.4, and Section 7.5 has indicated how the various performance measures are interpreted.

A system's type and the related steady-state error information (see Section 7.5.1) may be determined directly from the polar plot. Figure 7.31 shows a number of polar plots for systems of different type number. The $GH(j\omega)$ locus for a type 0

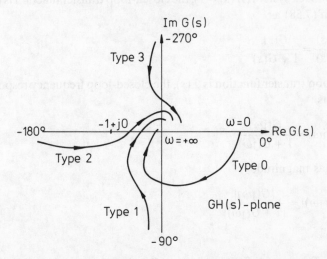

Figure 7.31 Polar plots for systems with different type number

system starts from a point on the $0°$ line, at $\omega = 0$. A type 1 system has a locus of infinite magnitude at $\omega = 0$ and starts from a line asymptotic to the $-90°$ axis. Type 2 and 3 systems start from lines asymptotic to the $-180°$ and $-270°$ axis, respectively. Incidentally, as $\omega \to \infty$ the magnitude ratio tends to zero and the phase angle to $-90R°$, for a minimum phase system (one with all poles and zeros in the left half s-plane). Here R is the rank of the system. If the closed-loop system is stable, the error in a type 0 system for a step input is given by

$$e(t)_{ss} = \frac{1}{1 + KGH(0)} \tag{7.70}$$

where K is any additional loop gain not included in $GH(s)$, and $GH(0)$ is the magnitude of the open-loop frequency response evaluated at zero frequency. The magnitude of $GH(0)$ may be measured directly from the polar plot, and indicates that, the further the starting point of the locus from the origin of the $GH(s)$ plane, the smaller the steady-state error.

For systems with higher type numbers the $GH(j\omega)$ locus becomes infinite at zero frequency. Therefore, if a particular forcing input produces a steady-state error, the magnitude of the error cannot be calculated directly from the polar plot. There are techniques for quantifying the error, using polar data, in which the location of the zero-frequency asymptote relative to the appropriate axis is found. The greater the distance between the axis and the asymptote, the greater the reduction in any steady-state error.

7.7.5   Open- and closed-loop polar plots

For a unity feedback system ($H(s) = 1$), the closed-loop transfer function is obtained from Equation (7.58) as

$$\frac{Y(s)}{R(s)} = \frac{G(s)}{1 + G(s)} \tag{7.71}$$

If the closed-loop transfer function is $T(s)$, the closed-loop frequency response may be expressed as

$$T(j\omega) = \frac{G(j\omega)}{1 + G(j\omega)} \tag{7.72}$$

Now, $T(j\omega)$ has magnitude

$$|T(j\omega)| = \frac{|G(j\omega)|}{|1 + G(j\omega)|} \tag{7.73}$$

and argument

$$\arg T(j\omega) = \arg G(j\omega) - \arg(1 + G(j\omega)) \tag{7.74}$$

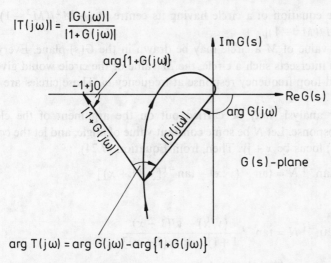

$$|T(j\omega)| = \frac{|G(j\omega)|}{|1+G(j\omega)|}$$

$$\arg\{1+G(j\omega)\}$$

$$\arg G(j\omega)$$

G(s)-plane

$$\arg T(j\omega) = \arg G(j\omega) - \arg\{1+G(j\omega)\}$$

Figure 7.32 Determination of the closed-loop frequency response from the $G(j\omega)$ locus

In the polar plot of the open-loop transfer function (see Fig. 7.32) the modulus of $G(j\omega)$ is the length of the vector from the origin of the plot to the point with value ω on the $G(j\omega)$ locus. The argument of $G(j\omega)$ is the angle this vector makes with the positive real axis. The modulus of $1 + G(j\omega)$ is the length of the vector from the $(-1 + j0)$ point to the point on the $G(j\omega)$ locus with value ω. Again, its argument is the angle this vector makes with the positive real axis.

By selecting a number of points on the $G(j\omega)$ locus and applying Equations (7.73) and (7.74), the closed-loop frequency response is obtained. With the use of a computer (see Program 7C in Chapter 11), even this simple procedure becomes unnecessary. However, it is worth exploring the method further since it illuminates the development of the Nichols chart.

The magnitude of the closed-loop frequency response at a given frequency is expressed as the ratio of the length of two vectors in Equation (7.73). Clearly, an infinite number of vectors could be found which would give the same magnitude value. Let the constant closed-loop magnitude have the value of M, and let the open-loop transfer function $G(j\omega)$ have the coordinate points $x + jy$. From Equation (7.73),

$$M = \frac{|x + jy|}{|1 + x + jy|} \tag{7.75}$$

Squaring both sides and rearranging yields

$$\left(x + \frac{M^2}{M^2 - 1}\right)^2 + y^2 = \frac{M^2}{(M^2 - 1)^2} \tag{7.76}$$

which is the equation of a circle having its centre at $(-M^2/(M^2-1),0)$ and a radius of $|M/(M^2-1)|$.

For each value of M a circle may be drawn in the $G(s)$-plane. Every time the $G(j\omega)$ locus intersects such a circle, the M value of the circle would give the gain of the closed-loop frequency response at frequency ω. These circles are known as M circles.

A similar analysis may be carried out on the argument of the closed-loop frequency response. Let N be some constant value of angle, and let the coordinates of the $G(j\omega)$ locus be $x+jy$. Then, from Equation (7.74),

$$\tan^{-1}N = \tan^{-1}(y/x) - \tan^{-1}[y/(1+x)] \tag{7.77}$$

or

$$\tan^{-1}N = \tan^{-1}\frac{(y/x)-y/(1+x)}{1+(y/x)[y/(1+x)]} \tag{7.78}$$

From which it may be deduced that

$$N = \frac{y}{x^2+x+y^2} \tag{7.79}$$

After further manipulation the above equation can be written in the form

$$\left(x+\frac{1}{2}\right)^2 + \left(y-\frac{1}{2N}\right)^2 = \frac{1}{4}\left(\frac{N^2+1}{N^2}\right) \tag{7.80}$$

which again is the equation of a circle, only this time with its centre at $(-1/2, 1/2N)$ and radius

$$\frac{1}{2N}\sqrt{(N^2+1)}$$

For each value of N a circle could be drawn in the $G(s)$-plane. Again, intersections of the $G(j\omega)$ locus with an N circle give the closed-loop phase angle.

Before dealing with the various closed-loop stability criteria, it is interesting to note that the Nichols chart (described in Section 7.4) is obtained by mapping M and N circles in polar coordinates, onto M and N loci in rectangular (dB) magnitude and phase coordinates.

The open-loop polar plot of the $G(j\omega)$ locus with superimposed M circles is used to determine the closed-loop peak magnification, bandwidth and cutoff rate. In Fig. 7.33, the $G(j\omega)$ locus touches the $M=2$ circle at frequency ω_p. This indicates that $M_p=2$, and the peak frequency is ω_p. Equation (7.34) may be used to approximate the effective damping ratio from M_p, which will be found to be 0.26. A further test would be to measure the phase margin, which in this case is 29°, and use the rule of thumb given in Section 7.5.2 to find an approximate value for the damping ratio, namely 0.29. The two methods thus give similar results.

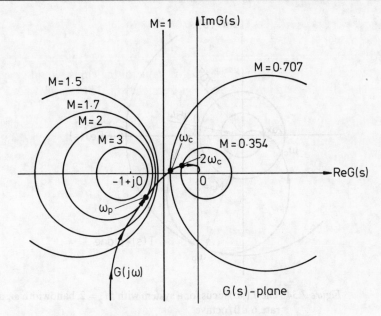

Figure 7.33 The $G(j\omega)$ locus with superimposed M circles

For a zero-frequency magnitude ratio of unity, the bandwidth may be determined from the frequency at which the $G(j\omega)$ locus intersects the $M = 0.707$ circle. In Fig. 7.33 the bandwidth is the value of the frequency at ω_c. The cutoff rate is the rate of decrease in the magnitude ratio; in Fig. 7.33 the bandwidth frequency doubles if the magnitude ratio is halved, which indicates the cutoff rate to be 6 dB/octave.

To produce the closed-loop polar plot of the $T(j\omega)$ locus, both the M and N circles would be required. In the $T(s)$-plane, M circles become circles of radius M with their centre at the origin of the plane, and the N circles become radial lines emanating from the origin of the $T(s)$-plane. The closed-loop frequency measures of bandwidth, peak magnification and cutoff rate may all be measured directly. Figure 7.34 shows the closed-loop $T(j\omega)$ locus corresponding to the open-loop $G(j\omega)$ locus in Fig. 7.33.

7.7.6   Inverse polar plots

This section provides a brief introduction to inverse polar plots. In principle, the inverse polar plot may be used to achieve the same objectives as direct polar plots. The Nyquist stability criterion, gain margin, phase margin, steady-state error checks and the various closed-loop frequency response measures all have their counterpart in the inverse plane.

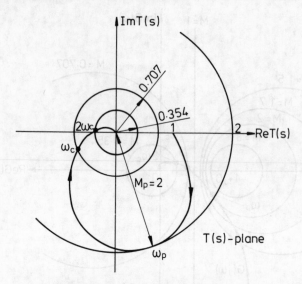

Figure 7.34 The $T(j\omega)$ locus for a system with $M_p = 2$, bandwidth ω_c and cutoff rate 6 dB/octave

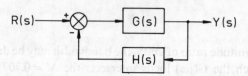

Figure 7.35 A closed-loop control system with feedback compensation

Consider the closed-loop system shown in Fig. 7.35, whose overall system transfer function is given by

$$\frac{Y(s)}{R(s)} = T(s) = \frac{G(s)}{1 + GH(s)} \qquad (7.81)$$

The inverse closed-loop transfer function for this system is

$$T^{-1}(s) = G^{-1}(s) + H(s) \qquad (7.82)$$

For feedback compensation, where the object is to design the dynamic characteristics of $H(s)$ for fixed plant dynamics $G(s)$, it is much easier to work with Equation (7.82) than with Equation (7.81). If direct polar plots are used, each change in $H(s)$ requires the rational polynomial $T(s)$ in Equation (7.81) to be evaluated. However, $T^{-1}(s)$ may be evaluated by using the fixed inverse polar plot $G^{-1}(s)$ of the plant and adding it vectorially to the direct polar plot $H(s)$ of the feedback compensator. In this way the effect of $H(s)$ on the closed-loop system is seen immediately.

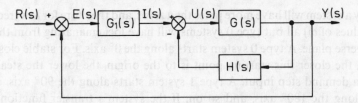

Figure 7.36 A control system containing a minor feedback loop

Another system for which it is advantageous to use the inverse notation is one in which a minor feedback loop is used. Figure 7.36 shows $G_1(s)$, a forward path compensator, and $H(s)$, a feedback compensator, for a plant with fixed dynamics $G(s)$. This type of control may have distinct advantages: the system may be easier to build, install and adjust than one with a more complex, single-loop controller. And although the designer will have more work to do, it is often possible to meet performance specifications which would otherwise be unobtainable. An example of the use of minor feedback loops is given in Section 8.3.2.

An iterative approach to the design of the compensators $G_1(s)$ and $H(s)$ is normally adopted. Typically, the inner loop is designed first: that is, the problem is reduced to the design of a feedback compensator similar to that shown in Fig. 7.35. For this stage of the design the inverse notation has a clear advantage. When the dynamics of $H(s)$ are fixed, the dynamics of the feedback compensated plant are fixed and given by

$$\frac{Y(s)}{I(s)} = \frac{G(s)}{1 + G(s)H(s)} \tag{7.83}$$

Bode and Nichols plots, or a Nyquist plot with M and N circles are then used to design the forward path compensator $G_1(s)$.

Figure 7.37 shows a number of inverse polar plots about the origin of the inverse plane. The general shape of the plots depends on the inverse transfer function.

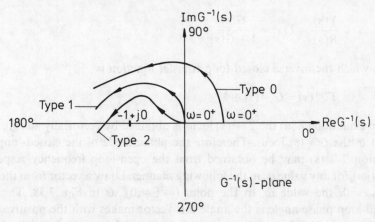

Figure 7.37 Inverse polar plots for type 0, 1 and 2 systems

Normally a system will have more poles than zeros, and therefore for low frequencies (small values of ω) all but type 0 systems will have loci emanating from the origin of the inverse plane. A type 0 system starts along the $0°$ axis. For stable closed-loop response, the closer this starting point is to the origin, the lower the steady-state error to a demand step input. A type 1 system starts along the $90°$ axis, a type 2 system along the $180°$ axis, and so on. If the system's transfer function has no zeros, the phase angle of the inverse locus increases with increasing ω. This produces an inverse locus which is a smooth curve in the anticlockwise direction. When zeros are present the locus still tends to rotate anticlockwise with increasing ω, but the curve may not be smooth. For high frequencies (large ω) the magnitude values become large. As $\omega \to \infty$ the magnitude ratio becomes infinite, and the phase angle $90R°$ for minimum phase systems. Again, R refers to the system rank. Given the system's transfer function, it is therefore possible to use the above argument to predict the general shape of the inverse locus.

The Nyquist stability criterion, when applied to the inverse transfer function, gives results similar to those obtained using direct polar plots in that the equation

$$Z = P - N \tag{7.84}$$

still holds (see Section 7.7.1). Again, Z is the number of zeros of the closed-loop characteristic equation in the right half s-plane, and for closed-loop stability Z must be zero. Likewise, N is the net number of encirclements of the $GH^{-1}(s)$ locus about the $(-1+j0)$ point in the $GH^{-1}(s)$ plane, anticlockwise encirclements being positive. However, P is now the number of zeros of the open-loop transfer function $GH(s)$ enclosed by the Nyquist contour, and not the number of poles in the right half s-plane, as with direct polar plots.

The open-loop gain and phase margin performance measures have been considered in Section 7.4. Closed-loop performance measures require the inverse equivalent of the M and N circles used with direct polar plots. The closed-loop transfer function of a unity negative feedback system is (see Fig. 7.35)

$$\frac{Y(s)}{R(s)} = T(s) = \frac{G(s)}{1 + G(s)} \tag{7.85}$$

from which the inverse closed-loop transfer function is

$$T^{-1}(s) = G^{-1}(s) + 1 \tag{7.86}$$

From Equation (7.86) the $T^{-1}(s)$ locus is obtained by vectorially adding 1 to each point in the $G^{-1}(s)$ locus. Therefore the phase angle of the closed-loop transfer function $T^{-1}(s)$ may be obtained from the open-loop frequency response plot $G^{-1}(j\omega)$, for any value ω, in the following manner. Draw a vector from the $G^{-1}(j\omega)$ locus, with the value ω, to the point $(-1+j0)$, as in Fig. 7.38. The required closed-loop phase angle is the angle this vector makes with the positive real axis. Similarly, a circle of constant magnitude ratio which is centred at the origin of the

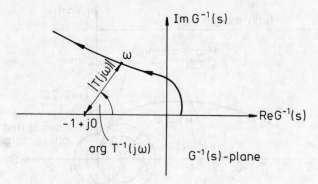

Figure 7.38 Closed-loop phase angle from inverse polar plot

$T^{-1}(s)$-plane is mapped onto a circle of identical radius in the $G^{-1}(s)$-plane, only centred at the $(-1+j0)$ point.

The above analysis indicates that lines of constant magnitude ratio in the closed-loop $T(s)$-plane map onto circles centred on the $(-1+j0)$ point in the $G^{-1}(s)$-plane. The inverse of the radius of the circle in the $G^{-1}(s)$-plane gives the closed-loop magnitude ratio. Also, lines of constant phase angle in the closed-loop $T(s)$-plane map onto radial lines centred on the $(-1+j0)$ point in the $G^{-1}(s)$-plane. In this mapping, the angle of any radial line undergoes a sign change.

Example 7.8

Draw the inverse polar plot of

$$G(s) = \frac{1}{s^4 + 6s^3 + 11s^2 + 6s}$$

and determine the gain margin, phase margin and closed-loop bandwidth. Also, comment on the M_p value.

Solution
In inverse notation,

$$G^{-1}(j\omega) = \omega^2(\omega^2 - 11) - j6\omega(\omega^2 - 1)$$

which is readily plotted on the $G^{-1}(s)$-plane, as shown in Fig. 7.39.

The gain margin is the magnitude of $G^{-1}(j\omega)$ when the imaginary part of $G^{-1}(j\omega)$ is zero. This occurs when $\omega = 1$ and the gain margin is 10 (see Fig. 7.39). From Fig. 7.39, the phase margin is measured to be 73°.

Since the open-loop system is type 1 with no right half s-plane poles, the gain and phase margins are sufficient to indicate closed-loop stability. The bandwidth is determined by drawing a circle of radius $1/0.707$ (or 1.414) centred at the

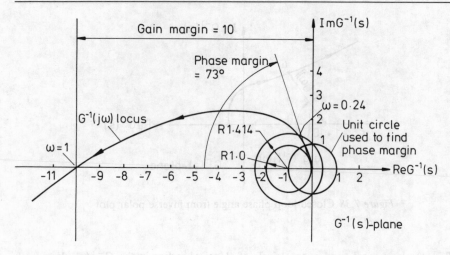

Figure 7.39 Inverse polar plot for $G(s) = 1/(s^4 + 6s^3 + 11s^2 + 6s)$

$(-1 + j0)$ point, as shown. This circle crosses the $G^{-1}(j\omega)$ locus at $\omega = 0.24$, and therefore the bandwidth is 0.24 rad/unit time.

The value of M_p is determined by drawing the circle of largest diameter centred on the $(-1 + j0)$ point which just touches the $G^{-1}(j\omega)$ locus without crossing it. However, the phase margin of 73° indicates an equivalent damping ratio of approximately 0.73, and therefore the peak magnification is likely to be at the zero frequency point (see Section 7.5.2). As expected, Fig. 7.39 shows that the largest circle with centre $(-1 + j0)$ which just touches the locus has unit radius.

7.7.7   Polar plots and time delays

The frequency response of a time delay for a Bode plot has been considered in Section 7.6.6. It was shown that the magnitude response is unity for all frequencies, and that the phase response is the product of the frequency and the time delay (see Equation (7.57)). The polar plot for a system consisting of a pure time delay

$$G(s) = e^{-\tau s} \tag{7.87}$$

shows the $G(j\omega)$ locus to be a circle of unit radius about the origin of the $G(s)$-plane (see Fig. 7.40).

When a time delay is added to a transfer function, the dominant feature of the plot is that the $G(j\omega)$ locus spirals into the origin of the $G(s)$-plane (see Fig. 7.41(b)). In general, the phase shift associated with a time delay will have a destabilizing effect, since it will reduce – or even eliminate – the system's margin of relative stability.

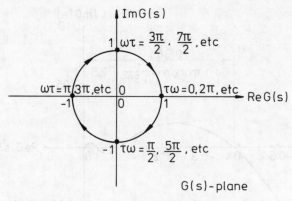

Figure 7.40 Polar plot of a pure time delay

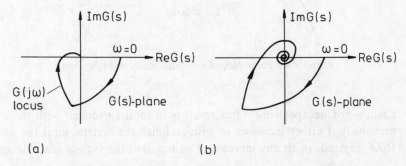

Figure 7.41 Effect of time delay on a polar plot of a type 0 system: (a) without delay, (b) with delay

7.7.8   Polar plots and conditionally stable systems

Conditionally stable systems were examined in the chapter on root loci, in Section 6.6. A polar plot of the example given in that section, namely

$$G(s) = \frac{K(s^2 + 2s + 4)}{s(s + 4)(s + 6)(s^2 + 1.4s + 1)} \qquad (7.88)$$

is shown in Fig. 7.42. For convenience, the gain K is set equal to 6. With this gain the $(-1 + j0)$ point is to the left of the $G(j\omega)$ locus, and the left-hand rule indicates that the system is stable. An increase in the gain affects only the magnitude ratio, and not the phase angle of the locus. The gain margin of the plot (see Fig. 7.42) is 2.6, and therefore the system is stable for all values of gain less than $2.6K$ (or 15.6), which agrees with the result obtained in Chapter 6. If the magnitude ratio is increased by a factor of 11.25, the locus again passes through the $(-1 + j0)$ point. Up to this point the response of the closed-loop system is unstable, and

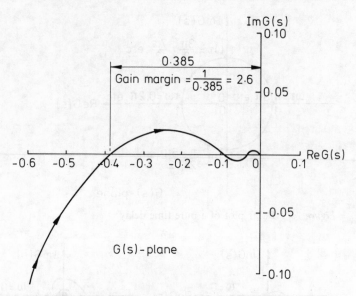

Figure 7.42 Polar plot of a conditionally stable system

again – not unexpectedly – this result is in total agreement with those obtained previously. Further increases in gain stabilize the system, until the loop gain is 163.6, beyond which any increase in gain makes the system unstable again.

7.8   Nichols charts

In principle, a Nichols chart plot may be used for adjusting a system's open-loop gain and phase margins, as well as modifying its closed-loop performance specifications. In practice, Bode plots are used to shape the open-loop response, and the Nichols chart plot to check the closed-loop characteristics and make any necessary gain adjustments.

A Nichols chart is essentially a transformation of the M and N circles on the polar plot into non-circular M and N contours on a plot of dB magnitude versus phase in rectangular coordinates (see Section 7.7.5 and Fig. 7.12). A plot of the open-loop frequency response function $GH(j\omega)$ on a Nichols chart yields the gain and phase margins directly (see Section 7.4).

Information on the gain and phase margins, together with the various closed-loop specifications, are summarized in Fig. 7.43. In this figure M_p, the peak maximum magnitude ratio, is obtained by finding the largest M circle which touches, but does not cross, the $GH(j\omega)$ locus. The frequency at M_p, namely ω_p, cannot be read directly but is usually readily found by interpolating between the $GH(j\omega)$ data points. The bandwidth of the closed-loop system is found from the intersection of the $GH(j\omega)$ locus with the -3 dB M contour.

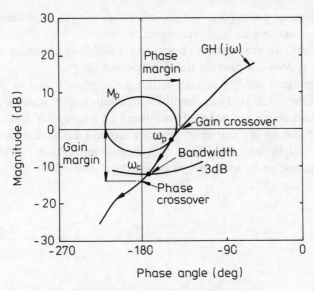

Figure 7.43 Summary of Nichols chart information

A change in gain does not alter the shape of the $GH(j\omega)$ locus on the Nichols chart, but shifts it vertically. Also, since the magnitude is expressed in dB form it is common practice to represent the open-loop transfer function $GH(j\omega)$ in Bode form (see Equation (7.40)). In this form the effect of a design change such as the inclusion or modification of a compensator is easily accommodated. The basic plot remains the same, and the modification becomes a problem in graphical addition. Unlike Bode plots, the Nichols chart does not allow asymptotic approximations to be made. Therefore data points must be calculated directly or transferred from the Bode to the Nichols chart.

7.9   Summary of Bode, polar and Nichols plots

The Bode, polar and Nichols plots are all paper-and-pencil methods of frequency design. Bode plots are by far the easiest to produce and modify, and provide information on a system's gain and phase margin. To obtain closed-loop information the data is normally transferred to a Nichols chart. Bode plots do not provide information on closed-loop stability; this must be tested for using the Nyquist or the Routh criterion. Gain and phase margin measures indicate stability only for a restricted class of systems (see Section 7.7.3).

Polar plots, together with the M and N circles, provide both open- and closed-loop frequency measures. When used with the Nyquist stability criterion, they also provide closed-loop stability information. Note that the Nyquist stability test is far more powerful than the Routh test, in that it indicates the adjustments required to improve stability and can deal with systems containing time delays. A major disadvantage of polar plots is that, after even simple modifications, the open-loop locus is unlikely to retain its orginal shape.

The Nichols chart also provides both open- and closed-loop frequency measures. Like Bode plots, it does not provide stability information.*

With computer-aided control system design, polar plots – and in particular inverse polar plots – for studying multi-input/multi-output systems are well established. Nichols charts are not easily produced on a computer and, as shown in Chapter 11, it is a simple matter to generate closed-loop data directly. Bode plots give such insight into the effects of system modification that they will undoubtedly continue to be used, but in conjunction with open-loop polar and closed-loop frequency plots.

Problems

7.1.   Determine the amplitude magnification $M(\omega)$ and the phase shift $\phi(\omega)$ for the following open-loop transfer functions:

(a) $G(s) = \dfrac{1}{s(s-1)}$

(b) $G(s) = \dfrac{K}{(\delta_1 s + 1)(\delta_2 s + 1)}$

(c) $G(s) = \dfrac{10(1+2s)}{s(1+0.02s)(1+0.5s)(1+s)}$

7.2.   Use a graphical technique to evaluate $M(\omega)$ and $\phi(\omega)$ at $\omega = 2$ rad/s for the transfer function

$$G(s) = \frac{0.2s+1}{(s+1)(0.5s+1)(0.25s+1)}$$

Check your solution by direct calculation of $M(\omega)$ and $\phi(\omega)$.

7.3.   For the transfer function

$$G(s) = \frac{10}{s(1+0.025s)(1+0.5s)(1+s)}$$

plot the polar, inverse polar and Bode diagrams. Also produce a plot of dB magnitude versus phase (a Nichols plot). From each of these plots, find the gain and phase margins. If the transfer function gain were reduced from 10 to 1, what would be the new gain and phase margins?

* Its main disadvantage is a corollary of the 'law of inanimate objects' which indicates that, if a special chart is required, one will not be available.

7.4.   A unity negative feedback system has an open-loop transfer function given by

$$G(s) = \frac{K(1 + 5s)}{s(1 + 10s)(1 + s)^2}$$

Draw the Bode diagrams and determine the loop gain K required for a phase margin of 20°. What is the gain margin?

7.5.   For the system of Problem 7.4, set the gain K to 1.52 and find the closed-loop transfer function. Draw the closed-loop Bode plot and determine M_p, ω_p, the bandwidth and the cutoff rate.

7.6.   Use the Nichols chart shown in Fig. 7.12 to obtain the closed-loop Bode plots from the open-loop Bode plots of Problem 7.4.

7.7.   A lag compensator (see Chapter 8, Section 8.4.1) with the transfer function

$$G_c(s) = \frac{1 + 10s}{1 + 50s}$$

is added to the system of Problem 7.4. Use Bode diagrams to find the reduction in steady-state error following a ramp change in the reference input, assuming the phase margin of 20° is maintained.

7.8.   A lead compensator (see Chapter 8, Section 8.4.2) with transfer function

$$G_c(s) = \frac{s + 0.1}{s + 0.5}$$

is added to the system of Problem 7.4. Use Bode diagrams to find:
(a) The new phase margin ($K = 1.52$).
(b) The increase in phase margin frequency and the gain K required if the phase margin is returned to 20°.

7.9.   Plot the Nyquist diagram for a plant having an unstable open-loop transfer function given by

$$G(s) = \frac{K(s + 0.4)}{s(s^2 + 2s - 1)}$$

Determine the range of gain K for which a closed-loop control system with unity negative feedback which incorporated this plant would be stable.

7.10.   An integrating controller having the transfer function

$$G_c(s) = \frac{K}{s}$$

is placed in the forward path of a control system with unity negative feedback whose open-loop transfer function is

$$G(s) = \frac{1}{(s+1)(2s+1)}$$

For this system:

(a) Determine the value of K that will give a gain margin of 2.5.

(b) Using this value of K, plot the full Nyquist diagram and hence prove that the closed-loop system is stable.

(c) Find the system's phase margin.

7.11.   A closed-loop control system with negative feedback has an open-loop transfer function given by:

$$G(s) = \frac{K}{(s+3)^3(s+1)}$$

From the polar plot, determine:

(a) The gain K that will make the closed-loop system marginally stable.

(b) The gain margin when the gain K is adjusted to give a phase margin of 45°.

(c) The phase margin when the gain K is adjusted to give a gain margin of 3.

7.12.   Figure 7.44 shows an M circle in the $G(s)$-plane. As usual, the circle has a radius $M/(M^2-1)$ and is centred at $M^2/(M^2-1)$ (see Section 7.7.5). A line $\overrightarrow{Oa}$ is drawn which is tangent to the M circle at a and passes through the origin of the $G(s)$-plane, as shown. Prove that:

(a) The angle ψ is such that $\sin\psi = 1/M$.

(b) A line perpendicular to the real axis which passes through the point of tangency intersects the real axis at -1. This is the line $\overrightarrow{ab}$ shown in Fig. 7.44.

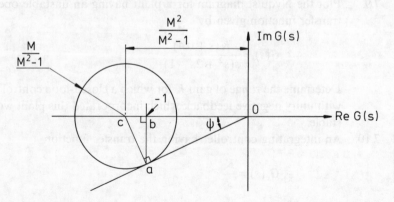

Figure 7.44

7.13. The relationships proved in Problem 7.12 are used to design control systems having a given closed-loop peak magnification M_p. The procedure used is first to draw a line from the origin of the $G(s)$-plane at angle ψ, where $\sin \psi = 1/M_p$. By trial and error a circle is found which is centred on the real axis and is tangent to both the $G(j\omega)$ locus and the line at angle ψ. From the tangency point of the line and circle, a line is drawn perpendicular to the real axis. Let this line intersect the real axis at $-T$. Now, if the closed-loop peak magnification is to be M_p, then the loop gain must be adjusted by an amount x, so that $Tx = 1$. Hence to produce the required closed-loop peak magnification, the loop gain must be adjusted by $1/T$.

Construct a polar plot for the system

$$G(s) = \frac{K}{s(1 + 0.5s)(6 + s)}$$

and determine the gain K that will give the closed-loop system a value of $M_p = 1.4$. With this gain, what are the system's gain and phase margins?

7.14. A frequency test on an open-loop system produced the following results:

ω, rad/s	2.5	3	3.5	4	4.5
$\operatorname{Re} G(j\omega)$	-2	-1.75	-1.5	-1.25	-1.0
$\operatorname{Im} G(j\omega)$	-2.5	-1.67	-1.17	-0.8	-0.5

Show that M_p has a value of 2.3 when $\omega = 4.7$. Determine the change in system gain required to give an M_p value of 1.5. What is the new resonant frequency? What is the new damped frequency?

7.15. Construct an inverse polar plot for the system

$$G(s) = \frac{K}{s(1 + 0.5s)(6 + s)}$$

and determine the gain K that will give the closed-loop system a value of $M_p = 1.5$. With this gain, what are the system's gain and phase margins? Also determine the closed-loop bandwidth.

CHAPTER 8

CONTROLLER AND COMPENSATOR DESIGN

8.1 Introduction

The normal procedure for designing a controller for single input/single output systems is an iterative one. Often the nature of the plant determines the structure of the controller, and the problem becomes one of selecting and tuning suitable compensating networks in order to meet the required performance specifications. Sometimes the controller's structure is not immediately evident; sometimes a preferred configuration proves impossible to implement. Such cases test the ingenuity of the designer. Also, the final control system design is invariably a compromise, not only between the various performance specifications but also between the feasible and the economically justifiable. There are no unique solutions. However, there are some simple controller structures and compensating elements which have proved useful, and it is a consideration of these which provides the basis for this chapter.

8.2 Controller design concepts

A control system consists of a plant with its actuators and sensors, and a controller. The actuator normally takes a low-energy signal, and transforms and amplifies it in order to produce a corresponding action which is applied to the plant's manipulable input. A sensor typically takes a high-energy signal from the plant's measurable output, and transforms it – ideally linearly – into an equivalent low-energy signal. The controller includes all the compensators, comparators, set points and paths required to complete the system.

In this section, rules of thumb are given for designing the controller and the control system configuration. These are not rigorous, but are intended to provide some insight into the design problem.

Some of the more common design problems are about the control of flow, level, temperature and pressure. Experience shows that satisfactory control may be achieved by using simple PID controllers (see Section 5.10).

Flow. Flow loops are notoriously noisy, and this precludes derivative action. The controller gain is rarely greater than one, and consequently, unless integral action is included, there is significant steady-state error.

Level. In most level control loops the actual liquid level is relatively unimportant, provided it is between some maximum and minimum value. The transfer function model of a level system will often be of type 1, so satisfactory control can be achieved with a simple proportional controller.

Temperature. In temperature loops, thermal lags (caused by heat transfer) and sensor lags make the loops relatively slow and noise-free, so full PID controllers are often used. Note that in the case study carried out on a temperature control loop (see Section 10.2), the sensor measures the temperature of a flowing air stream, so the loop acts like a flow loop.

Pressure. Pressure loops can vary considerably in their dynamic characteristics, from very fast loops almost like flow to slow averaging loops almost like level. Therefore, depending on the nature of the loop, a PI or P controller is used.

Compensators modifying low-energy signals are normally cheaper and easier to adjust than those for high-energy signals. If, for some reason, a forward path compensator has to be combined with the actuator, then it may be more economical to use feedback compensation. In general, though, forward path compensation provides the preferred configuration.

At this point it is worth noting that control is an important aspect of any plant design. If control is likely to be required, then dynamic considerations should be taken into account in the plant design. Various hardware configurations may be possible and some, for example those which avoid interactions, are easier to control than others. Allowance should be made in sizing and selecting equipment so that it can handle operational transients. Also, allowance should be made for the inclusion and optimal location of sensors and actuators so that they operate effectively.

In developing the controller, the following design points should be borne in mind:

1. Feedback paths should be designed so as to avoid time delays and lags. In practice this means that a sensor should be placed as close as possible to the source of the variable it is trying to control. For example, if the temperature of the steam from a boiler is required, then ideally the sensor should be in contact with the steam and located close to the boiler steam outlet point. Measurements at a point on the steam pipe away from the boiler would introduce a time delay, and if the sensor is not in contact with the steam a thermal lag would also be introduced into the feedback path.
2. Large, frequent measurable disturbances should be compensated for by feedforward control. Consider the single-loop control system shown in Fig. 8.1. The controller, consisting of the forward path compensator $G_c(s)$ and the comparator, is indicated, together with the actuator dynamics $G_1(s)$, the plant

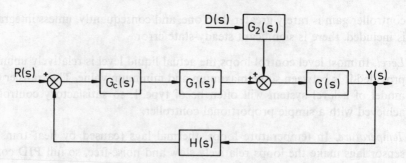

Figure 8.1 Single-loop control system with disturbance

dynamics $G(s)$, the sensor dynamics $H(s)$ and the disturbance input $D(s)$. This system could represent a valve-actuated system $G_1(s)$ controlling the flow of some liquid into the plant $G(s)$; $D(s)$ would then represent temperature disturbances (variations) affecting the viscosity, and hence the flow rate, of the fluid, and $G_2(s)$ would be a transfer function converting temperature changes into equivalent flow changes.

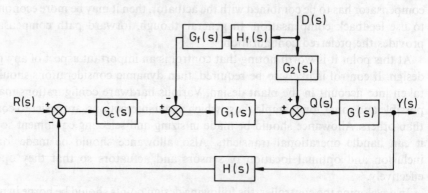

Figure 8.2 Single-loop control system with feedforward control

Figure 8.2 shows the same system with feedforward control. For the example given, $H_f(s)$ would be a temperature sensor and $G_f(s)$ a compensating network. For removal of the effect of the disturbance from $Q(s)$, it is easily shown that

$$G_f(s) = \frac{G_2(s)}{G_1(s)H_f(s)} \tag{8.1}$$

3. Where possible, minor disturbances should be eliminated by introducing a cascade controller (see Fig. 8.3). If, in the example given, the temperature changes were small but still produced flow disturbances, they could be compensated for by placing a cascade controller around the valve $G_1(s)$, as shown in Fig. 8.3. Here $H_{cc}(s)$ represents the flow sensor dynamics, and $G_{cc}(s)$

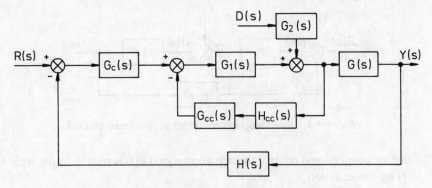

Figure 8.3 Single-loop control system with cascade controller

the cascade controller compensator. With this scheme the master controller is providing the set point value for the slave cascade control system.

4. Multiloop systems in which the operation of an external loop depends on the operation of an inner loop should be avoided. This is not always possible, as for example in cascade control for disturbance rejection. However, as a general rule loops should be independent.

5. The control structure should be such that the failure of a loop, or any element in the loop, does not tend to make the overall system unstable.

6. The manipulated variable should not be saturated. Aircraft control surfaces can deflect only by a certain amount; fluid flows are fixed between certain limits; motors have maximum torques. Once a loop saturates, the system becomes non-linear and there is often a significant recovery time before the loop becomes fully operative again. The saturation values provide design constraints.

7. A proposed design should be checked to ensure that it is effective during startup and shutdown conditions.

8.3   PID controllers

Control systems which have a forward path compensator, consisting of some PID combination, have proved suitable for most industrial applications. Since maintenance and failure problems associated with non-standard equipment can prove costly, it is normal practice to use a commercially available PID controller whenever possible.

The ideal PID controller has a transfer function given by

$$G_c(s) = K[1 + T_d s + 1/(T_i s)] \qquad (8.2)$$

where K is the gain of the proportional channel, T_d the derivative action time and T_i is the integral action time. In practice, the proportional channel will have a

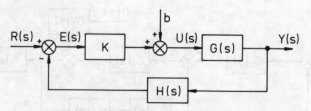

Figure 8.4 Closed-loop system under proportional control

bias: for proportional control the actuation signal $u(t)$ is related to the error signal $e(t)$ by the equation

$$U(s) = KE(s) + b \tag{8.3}$$

where b is the output bias.

Figure 8.4 shows the closed-loop block diagram of such a proportional controller. In this system the bias provides an offset for the zero error point. With no bias signal and an open-loop system transfer function of type 0, $E(s)$ can be zero in the steady state only if $R(s)$ and $Y(s)$ are zero. However, even if the bias is not zero, the zero steady-state $E(s)$ point still requires $R(s)$ and $Y(s)$ to be equal, but they will not be identically equal to zero. The bias therefore shifts the control system's operating point. Normally, the selected operating point is somewhere in the actuator's mid-range of operation so that it can respond to both positive and negative error signals.

Ideal derivative action cannot be achieved in practice, so a practical PID controller approximates derivative action via a lead compensator, whose action may be described by

$$T_d s \approx \frac{T_d s + 1}{\alpha T_d s + 1} \tag{8.4}$$

where typically, for a commercial device, α is fixed somewhere between $1/6$ and $1/20$. The lead compensator is placed in series with a PI controller. The commercial PID controller transfer function is therefore the product of those of a PI controller and a lead compensator:

$$G_c(s) = K\left(\frac{T_i s + 1}{T_i s}\right)\left(\frac{T_d s + 1}{\alpha T_d s + 1}\right) + b \tag{8.5}$$

For further details on lead compensation, see Section 8.4.2.

8.3.1   PID controller tuning

There have been many techniques proposed for the time domain tuning of PID controllers, but in practice the Ziegler and Nichols methods seem to have become the recognized standards in the process industries. Using their settings produces an under-damped response (see Section 5.10.1). For process control such responses are desirable since the oscillations tend to average out, and the overshoot helps to eliminate the effects of stiction and backlash in the valves and actuators. (Stiction is static friction, or Coulomb friction, the initial resistance to motion of contacting surfaces.) However, with servo control systems oscillations are not usually desirable and additional damping will therefore be required. In most cases this can be achieved by reducing the controller's proportional gain during on-line tuning.

The effect on a system's root loci plot of introducing a PID controller has been examined in Section 6.4. Proportional control is ideally suited to the root loci design method. With PI control the design procedure is normally an iterative one. However, a PI controller introduces a pole at the origin of the s-plane and a zero at $1/T_i$, and this requires the selection of two parameters: T_i, which defines the position of the zero in the open- and closed-loop transfer functions, and K, which locates the closed-loop pole positions. The pole at the origin of the s-plane slows down the response, so T_i is normally chosen so that its associated zero is close to the pole, so as to speed up the response. If derivative action is included in the controller, then a further pole–zero combination is introduced, the relative positions of which depend on the selection of α (see Equation (8.5)). In general, integral action is introduced to increase the system's type number and hence improve its steady-state tracking capabilities. Since most systems contain some noise, derivative action is seldom used; however, when it is used (see Section 8.2) the effect is to increase the system's speed of response.

For selecting the settings for a PID controller using frequency domain techniques, design guidelines have been developed. Typically, a proportional controller should satisfy a given gain margin criterion. For example, the Ziegler–Nichols setting for a proportional controller is based on a gain margin of 2. Checks are then carried out on the phase margin and M_p value to ensure satisfactory performance. With a PI controller the proportional gain would be adjusted to meet the M_p specification (this is effectively defining the closed-loop damping ratio). The introduction of integral action reduces the system's bandwidth, hence reducing the response speed of the closed-loop system. The zero introduced by the PI controller is then used to increase the bandwidth, therefore reducing any loss of response speed. One way of achieving this is to set the integral action time at

$$\frac{1}{T_i} = \frac{\omega_p}{5} \tag{8.6}$$

where ω_p is the resonant frequency, the closed-loop frequency corresponding to M_p. This places the zero in a position where it removes most of the integration phase angle lag at the point where the resonant peak occurs. The introduction of this combined PI action changes the closed-loop response, and consequently the proportional term must be readjusted to meet the bandwidth specification. If at this stage ω_p is essentially unaltered, the design is complete; otherwise the integral action time must be recalculated from Equation (8.6) and the process repeated.

The frequency design of a full PID controller gives three adjustable parameters. A typical procedure for selecting them is first to design a PI controller. The derivative action time T_d is then chosen so that the lead compensator network gives its maximum phase angle advance at the $-180°$ point on the PI controlled open-loop frequency response curve. Again, the closed-loop frequency response changes with the introduction of new terms, and the various design stages must be repeated until there is no appreciable change.

8.3.2   Velocity feedback

Derivative action has the attraction that it improves a system's rise time without producing an over-large overshoot or excessive oscillations. Despite this, it finds little practical application because of its intolerance to noise. With some systems, however, it is possible to measure the derivative of the output variable directly and so avoid the problems of noise. For historical reasons, the direct measurement of the derivative of an output for control purposes is known as velocity feedback.

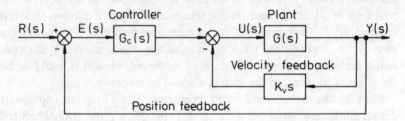

Figure 8.5 Velocity feedback system

A classic example of direct derivative measurement is the use of a tachogenerator to measure the angular velocity of rotational, position control machinery. Figure 8.5 shows the block diagram of such a system incorporating velocity feedback. The closed-loop transfer function for this system is given by

$$\frac{Y(s)}{R(s)} = \frac{G(s)G_c(s)}{1 + G(s)K_v s + G(s)G_c(s)} \tag{8.7}$$

where K_v is the gain associated with the tachogenerator. It is left as an exercise for the reader to show that, if $G_c(s)$ is a PI controller, then the above closed-loop characteristic equation is identical to that of the system under ideal PID control. This is true if

$$K_v = KT_d \tag{8.8}$$

where K and T_d are the gain and derivative action time of an ideal PID controller (see Equation (8.2)).

Although velocity feedback produces the same characteristic equation as a PID controller, PID action introduces a zero into the closed-loop system, whereas velocity feedback does not.

8.4   Simple compensators

When it is not possible to meet the desired performance specifications simply by adjusting the system's closed-loop gain, dynamic compensation is required. The simplest form of dynamic compensation is provided by the lag, lead and lag–lead compensators. With these devices, either singly or in combination, it is normally possible to adjust the system's response sufficiently to achieve the required performance. In this sense, the PID controller and its various derivatives are specific combinations of lead and lag compensators.

In the following subsections are considered the design of lag, lead and lag–lead compensators using root loci and frequency domain methods.

8.4.1   Lag compensation

A stable closed-loop design which has a satisfactory transient response, but too large a steady-state error, can often be improved by means of lag compensation. The transfer function of a lag compensator is

$$G_c(s) = \frac{1 + Ts}{1 + \alpha Ts}, \quad \alpha > 1 \tag{8.9}$$

The compensator's pole–zero map is shown in Fig. 8.6. Since the pole is located at $s = -1/\alpha T$, this compensator is a more general form of PI controller. A PI controller has its pole at the origin of the s-plane.

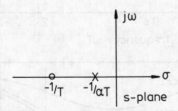

Figure 8.6 Pole–zero map of a lag compensator

Using the root loci design technique, the pole of the compensator is placed close to the origin of the s-plane in order to improve the steady-state error. Since the response is deemed to be satisfactory, the original shape of the root loci plot should remain unchanged. The change brought about by compensation can be minimized by placing the zero close to the pole. A useful rule of thumb is that the angle contributed by the compensator at the original dominant closed-loop poles should be less than $5°$.

The various frequency design methods enable closed-loop performance to be described in terms of M_p, ω_p and the static error coefficients. The usual design procedure is to fix M_p and then determine the corresponding frequency ω_p and loop gain K. Fixing M_p essentially defines the damping ratio, and from ω_p is found the undamped natural frequency ω_n; ω_n, together with the damping ratio, determines the response settling time.

The frequency response function of the lag compensator is

$$G_c(j\omega) = \frac{1 + j\omega T}{1 + \alpha j\omega T} \qquad (8.10)$$

Bode plots for this compensator are shown in Fig. 8.7. Phase lag compensation is normally applied to the very-low-frequency end of the open-loop frequency response. If the object is to reduce the steady-state error of a type 0 system, then α is selected to give the required increase in gain. Typically, for the same gain and phase margins as the uncompensated system, the loop gain of the compensated system has to be increased by α. There are practical limits on the magnitude of α,

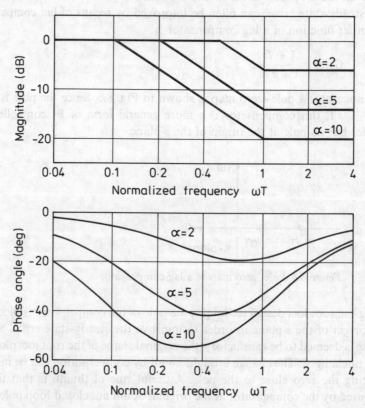

Figure 8.7 Bode plots for lag compensation

and normally α would not be greater than 10. Once α is selected, T is chosen so that the corner frequency of the zero is well away from the critical frequency range of the system. An acceptable value is

$$\frac{1}{T} = \frac{\omega_p}{10} \tag{8.11}$$

The above design procedure is intended only as a guide; in an actual study adjustments to these values are likely to be required before an acceptable solution is reached.

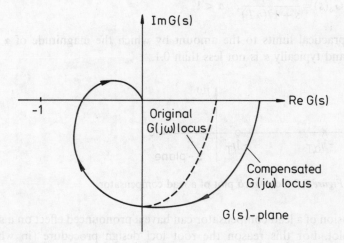

Figure 8.8 Effects of lag compensation on a system's polar plot

Figure 8.8 shows the effect of lag compensation on a type 0 system. In this figure the gain of the compensated system has been increased so that it has the same gain and phase margins as the original systems. As a corollary, if the steady-state error of the original design were acceptable, no increase in loop gain would be required and the gain margin of the compensated system would be increased. The increase in gain margin would be accompanied by an increase in phase margin.

In general, the effects of lag compensation are as follows:

1. The bandwidth is reduced.
2. The predominant time constant of the system is usually increased, producing a more sluggish system, as lag compensation introduces a pole close to the imaginary axis.
3. For a given relative stability the steady-state error is improved, and the system has better tracking capabilities.
4. For a given steady-state error the gain margin is improved.

8.4.2   Lead compensation

Lead compensation is normally used to improve the speed of response of stable systems of type 1 and higher. It can also be used to advantage in systems with time delays.

The compensator transfer function is given by

$$G_c(s) = \alpha \frac{1 + T_s}{1 + \alpha Ts} \tag{8.12}$$

or

$$G_c(s) = \frac{s + 1/T}{s + 1/(\alpha T)}, \quad \alpha < 1 \tag{8.13}$$

There are practical limits to the amount by which the magnitude of α may be decreased, and typically α is not less than 0.1.

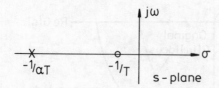

Figure 8.9 Pole–zero plot of a lead compensator

The inclusion of a lead compensator can have a pronounced effect on a system's root loci plot. For this reason the root loci design procedure (in which the compensator's pole–zero positions are fixed, see Fig. 8.9) is normally one of trial and error. However, an approach which can be used to good effect on systems of type 1 or higher is to choose the zero position so that it cancels the largest real pole of the open-loop system. (It is assumed that poles at the origin of the system are excluded, and that all other open-loop poles and zeros are in the left half s-plane.) The frequency response function of the lead compensator is

$$G_c(j\omega) = \alpha \frac{1 + j\omega T}{1 + \alpha j\omega T} \tag{8.14}$$

and its Bode plots are shown in Fig. 8.10.

In the frequency domain, lead compensation is commonly used to improve a system's phase margin. Improvements in phase margin are accompanied by increases in bandwidth and ω_p, and hence also by an increase in the closed-loop transient response speed. One method of selecting the parameters of the phase compensator is first to choose the desired phase margin, β. The frequency ω_B is then found at which the phase angle of the open-loop plant $GH(s)$ is given by

$$\arg GH(s) = -(180° - \beta + r\delta) \tag{8.15}$$

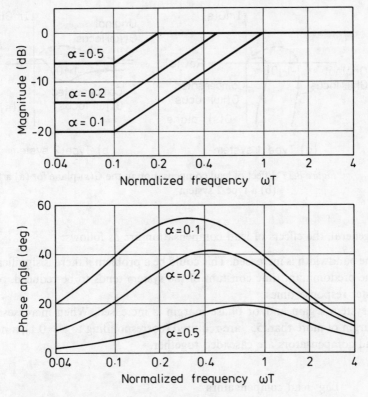

Figure 8.10 Bode plots for lead compensation

where

β = desired phase margin
r = system type number
δ = maximum phase advance of selected compensator

It may be shown that

$$\delta = \tan^{-1}\frac{1-\alpha}{2\sqrt{\alpha}} \tag{8.16}$$

If T is now chosen such that

$$T = \frac{1}{\omega_B\sqrt{\alpha}} \tag{8.17}$$

and the gain of the compensated system is adjusted to give a magnitude ratio of OdB at ω_B, the system will be found to have the desired phase margin β.

The effect of a phase advance compensator on the polar plots of type 1 and 2 systems is shown in Fig. 8.11.

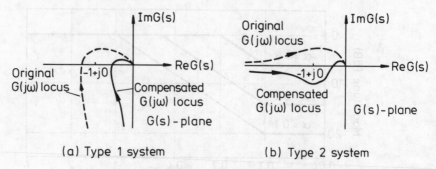

(a) Type 1 system   (b) Type 2 system

Figure 8.11 Effect of lead compensation in the $G(s)$-plane for (a) a type 1 and (b) a type 2 system

In general, the effects of lead compensation are as follows:

1. The bandwidth is increased. This could be a problem if there is significant noise.
2. The predominant time constant of the system tends to be reduced, producing faster response times.
3. A system's gain and/or phase margin is increased. When increases in phase margin of more than $55°$ are required (corresponding to $\alpha = 0.1$), a number of lead compensators are cascaded together.

8.4.3   Lag–lead compensators

If both the transient and steady-state responses of a system are unsatisfactory, it is usually more economical to use a combined lag–lead compensator instead of individual lag and lead elements. The transfer function of this type of compensator is

$$G_c(s) = \frac{(1 + T_1 s)(1 + T_2 s)}{(1 + \alpha T_1 s)(1 + T_2 s/\alpha)}, \quad \alpha > 1, \quad T_1 > T_2 \tag{8.18}$$

Figure 8.12 shows the compensator's pole–zero plot, and Fig. 8.13 a sketch of its Bode plots.

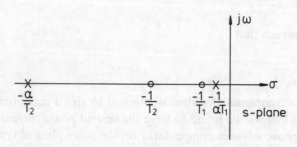

Figure 8.12 Pole–zero plot of a lag–lead compensator

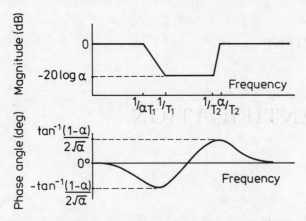

Figure 8.13 Bode plots of a lag–lead compensator

Design techniques for the lag–lead compensator combine the methods previously described for the lag and lead elements.

In general, a lag–lead compensator has all the advantages of both lag and lead compensation, and very few of their usually undesirable characteristics. The use of lag–lead compensators therefore makes it possible to meet many system specifications without incurring the penalties of excessive bandwidth or over-sluggish response.

Problems

8.1. A closed-loop control system with negative feedback has the open-loop transfer function

$$G(s) = \frac{K}{s^2(s+10)}$$

The closed system is to be stabilized by means of a forward path compensator. The required closed-loop performance specifications are a 5 per cent settling time of 5 s, and a subsidence ratio of about 5:1.

Select an appropriate compensator, and use the root loci method to find suitable values of α and T.

8.2. Repeat Problem 8.1 using frequency domain methods.

CHAPTER 9

IDENTIFICATION

9.1   Introduction

By using fundamental physical laws and the lumped parameter approach (described in Chapter 1), it is possible to model most engineering systems to a degree of accuracy sufficient for most controller design work. However, the models produced in this way are inevitably of high order, and possibly non-linear, and are consequently not in a form suitable for the application of the standard design techniques. The problems of reducing the order of the model (model reduction) and the associated problem of linearization (by considering small perturbations about some operating point) can often be more difficult than the original modelling exercise. For this reason an alternative approach is often used – that of identification. In this chapter techniques are developed which enable a low-order linear transfer function to be found, either directly from an existing plant or from some high-order model.

9.2   Identification defined

The purpose of identification is to determine the dynamics of a system by means of an experiment. The conditions under which the process operates, together with the physical constraints, determine the type of experiment which may be carried out. However, for any linear stable system the experiment will invariably be of one particular form. A well-defined bounded input is used to perturb the system from some initial operating condition, and measurements are then made of the resulting output. A boundary placed around this system may be thought of as containing a black box having a mathematical structure which, although unknown, is equivalent to the system's dynamics. The identification problem may therefore be considered the realization of a mathematical structure which produces the same input/output response as the system. If it is assumed to be linear, time-invariant and stable, with no external disturbances, then the relationship between the input $u(t)$ and the output $y(t)$ is, in Laplace transform form,

$$Y(s) = G(s)U(s) \tag{9.1}$$

where $G(s)$ is the unknown black-box transfer function. It is evident that, provided the input $u(t)$ is well defined, then the output

$$y(t) = \mathscr{L}^{-1}[Y(s)] = \mathscr{L}^{-1}[G(s)U(s)] \tag{9.2}$$

contains all the information necessary for determining the transfer function $G(s)$, and hence the system's dynamics.

9.3   Experimental procedures

Ideally, experiments are carried out on the open-loop system. If, however, drift in the output variable is apparent, or if the system is open-loop unstable, then the system must be operated closed-loop and the manipulable input and output signals measured. Both open- and closed-loop operation are considered here.

The initial phase of any identification procedure is to establish both the measurable output and the manipulable input that is to be used to control that output. This clearly requires some knowledge of the system and how it works. In this introduction to identification it is assumed that data obtained from this input/output pair is sufficient to produce the required transfer function.

Another assumption is that the system is operating linearly. Most industrial systems have some inherent non-linearity, and therefore small perturbations must be assumed; for small perturbations (disturbances) most systems may be adequately described by linear equations. However, this raises the question of how small is small. In practice most signals obtained from industrial equipment will be contaminated with noise. If the input signal is too small the output response could be lost in the noise, whereas if the output is too large the non-linear effects are likely to become significant. Normally the plant would be run at the selected steady-state operating conditions and a small step change made to the manipulable input. Here a small change would be, say, 5 per cent of the total possible adjustment. The resulting step response would be recorded. After returning the plant to its initial steady-state condition, a step change of 5 per cent would be made to the manipulable input, in a direction opposite to that originally chosen. The negative of this response about its original steady-state condition (the physical inversion of the response) is recorded and compared with the previously recorded response. If the two responses are identical – that is, if they have the same steady-state value and transient response characteristics, then the plant's response could be assumed to be linear about the selected operating point; otherwise, the amplitude of the forcing step input must be reduced and the experiment repeated. Increasing the input step until a change in the responses is detected indicates the range over which any derived linear model is valid.

If during this initial testing phase noise becomes significant, it must be accounted for in some way. Allowance may be made for noise by using one of the statistical methods of identification, or else the same experiment may be repeated several times and the average response used. Both methods use statistics, although in the former the techniques have their basis in statistics, and in the latter the methods are deterministic. The statistical methods are not dealt with here; suffice it to say that when there is a small signal-to-noise ratio, or if for some reason plant disturbances must be of extremely low amplitude, then these methods provide good practical solutions. Most commercially available equipment for providing frequency response and impulse response data uses statistical methods.

With step or pulse testing the effect of noise may be reduced by taking a number of responses, all from the same initial steady-state condition and all using the same forcing input. Averaging these responses will tend to remove the noise from the signal. The number of tests required depends on the signal-to-noise ratio and also on the desired level of noise reduction. A useful estimate of the number of experiments required is

$$n = 400/x^2 \tag{9.3}$$

where n is the number of experiments required to yield a response with a standard deviation of 5 per cent or less, and x is the signal-to-noise ratio.

In practice, a response with a 5 per cent standard deviation, or better, is adequate for the procedures to be outlined in this chapter. The signal-to-noise ratio may be estimated by comparing the amplitude changes in the measured steady-state output (the noise level) with the average change between the steady-state levels caused by a forcing step input (the signal level). Equation (9.3) indicates that if the signal-to-noise ratio is 20:1, or better, then a single experiment is adequate, and if it is 3:1 then 45 experiments would be required. Also, if the system's dominant time constant is T, then steady state conditions would be achieved in a time of approximately $5T$, and the duration of n tests would be about $5nT$. Clearly, for any given problem there is a point beyond which this deterministic approach is impractical.

In addition to step and pulse testing, the other common form of testing is frequency response testing. The frequency range must be established, and tests carried out at discrete frequencies within that range. Noise is an added complication in that, as in the other types of test, it increases the duration of each test. However, since the measured steady-state harmonic response is continuous, the averaging process is carried out on a continuous record rather than a number of discrete records. This must be done with care since it involves dividing a continuous record into a number of discrete records, and any error introduced will be cumulative. Under steady-state conditions the frequency of the response will be the same as the frequency of the forcing input, and time for one complete cycle can therefore be established. Sequential output recordings, each covering one (or more) cycle, could be averaged and, provided the position of the forcing input relative to the start of this sequence is known, the required magnification ratio and phase shift can be established.

The remainder of this chapter will consider step, pulse and harmonic testing with the object of establishing a system transfer function.

9.4   Step response data

For open-loop plant a step response is particularly useful. The following list indicates the information it can provide (see also Program 8 in Chapter 11):

1. *Time delays.* Most methods of transfer function identification are based on the assumption that the response data is instantaneous. If, for a step input initiated at time $t = 0$, there is a delay before the system begins to response, at time $t = \tau$, identification must be used on the input $u(t)$ and the shifted output $y(t - \tau)$. This extracts the time delay from the response and will produce a transfer function $G(s)$. Hence, if the time delay is included, the final transfer function in the s-domain becomes

$$e^{-\tau s}G(s) \tag{9.4}$$

   This approach works for a linear process having a time delay in the forward path. However, the extraction of any information from a response and its subsequent introduction to the model should always be undertaken with care.

2. *Absolute stability.* By estimating the upper bound of the response it is possible to determine the dominant pole position(s). This is useful for several reasons, not the least of which is that it indicates the frequency range over which frequency response testing should be carried out.

3. *Non-minimum phase.* The presence of non-minimum phase zeros is usually demonstrated by the initial time response moving in the opposite direction to that of the input. Knowledge of this characteristic is useful when applying Bode's decomposition method (see Section 9.7.1).

4. *Predominantly first-order process.* If the step response transient leaves its steady-state position at an angle of maximum slope, then the process is predominantly first order and may be represented by a transfer function having one more pole than zero. The process usually has a dominant first-order pole, the location of which may be estimated by using the 63 per cent rise time criterion, defined as follows. If a unit step is applied to a stable first-order system, the output $y(t)$ is given by

$$y(t) = 1 - e^{-pt}$$

   and when t is equal to $1/p$ the output $y(t)$ is approximately equal to 0.63. Hence the negative of the reciprocal of the time taken by the step response to reach 63 per cent of its final value may be used to approximate the position of the dominant pole.

5. *Dominant complex poles.* These are usually indicated by an oscillatory response.

6. *System gain.* This is defined as the change in the steady-state output divided by the magnitude of the input step response.

One identification technique is to find an approximate transfer function which will fit the step response. Such models are easily found and often form the basis for many of the on-line tuning techniques. For more detailed design studies the method is limited since it is difficult to distinguish between systems of second and third, or higher, order.

Closed-loop step responses are less useful, unless of course a model of the closed-loop system is required. In general, such closed-loop models cannot be manipulated to yield the system's open-loop transfer function. The only information likely to be extracted from the closed-loop response is the open-loop steady-state gain, and possibly any time delay. A simple analysis of an open-loop system with time delay indicates that the closed-loop response will have the same time delay.

9.5   Pulse response data

Pulse testing is one of the oldest techniques for characterizing a system's dynamics. Ideally, the open-loop system is excited by a forcing impulse and, from Equation (9.2), the resulting time response is the inverse Laplace transform of $G(s)$, namely the required transfer function.

Unfortunately an impulsive test signal can inflict damage on the control system and its hardware, particularly if the experiment has to be repeated to eliminate the effects of noise. For this reason direct impulsive testing tends not to be used on engineering plant. An open-loop impulse response can be obtained from an open- or closed-loop system by using statistical methods, the most common of which is the injection of a pseudo-random binary signal into the plant, followed by cross-correlation of the input and output signals. Details of correlation techniques are not given in this text.

An alternative to impulse testing is pulse testing. Here a pulse of known but otherwise fairly arbitrary shape is injected into the process and the response recorded. This data can be analysed to produce the system's open-loop frequency response (see Section 9.6). Pulse testing can be applied to both open- and closed-loop systems. In theory, for noise-free data just one pulse on the systems manipulable input is sufficient to generate the entire frequency response curve. In practice, even for noise-free data, several pulses are usually required simply to establish the required size and duration of the input pulse. The approximate width of the input pulse is probably the method's limiting factor. A large pulse width would tend to lose high-frequency information. As a general rule, the width of the pulse should be less than half the smallest time constant of interest. Once the width is fixed, the pulse height, which should be within the linear range, determines the disturbance in the system's response. If the signal-to-noise ratio is low, then again several tests will have to be averaged to produce a usable response.

9.6   Time and frequency response correlation

An aperiodic function $y(t)$ in the time domain may be transformed into a corresponding function $g(j\omega)$ in the frequency domain by Fourier transformation:

$$g(j\omega) = \int_{-\infty}^{\infty} y(t)e^{-j\omega t}\,dt \qquad (9.5)$$

Here it is assumed that the function $y(t)$ is analytic, but in experimental work $y(t)$ is likely to be discrete – to consist of a set of amplitude and time coordinates. The discrete form of Equation (9.5) is developed for Program 8 in Chapter 11. Since the discrete equivalents are easily derived, the arguments in this section are developed for the continuous form only.

From Equation (9.1) and by definition, the unknown black-box transfer function of the system to be identified is

$$G(s) = \frac{Y(s)}{U(s)} \qquad (9.6)$$

where $Y(s)$ is the system's measured output and $U(s)$ the manipulable forcing input. The frequency response of $G(s)$ may be obtained by setting $s = j\omega$ as

$$G(j\omega) = \frac{Y(j\omega)}{U(j\omega)} \qquad (9.7)$$

Now, $Y(j\omega)$ could be found from the Fourier transform of the measured output signal $y(t)$, and $U(j\omega)$ from the Fourier transform of the forcing input $u(t)$. If $u(t)$ is impulsive, then $U(j\omega)$ is simply the power of the impulse (a unit impulse, for example, has a Fourier transform of unity). If $u(t)$ is a step function, then its Fourier transform is

$$U(j\omega) = \frac{A}{j\omega} \qquad (9.8)$$

where A is the amplitude of the forcing step input. When the input $u(t)$ takes one of these simple forms, the frequency response of the system, $G(j\omega)$, can be derived directly from the output signal $y(t)$. Program 8 in Chapter 11 gives the discrete equations for both an impulsive and a step input.

With a pulse input, Fourier transforms of both the output $y(t)$ and input $u(t)$ are required. On transformation, let

$$Y(j\omega) = A + jB \qquad (9.9)$$

and

$$U(j\omega) = C + jD \qquad (9.10)$$

from which it can be shown that

$$G(j\omega) = \frac{(AC + BD) - j(AD - BC)}{C^2 + D^2} \qquad (9.11)$$

With analytic functions for $y(t)$, and $u(t)$, the integrals A, B, C and D could, in theory, be evaluated from Equation (9.5). For discrete data there are the usual problems associated with numerical integration, further compounded by the harmonic nature of the exponential term. The net result is that, as the frequency increases, the transforms of both $y(t)$ and $u(t)$ will eventually break down.

9.7 Frequency response

In the preceding sections it has been indicated that a frequency response may be obtained directly from frequency response testing, or indirectly by means of the Fourier transform. Direct testing tends to be a time-consuming procedure; indirect testing, although offering many advantages including short experimental times, tends to become unreliable at higher frequencies. Nevertheless, both methods yield the system's frequency response over a frequency range which is usually sufficient for control purposes. The best approach is probably to use a Fourier transform method and then to check the results, particularly at higher frequencies, by a direct method.

Many of the design techniques for single input/single output plant use the system's frequency response. If the frequency approach is to be adopted then the derived frequency data may be used directly. Such data is readily stored and manipulated within a computer. If a transfer function is required, it is most easily obtained from the system's frequency response. Below are described two methods of determining a transfer function. The first is essentially a pencil-and-paper method, and the second a computational method.

9.7.1 Bode's decomposition method

For transfer function identification, the reverse procedure to the Bode asymptotic approximation is carried out on the plot of dB magnitude versus logarithmic frequency response. It consists of the following steps:

1. The dB magnitude curve is approximated by a series of piecewise straight lines, or asymptotes. The asymptotes are chosen such that their slope is given by an integer; a unit change in slope is defined as being a gradient change of 20 dB/decade.
2. The asymptotes are decomposed into several components.
3. The corner frequencies are noted.
4. The corresponding transfer function is written in pole–zero form.

In using step 4, particular attention must be paid to the plot of phase versus frequency response. For a minimum phase process (one having poles and zeros with negative real parts), the magnitude and phase frequency responses are interdependent: a pole will tend to cause a $-90°$ phase shift, and a zero will tend to cause a $+90°$ phase shift. For a non-minimum phase process the converse is true: a non-minimum phase pole will tend to cause a $+90°$ phase shift, and a

Table 9.1

Frequency ω (rad/s)	Amplitude ratio	Phase
0.01	0.1585	81°
0.05	0.7499	62°
0.10	1.122	45°
0.20	1.334	25°
0.50	1.585	10°
1.0	1.585	3°
5.0	1.585	−15°
10.0	1.496	−26°
20.0	1.122	−45°
50.0	0.5957	−68°
100.0	0.3162	−78°

non-minimum phase zero will tend to cause a $-90°$ phase shift (see Section 7.2.1). Since the system is assumed to be stable, all the poles will have negative real parts. However, the sign of the zero must be selected by reference to the curve itself.

If the identification experiment is carried out in the frequency domain, the Bode plots may be used to determine the presence of a time delay in the process response, since the high-frequency phase plot decreases rapidly to minus infinity. The value of the time delay can be determined by plotting phase against frequency (not log of frequency), and measuring the negative slope of the resulting straight line. This is proved simply by considering the frequency response of a time delay,

$$\mathrm{e}^{-st} = \mathrm{e}^{-j\omega\tau} = 1 \arg(-\omega\tau) \qquad (9.12)$$

When the frequency response is to be calculated using the Fourier transform, any time delay should be removed from the output $y(t)$ before transformation takes place.

Example 9.1

Table 9.1 lists the experimentally obtained harmonic responses for a component of a system. Find the component's transfer function, explaining clearly how the result is obtained.

Solution

A transfer function may be obtained by plotting the data in Bode form and using the asymptotic approximation to determine the corner frequencies, and hence the pole–zero locations. To do this the data in Table 9.1 must be changed into Bode form, as in Table 9.2. Bode plots may now be produced using standard log–linear graph paper, as shown in Fig. 9.1.

The Bode gain plot has a final slope of -20 dB/decade, and the phase plot

Table 9.2

Frequency ω (rad/s)	Gain (dB)	Phase
0.01	−16.0	81°
0.05	−2.5	62°
0.1	1.0	45°
0.2	2.5	25°
0.5	4.0	10°
1.0	4.0	3°
5.0	4.0	−15°
10.0	3.5	−26°
20.0	1.0	−45°
50.0	−5.5	−68°
100.0	−10.0	−78°

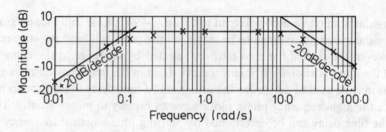

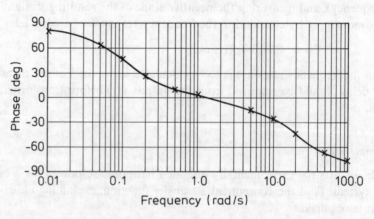

Figure 9.1 Bode plots from the data of Table 9.1

tends to −90°. This suggests that the component's transfer function is likely to be minimum phase with no time delay. It is worth noting that, although the final slopes of the magnitude and phase plots may indicate minimum phase, the system may still be non-minimum phase. However, this would require the transfer function to have the same number of right half s-plane poles as zeros, which in turn implies that the system is unstable.

Initially the gain plot has a slope of $+20$ dB/decade and a phase of $+90°$, so the system has a zero at $s = 0$. Two corner frequencies can be identified at 0.1 and 20 rad/s, and in both cases the gain plot changes slope by -20 dB/decade accompanied by a fall in phase. Combining this information produces a transfer function of the form

$$G(s) = \frac{Ks}{(1 + s/0.1)(1 + s/20)}$$

The Bode gain K may be determined from the dB magnitude expression for $F(s)$,

$$20 \log(G(j\omega)) = 20 \log K + 20 \log(\omega) - 20 \log \sqrt{\left[1 + \left(\frac{\omega}{0.1}\right)^2\right]}$$

$$- 20 \log \sqrt{\left[1 + \left(\frac{\omega}{20}\right)^2\right]}$$

At a frequency of 1 rad/s the system gain is 4 dB, so

$$\frac{4}{20} = \log K + 0 - 1.0022 - 0.0005$$

or

$$K = 10^{1.203} = 16$$

Consequently, the complete transfer function becomes

$$G(s) = \frac{16s}{(1 + 10s)(1 + 0.05s)}$$

9.7.2   Levy's method

Levy's method (named after E. C. Levy, who proposed it in 1959) is a complex curve-fitting technique designed for computer implementation. It uses a modified least-squares method to fit the frequency response curve of a system to the linear expression

$$G(s) = \frac{b_0 + \sum\limits_{i=1}^{q} b_i s^i}{1 + \sum\limits_{i=1}^{r} a_i s^i} \quad r \geqslant q \tag{9.13}$$

Since details of the method are given in Program 9 of Chapter 11, this section concentrates on its implementation.

For exact data the method gives good results provided a suitable data range is used. However, when attempting to identify noisy, high-order systems having a frequency range extending over several decades, the results – as might be expected

– slowly degenerate with increasing range or noise. Levy's method applied to such data yields a system of linear equations in which the lower-frequency values have very little influence. Since these lower frequencies contain the information relating to the dominant poles and zeros, the resulting transfer function will give a very poor representation of the system. For most control work where the main interest is in identifying the dominant poles and zeros, this usually involves neglecting much of the high-frequency data, and hence the importance of selecting a suitable data range.

The method does not identify systems which have poles at the origin of the s-plane. However, the presence of such poles is easily identified from a Bode or polar plot, and their contribution to the system's frequency response easily removed.

When using Levy's method it is normal practice first to produce a frequency response plot. Any time delay or pole at the origin of the s-plane should have its contribution to the system's frequency response removed. The modified frequency response is then plotted in order to determine a suitable frequency range over which the transfer function should be fitted. The frequency plot is used to establish the rank of the system (the difference between the number of poles and zeros in the system model). Levy's complex curve-fitting technique may now be used.

Levy's method requires the number of poles and zeros in the transfer function to be specified. The normal procedure is to let the number of poles equal the rank of the system; the first trial model then has no zeros. After a number of iterations a model will be produced, which should be checked for its stability and dynamic response characteristics. Specifically, does the model produce time and frequency responses compatible with those of the system? Only if the model's dynamic characteristics prove inadequate should a model of higher order be considered.

Finally, a useful tip is to normalize the frequency response data so that at zero frequency the system's gain is unity. Any constant gain adjustments can, of course, be reintroduced once the dynamics of the model have been identified. The most noticeable advantage is that the term b_0 in Equation (9.13) will tend to iterate to a value of unity. Any other value indicates too large a model, or a problem with the selected frequency range.

Problems

9.1.   A frequency response test on an asymptotically stable system produced the data given in Table 9.3. Use Bode plots to determine the system's transfer function.

9.2.   Repeat Problem 9.1, using Levy's method to find the system's transfer function.

Table 9.3

Frequency ω (rad/s)	Amplitude ratio (dB)	Phase
0.2	13.81	−102.5°
0.369	8.09	−112.4°
0.682	1.64	−128.2°
1.260	−6.20	−148.7°
2.327	−15.64	−169.8°
4.298	−26.29	−190.2°
7.937	−38.12	−211.3°
14.659	−51.65	−231.8°
27.073	−66.51	−247.6°
50.0	−82.11	−257.5°

9.3.   Use Program 2B of Chapter 11 to find the step response of a plant having the transfer function

$$G(s) = \frac{1.5(s+4)}{(s+1)(s+2)(s+3)}$$

Convert the step response data into the frequency domain using Program 8 of Chapter 11. Use a logarithmic scale for the frequency data from 0.2 to 8 rad/s.

Check the solution by using Program 7A of Chapter 11 to calculate the exact frequency values.

Finally, use Levy's method, Program 9 in Chapter 11, to find the transfer function $F(s)$. Try system orders of numerator 1 and denominator 3, and numerator 0 and denominator 2. Comment on the results.

CHAPTER 10

CASE STUDIES

10.1   Introduction

This part of the text deals with a number of case studies which use the theory developed in Chapters 1 to 9, and the computer programs presented in Chapter 11. The object of these studies is to demonstrate how the theory that has been developed may be applied to standard pieces of current, commercial laboratory equipment. Although there is inevitably some overlap between the studies, each is used to illustrate a different aspect of the theory.

10.2   Feedback Process Trainer PT326

The Feedback PT326, together with its PID controller (part of the PCS 327 process control simulator), is shown in Fig. 10.1, and a diagram of the temperature process in Fig. 10.2. In the system, air drawn through a variable orifice by a centrifugal blower is driven past a heater grid and through a length of tubing back to the atmosphere. The process consists of heating the air flowing in the tube to some desired temperature. The set value control knob can be used to raise the process air temperature to 60°C. By operating the switch SW, a step change in set value may be obtained. The detecting element consists of a bead thermistor fitted to the end of a probe inserted into the air stream 28 cm from the heater. This probe forms one arm of a d.c. bridge which is in balance at 40°C. The bridge output voltage is supplied to a d.c. amplifier and produces a voltage ranging from 0 to +10 V for an air temperature change from 30 to 60°C. The output from the measuring element can be monitored at point x.

10.2.1   Design objectives

In this study the control objectives were to maintain the process temperature at a given value and to respond rapidly to a step change in set value. From observation, it was noted that the plant was non-linear, had significant time delay and was noisy. In effect, the behaviour of the process was more like that of a flow loop

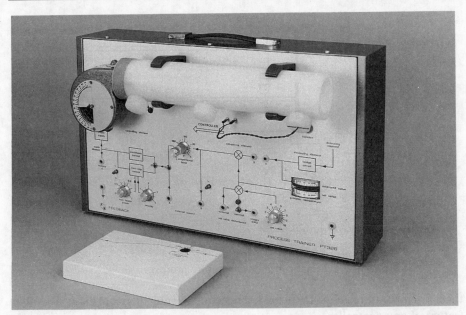

(a)

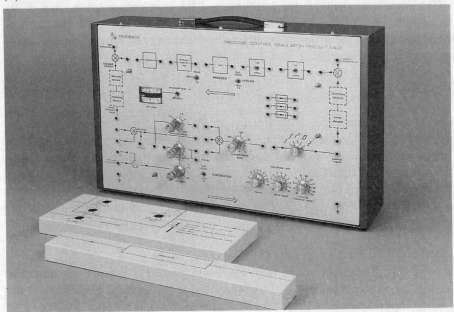

(b)

Figure 10.1 Photographs of the Feedback PT326, together with the PID
controller, part of the PCS327 process control simulator. As
shown in figures (a) and (b) respectively. (Courtesy of Feedback
Instruments Ltd.)

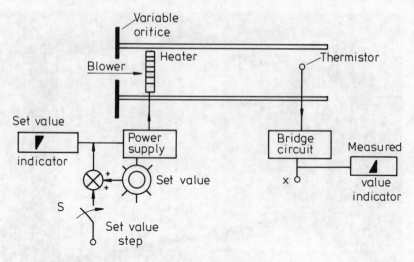

Figure 10.2 Schematic of the temperature process

than a temperature loop. Disturbance rejection would therefore be a desirable feature of any controller.

Hardware was available for imposing a PID type controller, or one of its derivatives, on the process. Such a controller should meet all the control requirements, and the problem became one of selecting the proportional, integral and derivative settings, K, T_i and T_d.

10.2.2   Model identification and validation

Model identification consisted of three distinct stages: experiment, data manipulation and transfer function identification. These are now described in turn.

Step response testing was used since there was no restriction on the availability or on the type of signal that could be injected into the process. On examination of the measured step response output, it was observed that the signal-to-noise ratio was approximately 12:1. To obtain an average step response with a standard deviation of 5 per cent or less, the number of tests needed is given approximately by

$$n = 400/x^2$$

where x:1 is the signal-to-noise ratio (see Equation (9.3)). Figure 10.3 shows the average response obtained from three tests. Using this average response and the demand input signals, the time delay and gain of the system are easily seen to be 0.2 s and 1.24, respectively. Time and amplitude data, for a total of 30 points, were obtained from the step response. A variable sampling rate was used, with more samples taken early in the response where the slope was larger.

Program 8 in Chapter 11 was used to convert the step response data into the frequency domain. Before entering the step data into this program, the time delay

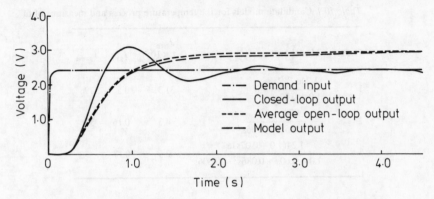

Figure 10.3 Step responses for process and second-order model

was removed by subtracting 0.2 s from all recorded time values. The program scales all amplitudes to give unity gain, before converting the data into the frequency domain over a chosen range of frequencies, Scaling is, strictly speaking, not necessary, although it is useful as a check on both the transformation into the frequency domain and in transfer function generation. Once a frequency range was obtained, it was modified until it contained the dominant corner frequencies, as seen from Bode diagrams in Fig. 10.4. The phase diagram indicates that the model structure should have at least two more poles than zeros. Both the gain and phase plots suggest that the frequency range should extend from about 0.2 to 9.5 rad/s.

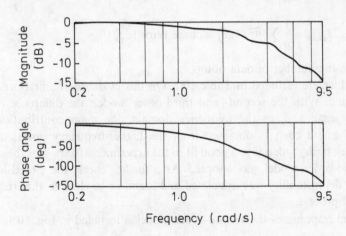

Figure 10.4 Bode plots for process frequency response, from average step data

The final step in the modelling process was transfer function identification using Levy's method (Program 9 in Chapter 11). Three candidate models were produced: see Table 10.1. The first-order model was included to see whether it would make

Table 10.1 Candidate models for the temperature process and measures of fit

Order	Model	J_{step} ($\times 10^3$)	J_{gain}	J_{phase}
1	$\dfrac{1.24e^{-02s}}{1.0 + 0.52s}$	31.3	0.5	579
2	$\dfrac{1.24e^{-0.2s}}{1.0 + 0.51s + 0.05s^2}$	4.3	0.16	32
3	$\dfrac{1.24(1.0 + 0.07s)e^{-0.2s}}{1.0 + 0.60s + 0.086s^2 + 0.006s^3}$	2.5	0.17	5.1

a reasonable approximation, although the frequency data of the temperature process is clearly of higher order.

In the model validation, the dynamics of each model were compared with those of the temperature process. To this end, step and frequency response data were generated using Programs 2 and 7, respectively (Chapter 11). Mean-square measures of fit were calculated, defined as follows:

$$J_{\text{step}} = \frac{1}{n} \sum_{i=1}^{n} \tilde{y}_i^2, \quad \tilde{y}_i = \text{step amplitude error (V)}$$

$$J_{\text{gain}} = \frac{1}{n} \sum_{i=1}^{n} \tilde{G}_i^2, \quad \tilde{G}_i = \text{gain error (dB)}$$

$$J_{\text{phase}} = \frac{1}{n} \sum_{i=1}^{n} \tilde{\theta}_i^2, \quad \tilde{\theta}_i = \text{phase error (deg)}$$

where n is the number of data points.

The values are included in Table 10.1. On this evidence, the first-order model was rejected. With the second- and third-order models the difference in J_{step} is relatively small and, in the frequency domain, the main contribution to the measured phase costs is obtained from the higher-frequency points. At lower frequencies, both models give a good fit to the experimental data. For these reasons the second-order model was adopted. As a further check, the closed-loop gain giving model instability was calculated and found to be close to that required by the actual process.

The step response of the second-order model is included in Fig. 10.3.

10.2.3   Controller design

A time delay indicates that a frequency response method rather than the root loci method of design should be selected. Since the process was noisy, it was decided not to include the derivative term in the PID controller. The open-loop transfer function, including the PI controller, was therefore

$$K\left(1 + \frac{1}{T_i s}\right)\frac{1.24e^{-0.2s}}{1 + 0.51s + 0.05s^2}$$

The performance specifications selected were a phase margin of 40° and a gain margin of about 6 dB, as these values should produce a reasonably fast closed-loop step response having a damping ratio of about 0.4.

The integral action time T_i was chosen as 0.5, which places an open-loop zero at $s = -2$. From Fig. 10.4, this zero should increase the bandwidth of the process. Adjusting the gain to $K = 1.1$ produced the required phase margin of 40°, and a gain margin of 5.3 dB. These values were sufficiently close to the desired performance specifications to merit connecting the PI controller with these settings to the process, resulting in the closed-loop step response shown in Fig. 10.3.

10.2.4   Discussion and conclusions

Since the process was noisy and incorporated a time delay, it could be considered typical of many industrial processes. The design stages consisted of step response testing, data processing, the development of candidate control models, validation, and controller design and testing.

A feature of this study is the attention given to modelling and validation. A good-quality model has been produced, giving increased confidence in the controller design. The designed phase margin of 40° should have produced a damping ratio of about 0.4 for a second-order plant. The achieved closed-loop step response shown in Fig. 10.3 has a damping ratio near this value. Modelling and validation is not a trivial exercise, and here the use of a computer has eased the problem. Without this advantage, it was likely that a first-order control model would have been used in the knowledge that feedback could compensate for some of the inaccuracies. It is clear from Table 10.1 that such a model would have been a poor approximation and, if used for control design, would have produced at best an over-optimistic gain margin. Consequently on-line adjustment of controller settings would then have been required.

10.3   Armfield PCT flow control experiment

The Armfield PCT (Process Control Technology) range of modular laboratory equipment was developed for process control training and teaching. Each experimental setup is built from a number of modules. In the current study the modules used were the PCT 10 electrical console, the PCT 9 process module and the PCT 15 flow control accessory (see Fig. 10.5).

A schematic of the equipment used in this case study is shown in Fig. 10.6. The PCT 10 electrical console contains power supplies and interfacing for all measurement and actuation signals. Also contained within the console is an

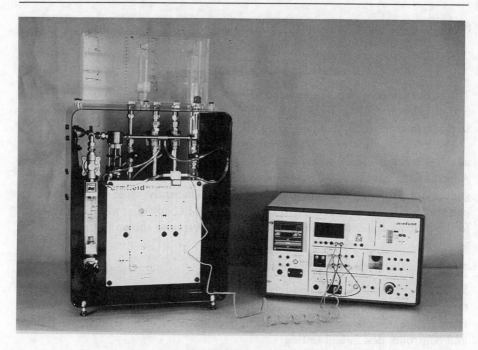

Figure 10.5 The Armfield Process Control Technology equipment. (Courtesy of Armfield Technical Education Co. Ltd.)

industrial PID controller. The flow system is contained within the PCT 9 process module. A constant-speed pump forces water along a pipe. Flow is controlled by means of a motorized valve, and changes in flow are detected by the PCT 15 flow sensor. In the closed-loop system shown in Fig. 10.6 flow is being controlled by the PID process controller.

10.3.1   Design objectives

The objectives were to identify a transfer function for the open-loop system, which consisted of a motorized valve, pipework and a flow sensor. The model was then to be used to design P and PI controller settings.

10.3.2   Model identification and validation

For the identification exercise the PCT 10 was used as the power supply for the constant-speed pump and motorized valve. It was also used for signal conditioning from the PCT 15 flow transducer. All other functions were carried out using an on-line IBM PC microcomputer.

An initial set of tests indicated that step response tests should be satisfactory for identification purposes. At the selected operating condition of 60 to 70 per cent

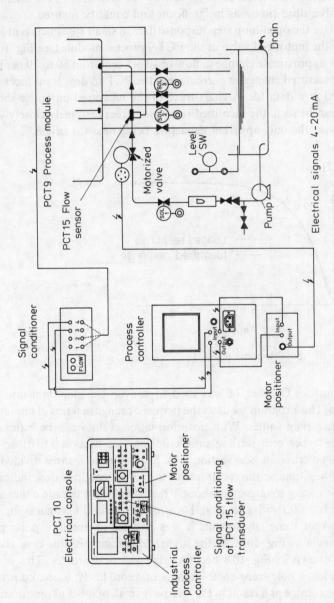

Figure 10.6 Armfield laboratory setup

of full flow, the measured response was relatively noise-free. This was not the case for all flow conditions, and therefore – as expected – derivative action would be precluded (see Section 8.2). Also, as suggested by the closeness of the sensor to the motorized valve, dead time was insignificant and could be ignored.

In order to obtain the open-loop step response data, a small step in current level was injected into the motorized valve of the PCT 9 process module (see Fig. 10.6). This produced an approximate change in flow of 10 per cent from 60 to 70 per cent of full flow. The measured change in current from the PCT 15 flow transducer was logged and stored in a data file. A chart trace of the measured output is shown in Fig. 10.7. As can be seen, the identified response agrees extremely closely with the logged response, the only apparent departure being near the origin.

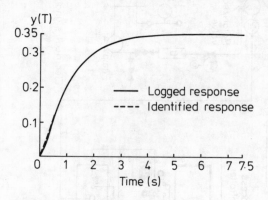

Figure 10.7 Step response plots

Again, Program 8 of Chapter 11 was used to convert the step data into the frequency domain. This program requests the frequency range in terms of minimum and maximum frequency values. With no information, values can be estimated only by using the 5 per cent settling time criterion (see Section 11.10) or the 63 per cent rise time criterion (see Section 9.4). From Fig. 10.7, three divided by the 5 per cent settling time or the reciprocal of the 63 per cent rise time indicates that the dominant corner frequency is about 1 rad/s. Taking a decade either side of this frequency, from 0.1 rad/s to 10 rad/s, provides a good first guess for the frequency range. After some adjustments it was possible to obtain a polar plot similar to that shown in Fig. 10.8. At the higher frequencies, in this case above 6 rad/s, the plot shown in Fig. 10.8 becomes different in character. This could indicate high-frequency pole–zero combinations or, more likely, a breakdown in the transformation routine as a result of the relatively small number of time domain sample points used, in this case 36.

A range of frequencies from 0.2 to 5 rad/s was deemed satisfactory, and the transfer function identification routine of Program 9 was used. This program requests the order of the transfer function denominator and numerator. From the

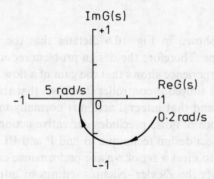

Figure 10.8 Polar plot of model

polar plot in Fig. 10.8 and from the various response data generated in arriving at this diagram, it is evident that there are three more poles than zeros. Therefore the simplest transfer function that could be used to fit the data would have a denominator of order 3, and a numerator of order 0. If higher-order systems are indicated, then it is likely that there would be approximate pole–zero cancellation, which would tend to reduce the system to third order. (It may be noted that systems of unnecessarily high order are often identified as being unstable.) Having selected the system's order, the transfer function coefficients are calculated and displayed. After several iterations the coefficients stabilized to produce a trial transfer function model of the open-loop flow system given by

$$G(s) = \frac{1}{1 + 1.38s + 0.43s^2 + 0.054s^3}$$

which has its poles at $s = -1$ and $s = -3.5 \pm j2.6$.

The model was validated by using the techniques described in the first case study (see Section 10.2.2), and the model proved to have good agreement with the various response data (see for example the model and plant step responses shown in Fig. 10.7).

A point not covered in the first case study is the need to specify the model's range of validity. In this case, frequency data was obtained between 0.2 and 5 rad/s. At the higher frequencies there are some discrepancies between the plant and the model, so without additional work it would be inadvisable to extrapolate the predicted model results beyond 5 rad/s. A polar plot obtained from the model over this range of frequencies is shown in Fig. 10.8. This indicates that the identified model may be used to test stability, and used to design P or PI controllers. However, it should not be used to design a PD or PID controller since frequency data with a phase content of less than $-180°$ has not been validated. From the design viewpoint, this is acceptable since flow loops are notoriously noisy, and normally a PI controller would be required.

10.3.3   Controller design

The Armfield equipment setup shown in Fig. 10.6 dictates that the flow loop controller must be of the PID type. Therefore the design problem reduced to one of finding appropriate settings. Experience shows that the gain of a flow controller is rarely greater than 1, so with a simple P controller it follows that there would be significant steady-state error, and that integral action is normally mandatory. As has been indicated, the presence of noise precludes derivative action.

This case study used the root loci design technique to find P and PI controller settings. Rather than attempting to meet a set of *s*-plane performance criteria, the root loci plots were used to modify the Ziegler–Nichols settings by adjusting the controller's proportional gain.

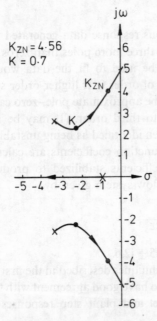

Figure 10.9 Root loci plot for system under P control

Figure 10.9 shows the root loci plot for the system with a simple proportional controller. This plot also indicates the Ziegler–Nichols controller setting K_{ZN}. With this gain the damping and decay ratio are relatively low. The magnitude of the loop gain is 4.56, which although greater than one is still going to produce a steady-state error of 18 per cent. Loop gains greater than ten would usually be required for simple proportional control. Reducing the gain to the value K indicated should produce a faster, less oscillatory response, but at the cost of increasing the steady-state error to approximately 59 per cent. The predicted responses for a unit demand step are shown in Fig. 10.10, and the corresponding plant responses in

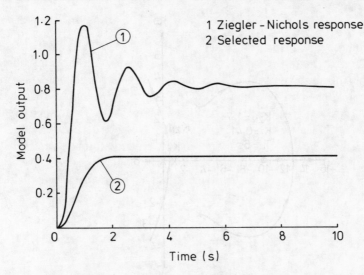

Figure 10.10 Step responses of model under P control

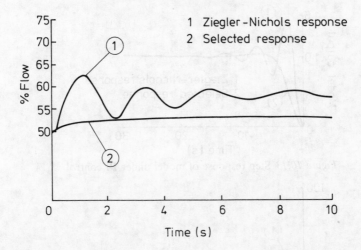

Figure 10.11 Step responses of equipment under P control

Fig. 10.11. Figure 10.10 shows that a significant offset is predicted with proportional control, and this is reflected in the equipment response shown in Fig. 10.11.

Figure 10.12 shows the root loci plot for the system under PI control. The integral action time was calculated from the Ziegler–Nichols rules and fixed at 6 s. The plot produced in the figure is the result of changing the proportional gain. The predicted responses to a unit step change for both the Ziegler–Nichols and selected settings are shown in Fig. 10.13; corresponding step changes to the equipment are shown in Fig. 10.14.

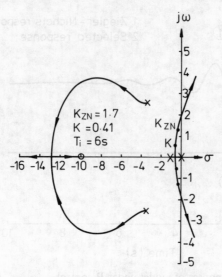

Figure 10.12 Root loci plot for system under PI control

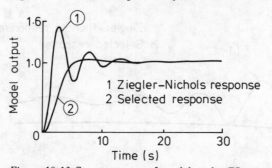

Figure 10.13 Step response of model under PI control

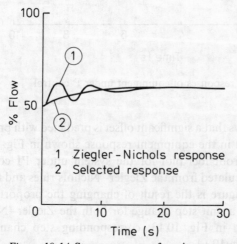

Figure 10.14 Step responses of equipment under PI control

10.3.4   Discussion and conclusions

No attempt has been made to optimize the design. Of the available design options, only the root loci technique has been used. However, sufficient results have been produced to demonstrate the high degree of correlation between the experimental and theoretically derived results.

A comparison of Figs. 10.13 and 10.14 is instructive. The selected gain should produce a damped response with a damping ratio of about 0.707. This value would be typical of that chosen for a positioning, or servo system. For process control a lower damping ratio is normal, partly to average out the output quantity and partly to avoid stiction in the actuation equipment. In Fig. 10.14 the selected response appears to have a much higher damping than that predicted. Since the model has been well validated, this increased damping is probably the result of stiction. It may be noted that this was a problem with the evaluation equipment used in this case study, which was from a preproduction run. All subsequent commercial equipment has been fitted with a modified valve.

10.4   TecQuipment CE8 coupled electric drives

This equipment, shown in Fig. 10.15, is representative of a recurring class of industrial problems. Figure 10.16 shows a strip of material wound from one spool

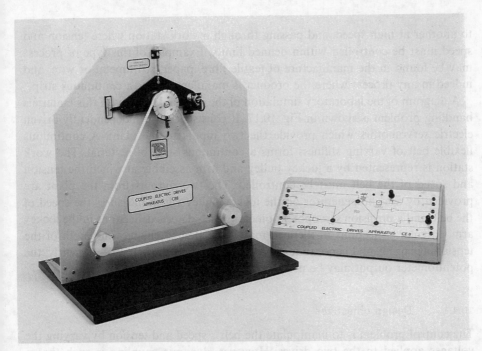

Figure 10.15 TecQuipment CE8 coupled electric drives. Designed by Dr Wellstead. (Courtesy of TecQuipment)

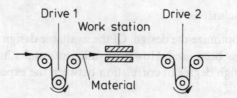

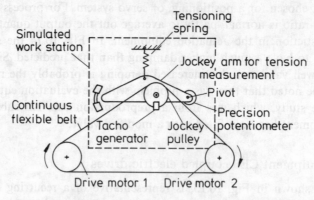

Figure 10.16 Material drive system

Figure 10.17 Laboratory simulation: coupled electrical drives

to another at high speed, and passing through a work station where tension and speed must be controlled within defined limits. Examples of this type of process may be found in the manufacture of textile fibre, paper, strip metal or wire, and indeed in any process where the product is manufactured as a continuous strip.

A diagram of the laboratory simulation of the essential features of this materials handling problem is shown in Fig. 10.17. It consists of two independently driven electric servomotors which provide the two inputs to the plant. A continuous flexible belt of varying stiffness forms a continuous strip of material. The work station is represented by a jockey pulley and arm assembly at which the tension and speed of the belt must be controlled. The two outputs from the plant are provided by a tacho generator on the jockey pulley which measures the speed of the belt, and a precision servopotentiometer which detects deflections in the jockey arm. Angular deflections in the arm may be related to changes in length of the tensioning spring, and hence to tension changes within the belt. Consequently, the potentiometer output may be used to measure the belt's tension.

10.4.1   Design objectives

The control problem is to manipulate the belt's speed and tension by varying the voltages applied to the two drives. However, dynamic coupling is such that a change in one drive motor voltage will affect both the speed and tension of the belt. Therefore independent control is not possible without some form of decoupling. The control problem is further complicated by significant plant non-linearity and the presence of noise.

The design objectives are therefore to design first a decoupler, and then two single-loop controllers, one for tension and the other for speed.

This case study uses material which goes beyond the scope of the rest of the book. However, it is included since it provides a link between classical control theory, the subject of this book, and more advanced control topics.

10.4.2   Additional theory

A multivariable plant is one with several inputs and outputs, and may be represented by a transfer function matrix $G(s)$. For the purpose of control, it is desirable to use one input to adjust a given output. This implies that $G(s)$ is a square matrix. Such systems are characterized by interactions, in that one input affects more than one output, and consequently classical single input/single output (SISO) control is not directly applicable. One approach is to design a precompensator which will eliminate the unwanted interactions. Such a device is referred to as a decoupler. Figure 10.18 shows the use of a decoupler $K(s)$ in a SISO control system for a two-input/two-output plant. The necessary decoupler equations are now developed.

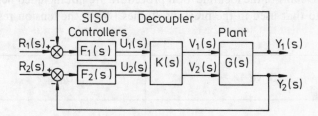

Figure 10.18 SISO control of multivariable plant

From Fig. 10.18 the open-loop cascaded plant and decoupler equation is

$$\begin{bmatrix} Y_1(s) \\ Y_2(s) \end{bmatrix} = \begin{bmatrix} G_{11}(s) & G_{12}(s) \\ G_{21}(s) & G_{22}(s) \end{bmatrix} \begin{bmatrix} K_{11}(s) & K_{12}(s) \\ K_{21}(s) & K_{22}(s) \end{bmatrix} \begin{bmatrix} U_1(s) \\ U_2(s) \end{bmatrix} \qquad (10.1)$$

If $K_{12}(s)$ and $K_{21}(s)$ are chosen so that

$$K_{12}(s) = -\frac{G_{12}(s)}{G_{22}(s)} K_{22}(s); \quad K_{21}(s) = -\frac{G_{21}(s)}{G_{22}(s)} K_{11}(s) \qquad (10.2)$$

then the plant has been decoupled since $G(s)K(s)$ is now a diagonal matrix. By substitution, the decoupler equation may be written as

$$\begin{bmatrix} 1 & -\dfrac{G_{12}(s)}{G_{11}(s)} \\ -\dfrac{G_{21}(s)}{G_{22}(s)} & 1 \end{bmatrix} \begin{bmatrix} K_{11}(s) & 0 \\ 0 & K_{22}(s) \end{bmatrix} \tag{10.3}$$

In this form, the first matrix describes the dynamics of the decoupler and is fixed. The terms $K_{11}(s)$ and $K_{22}(s)$ are combined with the single-loop controllers of Fig. 10.18 ($F_1(s)$ and $F_2(s)$) to produce acceptable closed-loop dynamics.

10.4.3   Plant identification

The decoupling compensator, defined by Equation (10.3), requires the identification of the plant transfer function matrix $G(s)$. The transfer function elements of this matrix may be obtained from step responses. A constant voltage of 3 V was applied to both drives, and positive and negative step changes of 0.8 V were then injected onto each drive in turn. The resulting open-loop responses are shown in Fig. 10.19. In this figure, responses to positive steps are indicated by a solid line, and the negative responses are inverted and shown as a broken line.

From these results it was clear that the plant was non-linear and that the tension responses were significantly noisy. However, examination of the speed responses suggested that the non-linearity was one of gain. This was confirmed by comparing the identified transfer functions obtained from the speed responses on drive 1 and also on drive 2. (Details of the identification procedure are not included here since it was identical to that used in the previous studies.) With the tension responses

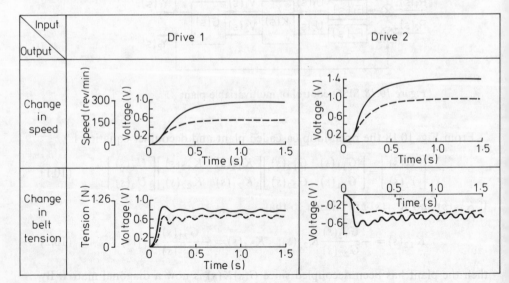

Figure 10.19 Open-loop step responses of plant

it was decided that, since the noise level was significant, a second-order transfer function would be fitted to the average of a series of responses. It was also assumed that the step-up and step-down dynamics were identical. The identified plant transfer function matrix, $G(s)$, is

$$G(s) = \begin{bmatrix} \dfrac{k_{11}}{s^3 + 31.8s^2 + 400s + 1450} & \dfrac{k_{12}}{s^3 + 30.1s^2 + 421s + 1480} \\ \dfrac{k_{21}}{s^2 + 26.1s + 4.6} & \dfrac{k_{22}}{s^2 + 22.3s + 274} \end{bmatrix}$$

(10.4)

where k_{ij} is a non-linear gain term.

10.4.4   Decoupler design

On comparing Equations (10.3) and (10.4) for the plant transfer function matrix and the decoupler matrix, it was clear that the dynamic elements in the decoupler would be of higher order, and furthermore would involve non-linear gain terms. It was therefore considered desirable to attempt to reduce the order of the decoupler dynamics, and then to adjust the gain terms on-line.

The first stage in reducing the order of the decoupler was to examine how its dynamics would modify a given input signal. Since the most severe signal likely to be experienced would be a step, unit step responses were obtained from both dynamic elements. The dynamics of element $G_{12}(s)/G_{11}(s)$ produced a response which varied between 0.96 and 1.05, before settling to a steady-state value of 1. Consequently, the dynamic effects for this element were insignificant compared with the non-linearities and could be ignored. A similar test on element $G_{21}(s)/G_{22}(s)$ produced a response which varied between 0.65 and 1.02 before settling to a steady-state value of 1. Since the speed of response was as fast as that of the fastest plant response, and also since elements $G_{21}(s)$ and $G_{22}(s)$ were derived from noise-contaminated measurements, it was again considered that these dynamics could be neglected.

The decoupler gains were then adjusted on-line to produce the matrix

$$\begin{bmatrix} 1 & -1 \\ 1 & 1 \end{bmatrix}$$

(10.5)

Note that the elements $K_{11}(s)$ and $K_{22}(s)$ in Equation (10.3) may be assumed to be unity. The open-loop step responses of the decoupled plant are shown in Fig. 10.20.

Owing to the non-linearities, the decoupling was not perfect. However, the amplitudes of the cross-coupled step responses were always smaller than those of the direct responses, and it may be shown that there is sufficient decoupling to permit the use of the SISO control on each loop.

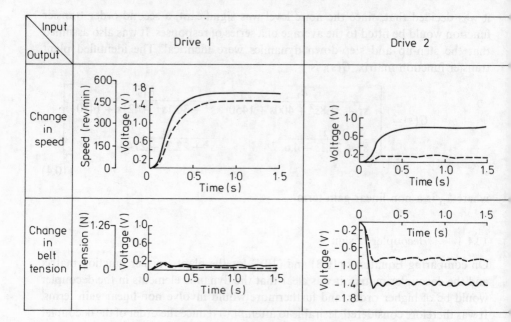

Figure 10.20 Open-loop precompensated step responses of plant for (a) $U_1(s)$, speed, and (b) $U_2(c)$, tension

It is interesting to consider the effect that this decoupler will have on the open-loop plant (see Fig. 10.18). A change in $U_1(s)$ will result in identical signals being applied to both drives. Since these are nominally the same, the result will be a change in belt speed with tension maintained constant. Similarly, a change in $U_2(s)$ will result in equal but opposite signals on both drives, and the result will be a change in belt tension, with speed maintained constant.

10.4.5   Controller design

Two feedback loops, one for speed and one for tension control, may now be introduced into the decoupled plant as indicated in Fig. 10.18. The control problem is to choose $F_1(s)$ and $F_2(s)$ so as to produce acceptable closed-loop responses for the two individual loops. (Since the design of single-loop controllers has been described for the previous case studies, it is not repeated here.) Figure 10.21 shows typical closed-loop responses.

10.4.6   Discussion and conclusions

The aim of this study was to describe how the theory and computer programs developed in this book for SISO systems could be used to produce controllers for multivariable plant. The method employed relies on the decoupling technique

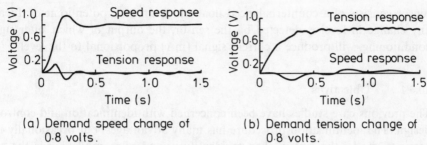

Figure 10.21 Closed-loop responses to demand change of 0.8 V for (a) speed and (b) tension

described in Section 10.4.2. This theory may be extended to systems having more than two inputs and outputs, but it has some limitations. When it is applied to non-minimum phase plant, stability problems can result. The technique is also known to produce high-order precompensators with associated realization problems. However, if used with care, useful results may be produced.

10.5   Armfield PCT level control experiment

The Armfield PCT 11 level controller is an accessory to the Armfield process control technology range. It consists of a level sensor used to detect a change in water level in the process tank of the PCT9 process module (see Fig. 10.22). The

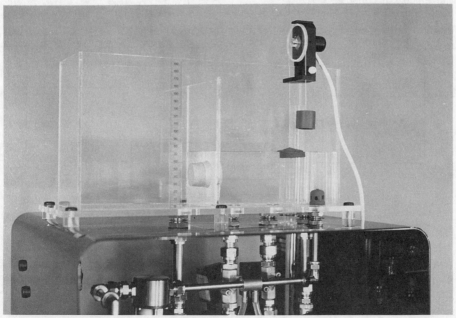

Figure 10.22 Armfield PCT 11 level sensor in process tank. (Courtesy of Armfield Technical Education Co. Ltd.)

sensor consists of a counterbalanced float which drives a potentiometer. Ideally, any change in level is detected by the sensor, the output of which, after signal conditioning, will produce a current signal (mA) proportional to the level.

10.5.1   Objectives

The previous case studies have been concerned with identification and controller design. The identification procedure has many advantages, but occasionally it is easier to model the system than to identify it, and often modelling is the only approach available. In this study the object is to model the level control loop.

10.5.2   Modelling the level loop

The level loop could be considered as consisting of a number of interacting subsystems, as shown in Fig. 10.23. In this figure are indicated the various Armfield modules. It is assumed that flow from the tank is a disturbance input, independent of any changes in head. It will be shown that this assumption, although not valid, is adequate and will produce a satisfactory control model.

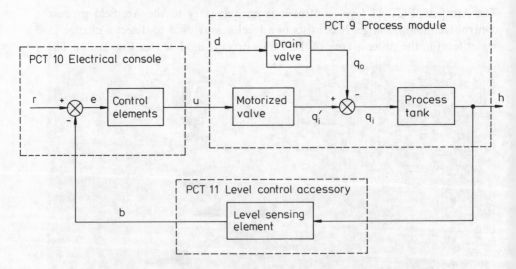

Figure 10.23 Signal flow diagram for the closed-loop level system

A simple lumped parameter technique is used to model the process tank (see Section 1.4.1), namely that volumetric flow into the tank less volumetric flow out equals the time rate of change of volume. Hence

$$q'_i - q_o = \frac{dV}{dt} \tag{10.6}$$

where

q_i' = volumetric flow into the tank, which may be controlled via motorized valve
q_o = volumetric flow out of tank
V = volume of liquid in tank

Clearly,

$$V = Ah \tag{10.7}$$

where

A = effective cross-sectional area of process tank
h = height of liquid

It is assumed that q_o is constant. This assumption is not strictly valid since q_o will be a function of h, but in this system the design is such that over the working region the outlet flow may be considered independent of the head.

Equation (10.6) may now be written as

$$q_i + (q - q_o) = A \frac{dh}{dt} \tag{10.8}$$

where $q_i' = q_i + q$, and if q is set equal to q_o, then

$$q_i = A \frac{dh}{dt} \tag{10.9}$$

This model indicates that the tank acts as an integrator, since any change in measured level h is given by

$$h = \frac{1}{A} \int q_i \, dt \tag{10.10}$$

The level sensor has some interesting characteristics. Close observation of the devices indicates that the element is in fact non-linear: a change in level changes the curvature of the water meniscus in contact with the float. Only when maximum curvature is reached (and this could be concave or convex, depending on the rise or fall of tank level) does the float move. A plot of water level h against sensor output b from any given steady-state value of level would take the form shown in Fig. 10.24. In this figure, the region of insensitivity between the maximum and minimum curvatures of the meniscus is a function of the displacement of the head from its steady-state position. A realistic linear lumped parameter description of this device, suitable for Laplace transform solution, is therefore difficult to develop. Either a digital simulation similar to that described in Program 1 of Chapter 11 must be used, or a simpler description of the non-linearity must be sought.

Although the region of insensitivity shown in Fig. 10.24 is small, it has two main effects which need to be considered.

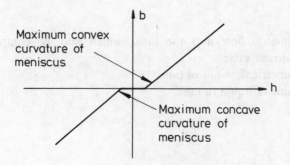

Figure 10.24 Sensor output as a function of level

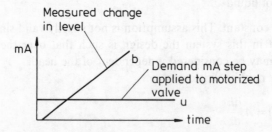

Figure 10.25 Typical open-loop step response plot from the PCT 9 process module

The first is the masking of the dynamics of the motorized valve on the feedback path b. An open-loop experiment in which a demand step was placed on $u(t)$, and the measurement $b(t)$ recorded, is shown in Fig. 10.25. The integral nature of the tank is clear from the constant slope of the measured level. The dynamics of the valve and motor would be expected to introduce a term of the form

$$q'_i + \tau_1 \frac{dq'_i}{dt} = Ku \tag{10.11}$$

where τ_1 is the time constant of the valve and motor. These additional dynamics would cause the measured response to be initially tangential to the time axis, and then to curve into the constant slope of the response shown in Fig. 10.25. Since this is not observed, the valve response must be faster than that of the sensor. Consequently, for closed-loop studies the valve dynamics may be ignored, and Equation (10.11) becomes

$$q'_i = Ku \tag{10.12}$$

The other effect of the sensor's non-linearity is to mask the self-regulation caused by the change in outflow from the tank caused by a change in head. In order to produce a measurable change in level, there will be a minimum step size which can be applied to the motorized value. Also, since the maximum change in head

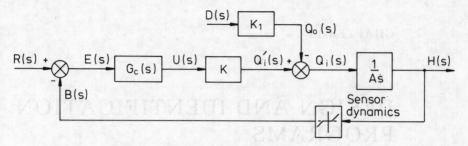

Figure 10.26 Closed-loop transfer function model

is relatively small, for all practical purposes about 10 cm, the feedback effect which will tend to alter the outflow may be ignored. Consequently, the relation between the disturbance input (drain valve adjustment) and the outflow may be simply modelled by an equation of the form

$$q_o = K_1 d \qquad (10.13)$$

The closed-loop transfer function model now takes the form shown in Fig. 10.26.

In practice only K and the cross-sectional area of the tank need be established, since in this model K_1 does not affect the loop dynamics. The analysis would be simplified if the sensor's region of insensitivity were assumed to be a pure time delay of constant duration. It is worth noting that a model identified from the response of Fig. 10.25 would be just such a model.

10.5.3   Discussion and conclusions

Throughout the analysis is was assumed that a closed-loop model was required. At each stage reference was made to the equipment, either to validate or to simplify parts of the model. Much has been made of the region of insensitivity within the sensor, and some of the assumptions made are a consequence of its effects. However, these effects are small, and satisfactory results have been obtained by ignoring sensor dynamics – that is, by making the further assumption that the duration of the assumed sensor time delay is zero. The resulting model is of a controlled type 1 system with a disturbance input. It is therefore not surprising that this is an excellent piece of equipment for demonstrating system type.

CHAPTER 11

DESIGN AND IDENTIFICATION PROGRAMS

11.1   Introduction

In this chapter are presented digital computer algorithms and programs which will assist in the design of single input/single output control systems. Most of the design problems with which control engineers are likely to be confronted consist of three stages:

1. *Identification and/or system modelling.* A mathematical model is produced on which further investigations may be carried out. A model may be obtained by measuring variables on a physical system, converting the data into a form suitable for analysis and then obtaining the Laplace transform model which best describes the system dynamics. Such a procedure is described in Chapter 9. Alternatively, a model could be obtained by directly formulating a mathematical description based on physical laws; this approach was introduced in Chapter 1.
2. *Controller design and system stability.* The mathematical model is used to develop a controller, probably based on some form of lead–lag or PID combination. The various design techniques that have been described in the book are then applied in a trial-and-error fashion to achieve desired stability and performance specifications. Factors such as the system's response characteristics, linearity and noise must be considered and will form constraints on any proposed design.
3. *Hardware problems.* These are the problems associated with the implementation of a given controller design. Typically, consideration will be given to the materials and energy sources available, and the possibility of modifying an existing system or of developing a new one which will achieve the specific purposes or tasks specified in the design.

The design process outlined above (and described in more detail in Chapter 1) is an iterative one. Since each stage is dependent on the others, a change in one stage usually has ramifications for the others. The process, calling as it does for much routine and repetitive calculation, is one for which a computer may be used, and there are available computer aided control system design (CACSD) packages.

There are no recognized standards for such packages, and educational institutions, companies and so on tend to produce their own routines for control work. With microcomputer technology this is almost a necessity, since unless a machine is designed as a clone of some other manufacturer's product, it is highly unlikely that it will run someone else's software.

In the sections that follow, CACSD programs are developed for use in the areas of design described in previous chapters. References to background reading are listed at the end of the chapter.

11.2   A CACSD package

When developing any CACSD package, consideration must be given to the various interfaces the computer is likely to have with the real world. This is particularly important with a microcomputer which, because of its physical portability, may well be required to interface with the plant, in either a data logging or a real-time control role.

Any CACSD design package will be required to accept user inputs. The main input devices are currently the keyboard, mouse and light pen (or some other form of digitizing pen and board); outputs include the VDU screen, printer and graph plotter. The package may be required to store and retrieve data captured from external equipment, or generated internally. Clearly there are many levels of sophistication, and it is likely that in the next few years the major advancements in package development will be in improving user interfaces. However, for most microcomputer packages the main interfaces are likely to be a keyboard, a VDU screen, possibly a printer capable of producing screen dumps, and some form of disc storage. This is the minimum system configuration suitable for interactive control system design work.

Most microcomputer packages are likely to be written in BASIC. However, the language has many dialects and so most BASIC programs are machine-dependent. Graphical and file handling statements are invariably machine-dependent. Some machines will accept array elements which start from 0, whereas others take the first element in the array as 1. Consequently, the dimension statement

```
10 DIM A(20)
```

could mean that the array A has 21 locations if the first element of the array is $A(0)$, but 20 locations if the first element is $A(1)$. Some machines accept both upper and lower case letters for variable assignment; others only upper case. Only the first two letters of any variable are significant with some machines, while others may accept six or more significant letters. For example, if only the first two letters in a variable were significant then the statements

```
10  FRED=20
20  FRANK=30
30  PRINT FRED,FRANK
```

would result in the output 30, 30, rather than 20, 30, the desired response.

The above points are enough to illustrate the problems associated with producing microcomputer programs. For this reason, all the programs presented in this chapter are developed for a hypothetical microcomputer which has a very restricted form of BASIC. All arrays use 1 as the first location, and all variables are restricted to having as significant the first two upper case letters only (note that more than two letters are often used to describe a variable in order to make the programs more legible). There are no graphics or file handling routines. Consequently, these programs should be suitable for any microcomputer having a memory size of about 12k. If graphics or file handling routines are introduced, then the minimum memory requirements will increase; the degree of added sophistication, however, will depend on both the user's requirements and the microcomputer's limitations.

A typical CACSD package is menu driven and contains some form of database. When a package is entered, a menu is displayed which indicates the various programs available. When one is selected, the appropriate routine is loaded; when that routine is exited the menu is displayed again. This is useful, as a design process may require many movements between several routines. A database which allows common information to be transferred between several different programs is therefore desirable. In all the programs presented here, a standard form of data entry is used, wherever possible, in order to facilitate the introduction of such a database.

There are nine programs. The first deals with the solution of system equations by numerical integration techniques. Program 2 uses Heaviside partial-fraction expansion techniques to find inverse Laplace transforms and transient responses. Programs 3, 4 and 5 deal with stability. Control system design techniques are covered in programs 6 and 7, which deal, respectively, with root loci and the various frequency-based design methods. Programs 8 and 9 are identification routines which may be used to find a transfer function from step or frequency response data.

11.3 Program 1: Numerical integration by the Euler–Cauchy method

The solution of ordinary differential equations on a digital computer calls for some form of numerical integration. These methods yield approximate numerical values of the solution, and may be applied to equations containing non linearities, or time delays, which cannot be solved exactly.

By far the simplest of the methods is the Euler–Cauchy algorithm (often called, more simply, the Euler algorithm). This method is named after the Swiss mathematician Leonhard Euler (1707–1783) and the French mathematician Augustin-Louis Cauchy (1789–1857). The method is best illustrated by means of an example.

Consider the second-order, linear differential equation

$$\frac{d^2x}{dt^2} + 2\zeta\omega_n\frac{dx}{dt} + \omega_n^2 x = \omega_n^2 y \tag{11.1}$$

where

$$y = \begin{cases} 0 & \text{for } t < 0 \\ 1 & \text{for } t \geqslant 0 \end{cases}$$

This definition represents a step forcing function on the system's input y.

To solve this equation it is necessary to reduce the second-order equation into two first-order equations. This may be achieved by defining new variables (see Section 2.2). Let

$$x_1 = x \tag{11.2}$$

$$x_2 = \frac{dx}{dt} \tag{11.3}$$

Substituting these variables into the differential equation (11.1) gives

$$\frac{dx_2}{dt} + 2\zeta\omega_n x_2 + \omega_n^2 x_1 = \omega_n^2 y \tag{11.4}$$

There are now two first-order differential equations to be solved, namely

$$\frac{dx_1}{dt} = x_2 \tag{11.5}$$

$$\frac{dx_2}{dt} = \omega_n^2 y - 2\zeta\omega_n x_2 - \omega_n^2 x_1 \tag{11.6}$$

It should be noted that, in general, this technique of reduction may be applied to any high-order system. However, with non-linear systems the procedure must be carried out with care.

Euler's algorithm may now be applied to each of the two first-order differential equations. To start the algorithm an initial time and starting point for each of the two variables is required; these are the initial conditions. For most transient response work it is assumed that the system is in a steady state, and that the forcing input occurs at time zero:

$$x_1(0) = 0 \quad \text{and} \quad x_2(0) = 0 \tag{11.7}$$

If the response is required to start from some other point in time, t say, or if the system is not in a steady state, then $x_1(t)$ and $x_2(t)$ must be defined.

On moving forward in time a small step h to a new time $t + h$, estimates of the new values of $x_1(t + h)$ and $x_2(t + h)$ are obtained by linear extrapolation from the initial rates of change of $x_1(t)$ and $x_2(t)$. The new values of x are approximately equal to the old ones, plus the product of the derivatives of x, times the step size:

$$x_1(t + h) = x_1(t) + h\frac{dx_1(t)}{dt} \tag{11.8a}$$

and

$$x_2(t + h) = x_2(t) + h\frac{dx_2(t)}{dt} \qquad (11.8b)$$

Since the derivatives of $x_1(t)$ and $x_2(t)$ are numerical algebraic expressions, the procedure of stepping forward in time and estimating new values may be repeated until the required solution is obtained. This procedure forms the basis of the Euler integration routine.

Program 1 is written specifically to solve the problem presented in this example. Clearly the method is extremely simple to program. (Note, for example, that lines 520 and 530 give the BASIC coding associated with updating $x_1(t)$ and $x_2(t)$.) However, as with all integration routines, the problems associated with accuracy, numerical stability and computational efficiency (speed) must be considered.

It is instructive to test Program 1 on a problem in which a number of runs are carried out using an ever-increasing step size. For example, consider a second-order system having a damping ratio of 0.1 and a natural frequency of 1 rad/s. For this system the results of such a test would be as follows. With a step size of 0.0001 computational efficiency is low – that is, the program runs very slowly. However, the results are good, with an accuracy of three to four significant figures. Comparisons may be carried out between this program and Program 2, which may be used to give an exact solution to the problem. Increasing the step size speeds up the program but reduces its accuracy. When the step size is 0.1 the results are probably good enough to show general response trends, although little confidence can be placed on individual points. For further increases the algorithm remains stable until a step size of about 1.8 is reached. At this point the algorithm may be thought of as being stable, since the response eventually settles to give a value of X equal to 1, and a value of $XDOT$ equal to 0. With steps greater than 1.8 a steady state cannot be achieved, and the algorithm is said to have become unstable.

The normal method of finding an acceptable step size is to start with a very low value, which is then used to get the program debugged and running. Tests are then carried out in which the step size is doubled on successive runs until the biggest step is found for which the algorithm gives sufficiently accurate results and is stable. With some low-accuracy microcomputers (those using only a few bits) a small step size may produce roundoff errors. If the number of significant figures is low, the error produced in taking the difference between two numbers which, because of small step sizes, are about the same can be significant. The golden rule in any simulation work is therefore to check and recheck the validity of any generated result.

The accuracy, computational efficiency and stability of numerical integration algorithms are all important considerations. The Euler algorithm has the advantages of being easily programmed, and having reasonable computational efficiency and stability. Its main disadvantage is its accuracy.

At this stage the reader may be wondering why an engineer should be satisfied with anything less than a very accurate algorithm. There are several reasons. In process plant, for example, pipe flows may occur in seconds whereas composition or temperature changes may take hours. The mathematical models describing such plant contain a mixture of fast and slow dynamics, and the resulting set of differential equations is referred to as being stiff. If, under these conditions, accurate integration is specified, then the step size on the fast loops becomes the limiting factor. The slow dynamics can tolerate a much larger step size and still produce accurate results. Since a process is normally dominated by its slow dynamics, it is often possible to select a step size for which the algorithm on the fast loops just remains stable, but produces quite accurate results on the slow loops. In stiff systems inaccuracies in the rapidly changing variables tend to have little effect on the accuracy of the slowly changing variables. Neither is accuracy critical if trends rather than absolute values are required: when there are large inaccuracies in the mathematical model, or when a qualitative assessment of a response is needed, for example.

Clearly, the Euler algorithm is good enough for most simulation problems. However, if better accuracy is required then the algorithm rather than the step size should be changed.

Probably the most popular numerical integration algorithm is the fourth order Runge–Kutta method (details of which may be found in most advanced engineering mathematics books) named after the two German mathematicians Carl Runge (1856–1927) and Wilhelm Kutta (1867–1944). At each step of the algorithm four auxiliary equations are calculated for each differential equation, so programming is more complicated than for the Euler algorithm. However, for the same computing time (the step size being four times greater than on the Euler algorithm) the Runge–Kutta algorithm is more accurate. It should be noted that if accuracy is not important, then the Euler method is the superior of the two. Both algorithms have a maximum step size for stability of about the same value. Consequently, when large step sizes are used the Euler algorithm will run four times as fast as the Runge–Kutta algorithm.

For control work using microcomputers, the Euler and the fourth-order Runge–Kutta algorithms are probably the only two numerical integration routines that need be considered. Variable step size routines, of which Chai's modified Merson's algorithm[1] is a well-tried method, do not show their advantages on smaller machines. The idea behind variable step size routines is that small steps must be taken while the process is changing rapidly, but large steps may be taken when the variables are changing slowly. However, when the fastest differential equation is running at its numerical stability limit, the step size cannot be increased no matter how slowly other variables are changing. Since the constant checking and adjustment of step size requires additional calculation at each step, computational efficiency can be low.

11.3.1   Listing of Program 1

```
10 REM    **************     PROGRAM 1     **********************
20 REM    THIS PROGRAM USES THE EULER NUMERICAL INTEGRATION
30 REM    ROUTINE TO FIND THE FORCED RESPONSE OF A SECOND
40 REM    ORDER SYSTEM TO A UNIT STEP INPUT. NOTE THE PROGRAM
50 REM    IS EASILY CHANGED TO TAKE NON-LINEAR, HIGHER ORDER
60 REM    SYSTEMS HAVING WELL DEFINED TIME VARYING INPUTS.
70 REM
80 PRINT:PRINT
90 PRINT "INPUT THE SYSTEM'S DAMPING RATIO AND NATURAL FREQUENCY"
100 INPUT "THE DAMPING RATIO (ZETA) IS: ";Z
110 INPUT "THE UNDAMPED NATURAL FREQUENCY (OMEGA-N) IS: ";WN
120 PRINT:PRINT
130 PRINT "INPUT THE SIMULATION START AND END TIME AND THE"
140 PRINT "INTEGRATION STEP SIZE"
150 INPUT "THE SIMULATION START TIME IS:";TS
160 INPUT "THE SIMULATION END TIME IS: ";TE
170 INPUT "THE INTEGRATION STEP SIZE IS: ";DT
180 T=TS:PT=TS
190 PRINT:PRINT
200 PRINT "THE PRINT INTERVAL TIME IS THE ELAPSED TIME BETWEEN OUTPUTS"
210 INPUT "INPUT THE PRINT INTERVAL TIME";TP
220 PRINT:PRINT
230 IF TS=0 THEN 280
240 PRINT "INPUT THE INITIAL CONDITIONS AT TIME ";TS
250 INPUT "THE MEASURED OUTPUT OF THE SYSTEM IS: ";X1
260 INPUT "THE RATE OF CHANGE OF THE OUTPUT IS: ";X2
270 GOTO 300
280 PRINT "ALL INITIAL OUTPUT CONDITIONS ARE ASSUMED TO BE ZERO"
290 X1=0:X2=0:P1=100:P2=1000
300 PRINT:PRINT
310 REM
320 REM THE FORCING INPUT YI IS ASSUMED TO BE A UNIT STEP
330 REM
340 YI=1
350 REM
360 REM MULTIPLY THE INPUT BY OMEGA-N SQUARED
370 REM
380 Y=YI*WN*WN
390 REM
400 REM PRINT HEADINGS
410 REM
420 PRINT"                    OUTPUTS                    INPUT"
430 PRINT" TIME         X              XDOT            Y"
440 IF T<PT THEN 510
450 PRINT INT(T*P1+.5)/P1,INT(X1*P2+.5)/P2,INT(X2*P2+.5)/P2,YI
460 PT=PT+TP
470 REM 480 REM SET INTEGRATION LOOP
490 REM
500 IF T>TE THEN 550
510 T=T+DT
520 X1=X1+X2*DT
530 X2=X2+(Y-(2*Z*WN*X2+WN*WN*X1))*DT
540 GOTO 440
550                           END
```

Example 11.1

This example demonstrates a typical output response. It is assumed that the system has a damping ratio of 0.1 and a natural frequency of 2 rad/s.

Output from Program 1

```
INPUT THE SYSTEM'S DAMPING RATIO AND NATURAL FREQUENCY
THE DAMPING RATIO (ZETA) IS: ? .1
THE UNDAMPED NATURAL FREQUENCY (OMEGA-N) IS: ? 2


INPUT THE SIMULATION START AND END TIME AND THE
INTEGRATION STEP SIZE
THE SIMULATION START TIME IS: ? 0
THE SIMULATION END TIME IS: ? 10
THE INTEGRATION STEP SIZE IS: ? .001


THE PRINT INTERVAL TIME IS THE ELAPSED TIME BETWEEN OUTPUTS
INPUT THE PRINT INTERVAL TIME ? .25


ALL INITIAL OUTPUT CONDITIONS ARE ASSUMED TO BE ZERO
```

	OUTPUTS		INPUT
TIME	X	XDOT	Y
0	0	0	1
.25	.118	.913	1
.5	.432	1.527	1
.75	.847	1.725	1
1	1.259	1.502	1
1.25	1.57	.952	1
1.5	1.72	.232	1
1.75	1.688	-.474	1
2	1.499	-1.002	1
2.25	1.211	-1.247	1
2.5	.902	-1.177	1
2.75	.644	-.838	1
3	.495	-.337	1
3.25	.477	.195	1
3.5	.584	.631	1
3.75	.777	.878	1
4	1.003	.898	1
4.25	1.208	.706	1
4.5	1.344	.368	1
4.75	1.388	-.024	1
5	1.336	-.372	1
5.25	1.212	-.601	1
5.5	1.05	-.668	1
5.75	.891	-.573	1
6	.773	-.353	1
6.25	.72	-.072	1
6.5	.736	.199	1
6.75	.813	.398	1
7	.926	.485	1
7.25	1.045	.451	1

7.5	1.143	.315	1
7.75	1.198	.119	1
8	1.201	-.087	1
8.25	1.158	-.252	1
8.5	1.082	-.343	1
8.75	.994	-.346	1
9	.916	-.268	1
9.25	.865	-.135	1
9.5	.85	.017	1
9.75	.872	.151	1
10	.921	.236	1

11.4   Program 2: Inverse Laplace transforms and transient responses

This section presents two programs, one for calculating inverse Laplace transforms and the other for generating transient responses. As both programs have much in common, they are dealt with together.

The general procedure for obtaining an inverse Laplace transform or a transient response from a system's transfer function $G(s)$ is initiated by decomposing $G(s)$ in order to obtain the system's poles and zeros. A partial-fraction expansion is then carried out, the form of which depends on the type of pole. Partial-fraction expansion coefficients are normally determined by a procedure which has been credited to the electrical engineer Oliver Heaviside (1850–1925). (The so-called Heaviside techniques are dealt with in detail in Chapters 2 and 4.) Essentially, the method consists of using four formulae to calculate the partial-fraction coefficients and hence the inverse Laplace transforms. Each formula corresponds to one of the four possible types of factor that can arise in the denominator of $G(s)$:

1. a distinct linear factor, or real pole
2. a repeated linear factor, or multiple pole
3. a distinct quadratic factor, resulting from the product of a complex pole, and its complex conjugate
4. a repeated quadratic factor, obtained from the product of multiple complex poles and their complex conjugates.

In this context, multiple poles are poles that have the same real and imaginary parts – that is, they occupy the same s-plane locations. The four formula, or their derivatives, form the basis for most inverse Laplace transform and transient response programs.

Several algorithms for evaluating Laplace transforms are available[2,6]. Most are based on partial-fraction expansions and differ only in the procedure used to determine the coefficients in the expansion. Chen and Haas[2] use a graphical interpretation of the Heaviside theorems. (Details of this type of graphical approach are given in Chapter 4.) The main weakness of the method is in the evaluation of multiple poles of order higher than two. Since the pole–zero locations of most real systems are distinct, this constraint does not necessarily pose any problem. A

FORTRAN listing is provided by Chen and Haas. It is valid only for distinct real or complex poles, but is readily translated into BASIC for use on microcomputers.

Vernon's method of solution[3] is to use the conventional Heaviside technique, as described in Chapter 2. The method is capable of computing the inverse Laplace transform of a transfer function having multiple real and complex poles. A FORTRAN listing of the program is given but its translation into BASIC must be carried out with care. Also, on a microcomputer a transfer function having a number of multiple poles can take several minutes to calculate. Errors are also likely to occur when calculating high-order poles.

Computer listings are not given in References 4-6, but the methods described all have some attraction from the point of view of microcomputer implementation. Watkins' algorithm,[4] like that of Chen and Haas, considers only distinct poles. Its main virtue is that it utilizes only real numbers, and thus avoids the complex arithmetic associated with many algorithms developed for larger machines.

Pottle's algorithm[5] deals with repeated poles; it reduces to the normal Heaviside residue evaluation when the poles are not repeated. An advantage of the method is that it provides an indication of its performance. However, it is reported that although the technique has good accuracy, it – in common with others – may suffer from large errors when applied to high-order poles.

N-Nagy and Al-Tikriti[6] restrict themselves to multiple poles, assuming that distinct poles are first extracted from the transfer function and their partial fraction coefficients found using conventional Heaviside techniques. Their proposed approach, unlike that of Heaviside, does not require repeated differentiation to evaluate the coefficients associated with multiple roots. Instead, a process involving repeated matrix multiplication is carried out until all multiple poles have been considered. It is claimed to be quicker than other methods, which makes it attractive from the point of view of microcomputer implementation.

The algorithm used in Program 2 is similar to that of Chen and Haas in that it is based on a graphical interpretation of the Heaviside method. (Full details of the algorithm used are given in Section 4.2.) The program consists of a main section, lines 10 to 1270, and two alternative endings. The first of these, Program 2A, gives the system's inverse Laplace transform, and the second, Program 2B, produces a transient response of the output in terms of time and amplitude coordinates.

11.4.1   Listing of main program (Program 2)

```
10 REM    ************     PROGRAM 2      ******************
20 REM    THIS IS THE MAIN SECTION OF THE INVERSE LAPLACE
30 REM    TRANSFORM AND TRANSIENT RESPONSE PROGRAM. IT USES
40 REM    TRANSFER FUNCTION DATA GIVEN IN POLE-ZERO FORM. IT WILL
50 REM    ONLY ACCEPT DISTINCT REAL OR DISTINCT COMPLEX POLES.
60 REM    **** NOTE POLES MAY BE IN ANY ORDER BUT COMPLEX POLE
70 REM            AND ZERO PAIRS MUST BE ADJACENT. *********
80 REM
90 DIM ZR(20),ZI(20),PR(20),PI(20),G(20)
100 PRINT:PRINT"INPUT NUMBER OF POLES"
```

```
110 INPUT NP
120 PRINT"INPUT NUMBER OF ZEROS"
130 INPUT NZ
140 IF NZ>NP THEN PRINT"NUMBER OF ZEROS IS GREATER THAN NUMBER OF
    POLES":END
150 REM
160 REM INPUT POLE AND ZERO LOCATIONS
170 REM
180 IF NZ=0 THEN 350
190 PRINT:PRINT"INPUT ZERO LOCATIONS"
200 M=0
210 FOR I = 1 TO NZ
220 INPUT "REAL,IMAGINARY ";ZR(I),ZI(I)
230 IF ABS(ZI(I))<.0001 THEN M=0:GOTO 330
240 IF M=0 THEN M=1:GOTO 330
250 IF ABS(ZR(I)-ZR(I-1))>.0001 THEN 320
260 IF ABS(ZI(I-1)+ZI(I))>.0001 THEN 320
270 IF ZI(I)<0 THEN M=0:GOTO 330
280 ZI(I-1)=ZI(I)
290 ZI(I)=-ZI(I)
300 M=0
310 GOTO 330
320 PRINT"COMPLEX ZEROS MUST BE ENTERED AS CONJUGATE PAIRS.":END
330 NEXT I
340 M=0
350 PRINT:PRINT"INPUT POLE LOCATIONS"
360 FOR I=1 TO NP
370 INPUT "REAL,IMAGINARY ";PR(I),PI(I)
380 IF ABS(PI(I))<.0001 THEN M=0:GOTO 480
390 IF M=0 THEN M=1:GOTO 480
400 IF ABS(PR(I)-PR(I-1))>.0001 THEN 470
410 IF ABS(PI(I-1)+PI(I))>.0001 THEN 470
420 IF PI(I)<0 THEN M=0:GOTO 480
430 PI(I-1)=PI(I)
440 PI(I)=-PI(I)
450 M=0
460 GOTO 480
470 PRINT"COMPLEX POLES MUST BE ENTERED AS CONJUGATE PAIRS.":END
480 NEXT I
490 REM
500 REM  CALCULATE THE STATIC GAIN AND TEST TO SEE IF IT IS REAL
510 REM
520 RR=1:M=0:SG=0
530 FOR I = 1 TO NP
540 IF M=1 THEN M=0:GOTO 620
550 IF ABS(PR(I))>.00001 THEN 570
560 IF ABS(PI(I))<.00001 THEN 620
570 IF PI(I)<.0001 THEN 610
580 RR=RR*(PR(I)*PR(I)+PI(I)*PI(I))
590 M=1
600 GOTO 620
610 RR=RR*(-PR(I))
620 NEXT I
630 R1=1:M=0
640 IF NZ=0 THEN 750
650 FOR I = 1 TO NZ
660 IF M=1 THEN M=0:GOTO 740
670 IF ABS(ZR(I))>.00001 THEN 690
680 IF ABS(ZR(I))<.0001 THEN 740
```

```
690 IF ZI(I)<.0001 THEN 730
700 R1=R1*(ZR(I)*ZR(I)+ZI(I)*ZI(I))
710 M=1
720 GOTO 740
730 R1=R1*(-ZR(I))
740 NEXT I
750 GR=R1/RR
760 PRINT:PRINT"STATIC GAIN = ";GR
770 M=0
780 REM
790 REM                               ***************
800 REM
810 REM THIS IS THE MAIN LOOP FOR CALCULATING THE PARTIAL FRACTION
820 REM COEFFICIENTS AS DESCRIBED IN CHAPTER 4, SECTION 4.2.
830 REM
840 FOR I=1 TO NP
850 IF M=1 THEN M=0:GOTO 1240:REM A COMPLEX CONJUGATE POLE
860 IF I = NP THEN 880
870 IF PI(I)>0 THEN M=1
880 R1=1:I1=0:RR=1:II=0
890 IF NZ = 0 THEN 1000
900 REM
910 REM FIND THE PRODUCT OF ALL THE ZEROS TO POLE P(I)
920 REM
930 FOR J = 1 TO NZ:RH=R1
940 R1=R1*(PR(I)-ZR(J))+I1*(PI(I)-ZI(J))
950 I1=I1*(PR(I)-ZR(J))-RH*(PI(I)-ZI(J))
960 NEXT J
970 REM
980 REM FIND THE PRODUCT OF ALL OTHER POLES TO POLE P(I)
990 REM
1000 FOR J = 1 TO NP
1010 IF J=I THEN 1050
1020 RH=RR
1030 RR=RR*(PR(I)-PR(J))+II*(PI(I)-PI(J))
1040 II=II*(PR(I)-PR(J))-RH*(PI(I)-PI(J))
1050 NEXT J
1060 REM
1070 REM DIVIDE THE ZERO PRODUCT BY THE STATIC GAIN AND THE
1080 REM REMAINING POLE PRODUCT.
1090 REM
1100 DD=GR*(RR*RR+II*II)
1110 RG=(R1*RR+I1*II)/DD
1120 IG=(RR*I1-II*R1)/DD
1130 IF M=1 THEN 1210
1140 REM    ROUTINE FOR REAL POLES.
1150 G(I)=RG
1160 SG=SG+RG
1170 GOTO 1240
1180 REM
1190 REM ROUTINE FOR COMPLEX POLES
1200 REM
1210 G(I)=2*RG
1220 SG=SG+2*RG
1230 G(I+1)=2*IG
1240 NEXT I
1250 REM
1260 REM END OF PARTIAL FRACTION LOOP ***************
1270 REM
```

11.4.2   Listing for inverse Laplace transforms (Program 2A)

```
1280 REM    ************     PROGRAM 2A     *****************
1290 REM    WHEN COMBINED WITH THE MAIN PART OF PROGRAM 1
1300 REM    THIS PROGRAM GIVES THE INVERSE LAPLACE TRANSFORM.
1310 REM
1320 PRINT:PRINT
1330 PRINT"THE INVERSE LAPLACE TRANSFORM IS OBTAINED"
1340 PRINT"BY ADDING ALL THE FOLLOWING TERMS."
1350 N=1
1360 IF (NP-N)<0 THEN 1460
1370 IF (NP-N)=0 THEN 1430
1380 IF PI(N)<.0001 THEN 1430
1390 PRINT:PRINT"EXP(";PR(N);"*T)"
1400 PRINT"      *(";G(N);"*COS(";PI(N);"*T)+";G(N+1);"*SIN(";PI(N)"*T))"
1410 N=N+2
1420 GOTO 1360
1430 PRINT:PRINT G(N);"*EXP(";PR(N);"*T)"
1440 N=N+1
1450 GOTO 1360
1460                                    END
```

11.4.3   Listing for transient response (Program 2B)

```
1280 REM    ************     PROGRAM 2B     ************
1290 REM    WHEN COMBINED WITH THE MAIN PART OF PROGRAM 1
1300 REM    THIS PROGRAM WILL PRODUCE A TRANSIENT RESPONSE
1310 REM    IN TERMS OF TIME AND AMPLITUDE POINTS.
1320 REM
1330 AT=SG:PRINT
1340 IF NZ=NP THEN AT=AT-1/RG
1350 IF ABS(AT)>.0001 THEN PRINT"DYNAMIC GAIN AT T=0 IS ";AT:PRINT
1360 PRINT "INPUT START TIME, END TIME, AND NUMBER OF TIME INCREMENTS"
1370 INPUT TS,TE,TI
1380 TD=(TE-TS)/TI
1390 TT=TS
1400 PRINT:PRINT "TIME            RESPONSE"
1410 IF TT >TE THEN END
1420 R=0:N=1
1430 IF (NP-N)<0 THEN 1520
1440 IF (NP-N)=0 THEN 1490
1450 IF PI(N)<.0001 THEN 1490
1460 R=R+EXP(PR(N)*TT)*(G(N)*COS(PI(N)*TT)+G(N+1)*SIN(PI(N)*TT))
1470 N=N+2
1480 GOTO 1430
1490 R=R+G(N)*EXP(PR(N)*TT)
1500 N=N+1
1510 GOTO 1430
1520 PRINT TT,R
1530 TT=TT+TD
1540              GOTO                    1410
```

Example 11.2

This example demonstrates the use of both program endings. Consider the system

$$Y(s) = G(s)R(s)$$

where $Y(s)$ is the system's output, $G(s)$ the plant transfer function and $R(s)$ a unit demand step input; and assume the plant transfer function to be given by

$$G(s) = \frac{K(s+1)}{(s^2 + 2s + 26)(s+3)(s+6)}$$

Both programs give the static gain due to the pole–zero distribution alone. However, the output data is then normalized by adjusting K to give a total static gain of unity. The static gain is the product of all the constant terms in the transfer function when the coefficient of each s term is unity. In this example, the static gain is $K \times 1/(26 \times 3 \times 6)$, or $0.002136752K$.

The required input for the program data is then that there is one zero, at

$$s = -1 + j0$$

and five poles, four from the plant transfer function at

$$s = -1 + j5, \quad s = -1 - j5, \quad s = -3 + j0, \quad s = -6 + j0$$

and one from the input $R(s)$, at

$$s = 0 + j0$$

The output corresponding to this input data is shown below.

Output from Program 2

```
INPUT NUMBER OF POLES
? 5
INPUT NUMBER OF ZEROS
? 1

INPUT ZERO LOCATIONS
REAL,IMAGINARY ? -1,  0
INPUT POLE LOCATIONS
REAL,IMAGINARY ? 0,  0
REAL,IMAGINARY ? -1,  5
REAL,IMAGINARY ? -1, -5
REAL,IMAGINARY ? -3,  0
REAL,IMAGINARY ? -6,  0

STATIC GAIN =  2.136752E-03
```

Output from Program 2A

```
THE INVERSE LAPLACE TRANSFORM IS OBTAINED
BY ADDING ALL THE FOLLOWING TERMS.

 1 *EXP( 0 *T)

EXP(-1 *T)
     *(-1.986207 *COS( 5 *T)+ -1.365517 *SIN( 5 *T))

 3.586207 *EXP(-3 *T)

-2.6 *EXP(-6 *T)
```

Output from Program 2B

```
INPUT START TIME, END TIME, AND NUMBER OF TIME INCREMENTS
? 0,  8,  32
TIME            RESPONSE
0               0
.25             .6168938
.5              2.140207
.75             2.487635
1               1.446546
1.25            .5271375
1.5             .6000981
1.75            1.140015
2               1.334956
2.25            1.090797
2.5             .8467378
2.75            .8721623
3               1.031356
3.25            1.093498
3.5             1.027167
3.75            .9567587
4               .9623435
4.25            1.00776
4.5             1.026664
4.75            1.008395
5               .9879536
5.25            .9889927
5.5             1.001901
5.75            1.00761
6               1.002585
6.25            .9966538
6.5             .9967888
6.75            1.000452
7               1.00217
7.25            1.000791
7.5             .9990726
7.75            .9990644
8               1.000103
```

11.5   Program 3: Polynomial root solving

Finding the roots of a polynomial is one of the more difficult computational problems for a microcomputer. There are formulae for finding the roots of second-, third- and fourth-order polynomials, but the cubic solution is rather long, and the quartic is very complicated. Consequently it is normal to use a numerical method of solution for polynomials of third or higher order.

There are many numerical algorithms,[7] but none offers a satisfactory solution. For microcomputer work there are probably two main methods, namely Newton's method and the Lin–Bairstow[8,9] methods. These methods will usually be adequate for most problems, but in certain situations calculation times can become excessive, and occasionally the algorithms will fail to converge.

The method originally proposed by Sir Isaac Newton (1642–1727) requires an approximate value of a root x_0 (real or complex) for the function

$$f(x) = 0 \tag{11.9}$$

where $f(x)$ is a polynomial of degree n in x which may be differentiated to give $f'(x)$. A better solution x_1 may then be obtained as

$$x_1 = x_0 - (f(x_0)/f'(x_0)) \tag{11.10}$$

Successive approximations will be closer and closer to the true value of the root.

One of the main problems encountered in coding Newton's algorithm is that of ensuring that the initial guess is sufficiently close to the true value, otherwise convergence cannot be guaranteed. Also, depending on the nature of the polynomial, it may take many iterations before a satisfactory solution is obtained. Once a root is found it should be removed from the polynomial and the procedure repeated until all the roots are accounted for. In general, as a means of finding the roots of polynomials Newton's algorithm is inferior to the Lin–Bairstow method.

There are many variations on the Lin–Bairstow method (see e.g. References 7, 10, 11), but they all suffer from occasional convergence problems or protracted solution times. Nearly all of them try to find a quadratic factor of the original polynomial by means of synthetic division (described in Reference 7). The division procedure produces a quotient polynomial, a remainder polynomial and a potential quadratic factor. Various schemes are then used to adjust the terms of the quadratic until convergence is obtained. Once this is achieved (after numerous divisions and adjustments to the quadratic, the coefficients of the remainder polynomial are close to zero), the quotient polynomial replaces the original polynomial and the procedure is repeated. At each iteration the roots of the resulting quadratic factors give the roots of the original polynomial.

Hovanessian[7] provides a BASIC listing of a Lin–Bairstow method, and Melsa[10] gives a FORTRAN subroutine for a modified Bairstow method. When translated into BASIC for use on a microcomputer, Melsa's routine is probably the better of the two. Borrie[11] gives an outline of a Bairstow–Hitchcock algorithm which, when coded, is probably as efficient as that proposed by Melsa. The BASIC listing given in Program 3 below is based on the Bairstow–Hitchcock method.

11.5.1   Listing of Program 3

```
10 REM     ****************   PROGRAM 3   ******************
20 REM     THIS PROGRAM FACTORIZES A POLYNOMIAL OF UP TO ORDER
30 REM     9 USING A MODIFIED BAIRSTOW METHOD.
40 REM
50 DIM A(10),B(10),C(10)
60 PRINT:PRINT
70 REM
80 REM     INPUT DEGREE OF POLYNOMIAL AND COEFFICIENTS
90 REM
100 INPUT "INPUT DEGREE OF POLYNOMIAL";ND
110 IF ND<10 THEN 150
120 PRINT"THE MAXIMUM ORDER PERMITTED IS 9"
130 PRINT
140 GOTO 100
```

```
150 N= ND+1
160 PRINT:PRINT
170 PRINT"POLYNOMIAL COEFFICIENTS ARE ASSUMED TO BE"
180 PRINT"OF THE FORM:"
190 PRINT
200 PRINT"A(0)*S^N+A(1)*S^(N-1)+.......A(N-1)*S+A(N)=0"
210 PRINT:PRINT
220 FOR I=1 TO N
230 PRINT"INPUT COEFFICIENT A(";I-1;")"
240 INPUT A(I)
250 NEXT I
260 PRINT:PRINT
270 PRINT"THE REQUIRED ROOT LOCATIONS ARE"
280 PRINT:PRINT"REAL            IMAGINARY"
290 PRINT
300 REM
310 REM    TEST FOR ROOTS AT S EQUALS ZERO
320 REM
330 IF A(N)<>0 THEN 440
340 PRINT 0,0
350 N=N-1
360 REM
370 REM    TEST IF ALL ROOTS HAVE BEEN FOUND
380 REM
390 IF N=1 THEN 1070
400 GOTO 330
410 REM
420 REM    TEST FOR A REMAINING REAL ROOT
430 REM
440 IF N=2 THEN 1060
450 REM
460 REM    ESTIMATE REMAINDER POLYNOMIAL COEFFICIENTS
470 REM
480 IF A(N-2)<>0 THEN 510
490 P=0:Q=0
500 GOTO 530
510 P=A(N-1)/A(N-2)
520 Q=A(N)/A(N-2)
530 IF N=3 THEN 820
540 REM
550 REM    EVALUATE QUOTIENT POLYNOMIAL B AND PARTIAL
560 REM    DERIVATIVES OF B.
570 REM
580 B(1)=A(1)
590 C(1)=B(1)
600 B(2)=A(2)-P*B(1)
610 C(2)=B(2)-P*C(1)
620 FOR I=3 TO N-1
630 B(I)=A(I)-P*B(I-1)-Q*B(I-2)
640 C(I)=B(I)-P*C(I-1)-Q*C(I-2)
650 NEXT I
660 B(N)=A(N)-P*B(N-1)-Q*B(N-2)
670 DV=C(N-2)*C(N-2)-(C(N-1)-B(N-1))*C(N-3)
680 IF DV = 0 THEN DV =E-20
690 P1=(B(N-1)*C(N-2)-B(N)*C(N-3))/DV
700 Q1=(C(N-2)*B(N)-(C(N-1)-B(N-1))*B(N-1))/DV
710 REM UPDATE THE TRIAL QUADRATIC FACTOR.
720 P=P+P1
730 Q=Q+Q1
```

```
740 REM
750 REM    TEST THE ACCURACY OF THE QUADRATIC FACTOR.
760 REM
770 IF ABS(P1)>.000001 THEN 600
780 IF ABS(Q1)>.000001 THEN 600
790 REM
800 REM    CALCULATE THE ROOTS OF THE QUADRATIC FACTOR.
810 REM
820 G=P*P-4*Q
830 PR=-P/2
840 IF G<0 THEN 890
850 PG=SQR(G)/2
860 PRINT PR+PG,0
870 PRINT PR-PG,0
880 GOTO 960
890 PG=SQR(-G)/2
900 PRINT PR,PG
910 PRINT PR,-PG
920 REM
930 REM    REPLACE THE ORIGINAL POLYNOMIAL WITH THE
940 REM    QUOTIENT POLYNOMIAL.
950 REM
960 N=N-2
970 IF N=1 THEN 1070
980 FOR I= 1 TO N
990 A(I)=B(I)
1000 NEXT I
1010 REM
1020 REM    TEST ORDER OF POLYNOMIAL. IF IT IS <3 THEN
1030 REM    SOLVE FOR ROOTS DIRECTLY, OTHERWISE REPEAT.
1040 REM
1050 GOTO 390
1060 PRINT -A(N)/A(N-1),0
1070 PRINT
1080 END
```

Example 11.3

Show that the roots of the fifth-order polynomial

$$s^5 + 8s^4 + 33s^3 + 88s^2 + 122s + 60 = 0$$

are

$$s = -1, \quad s = -2, \quad s = -3, \quad s = -1+j3, \quad s = -1-j3$$

Output from Program 3

INPUT DEGREE OF POLYNOMIAL 5

POLYNOMIAL COEFFICIENTS ARE ASSUMED TO BE
OF THE FORM:

A(0)*S^N+A(1)*S^(N-1)+.......A(N-1)*S+A(N)=0

```
INPUT COEFFICIENT A(0)
? 1
INPUT COEFFICIENT A(1)
? 8
INPUT COEFFICIENT A(2)
? 33
INPUT COEFFICIENT A(3)
? 88
INPUT COEFFICIENT A(4)
? 122
INPUT COEFFICIENT A(5)
? 60


THE REQUIRED ROOT LOCATIONS ARE

REAL          IMAGINARY
-.9999999        0
-1.999999        0
-.9999997        3
-.9999997       -3
-3.000002        0
```

11.6   Program 4: The Routh array

A necessary and sufficient condition for the asymptotic stability of a linear time-invariant system is that all the zeros of its characteristic polynomial have negative real parts. This means that all the dynamic modes in the system will, after excitation, decay to zero. This condition is also sufficient to ensure the stability of the forced system in situations for which the forcing input is bounded. Stability could therefore be tested directly by using a root solving routine to establish the root locations. However, as discussed for Program 3, root solving routines tend to suffer from occasional convergence problems, and a solution cannot always be guaranteed. It is therefore useful to establish stability without actually computing root locations.

The Routh array (see Section 5.7) provides just such a means for determining the stability of a system. It is based on the continued fraction algorithm, details of which may be found in Reference 12. If the array is being used simply to test the stability of a system, then coding is straightforward. However, if information on the exact number of roots having negative, positive or zero valued real parts is required, then the coding is more complicated, particularly if the array produces an all-zero row or has left-column zeros.

If none of the elements in the first column of the Routh array is zero (a first-column zero would normally cause premature termination of the array), then the number of sign changes in the first column is equal to the number of roots of the characteristic polynomial having positive real parts. All the other roots will have negative real parts. A program which tests for stability must terminate once a sign change or zero element occurs in the first row, since a zero element in the first row always indicates instability or, at best, marginal stability. Program 4, which simply tests for stability, prints out all the elements in the Routh array until the array is complete, or premature termination occurs, and then comments on the system's stability.

The appearance of an all-zero row indicates the presence of a divisor polynomial whose roots are all symmetrically placed about the origin of the real and imaginary axes. This symmetry of roots means that the divisor polynomial will always be of even order. The normal procedure for dealing with an all-zero row is to form an auxiliary equation from the row immediately above the all-zero row (the method is described in Section 5.7.1). It is easily shown that this equation is the required divisor polynomial. The auxiliary equation is then differentiated with respect to s, and the coefficients of the resulting equations are substituted into what was the all-zero row. This substitution enables the array to be completed in the normal way. Hence, from the row immediately above what was the all-zero row, each change in sign in the first column of the array indicates one root in the right half s-plane and one root in the left half s-plane. From the order of the auxiliary equation, the number of complex roots on the imaginary axis may then be found. The root distribution up to the all-zero row may be found in the normal way, and the root distribution of the characteristic equation will thus have been determined. This procedure is readily programmed.

The procedures described in Section 5.7.1 for dealing with left-column zeros are not in general suitable for computer implementation. Shamash[13] outlines such a situation, in which a left-column zero precedes an all-zero row. A computationally efficient method of dealing with left-column zeros is given by Yeung,[14] although the method is less suitable than those given in Section 5.7 for pencil-and-paper solutions.

11.6.1   Listing of Program 4

```
10  REM     ***************   PROGRAM 4   ****************
20  REM     THIS PROGRAM TESTS THE STABILITY OF A CHARACTERISTIC
30  REM     POLYNOMIAL USING THE ROUTH STABILITY CRITERION
40  REM
50  DIM A(16,9)
60  PRINT:PRINT
70  REM
80  REM     INPUT DEGREE OF POLYNOMIAL AND COEFFICIENTS
90  REM
100 INPUT "INPUT DEGREE OF POLYNOMIAL";ND
110 IF ND>2 AND ND<16 THEN 150
120 PRINT"THE ORDER OF THE POLYNOMIAL MUST BE EQUAL"
130 PRINT"TO OR GREATER THAN 3,AND LESS THAN 16"
140 GOTO 60
150 N= ND+1
160 IF ABS(INT(N))<>N THEN 100
170 PRINT:PRINT
180 PRINT"POLYNOMIAL COEFFICIENTS ARE ASSUMED TO BE"
190 PRINT"OF THE FORM:"
200 PRINT
210 PRINT"A(0)*S^N+A(1)*S^(N-1)+.......A(N-1)*S+A(N)=0"
```

```
220 PRINT:PRINT
230 IC=0
240 REM
250 REM    TEST IF POLYNOMIAL IS OF ODD OR EVEN ORDER
260 REM    AND INPUT ITS COEFFICIENTS ACCORDINGLY.
270 REM
280 IF (INT(ND/2)*2)=ND THEN 440
290 REM
300 REM    DATA INPUT FOR POLYNOMIAL OF EVEN ORDER
310 REM
320 M=(ND+1)/2
330 FOR J=1 TO M
340 FOR I=1 TO 2
350 PRINT"INPUT COEFFICIENT A(";IC;")"
360 IC=IC+1
370 INPUT A(I,J)
380 NEXT I
390 NEXT J
400 GOTO 560
410 REM
420 REM    DATA INPUT FOR POLYNOMIAL OF ODD ORDER
430 REM
440 M=(ND+2)/2
450 L=M-1
460 FOR J=1 TO L
470 FOR I=1 TO 2
480 PRINT"INPUT COEFFICIENT A(";IC;")"
490 IC=IC+1
500 INPUT A(I,J)
510 NEXT I
520 NEXT J
530 PRINT"INPUT COEFFICIENT A(";ND;")"
540 INPUT A(1,M)
550 A(2,M)=0
560 PRINT:PRINT
570 REM
580 REM    FORM THE ROUTH ARRAY
590 REM
600 PRINT"THE ROUTH ARRAY IS:"
610 SC=ABS(A(1,1))/A(1,1)
620 PRINT
630 FOR J=1 TO 2
640 PRINT"ROW ";J;" : "
650 FOR I=1 TO M
660 A(J,I)=A(J,I)*SC
670 PRINT A(J,I);
680 NEXT I
690 PRINT
700 NEXT J
710 FL=0
720 FOR J=3 TO N
730 IF FL<>0 THEN 810
740 PRINT"ROW ";J;" : "
750 FOR I=1 TO M-1
760 A(J,I)=A(J-2,I+1)-(A(J-2,1)*A(J-1,I+1))/A(J-1,1)
770 PRINT A(J,I);
780 IF A(J,1)=0 THEN FL=J
790 NEXT I
800 PRINT
```

```
810 NEXT J
820 REM
830 REM    TEST FOR A LEFT COLUMN ZERO
840 REM
850 PRINT:PRINT
860 IF FL=0 THEN 1050
870 REM
880 REM    TEST IF THERE IS AN ALL ZERO ROW
890 REM
900 JJ=0
910 FOR I= 1 TO M
920 IF JJ<>0 THEN 940
930 IF A(FL,I)<>0 THEN JJ=1
940 NEXT I
950 IF JJ=1 THEN 990
960 PRINT"ROW ";FL;" IS AN ALL ZERO ROW - THIS INDICATES"
970 PRINT"THE SYSTEM IS UNSTABLE OR MARGINALLY STABLE"
980 GOTO 1120
990 PRINT"THE LEFT COLUMN ZERO, ROW ";FL;" INDICATES"
1000 PRINT"THE SYSTEM IS UNSTABLE"
1010 GOTO 1120
1020 REM
1030 REM    TEST FIRST COLUMN FOR STABILITY
1040 REM
1050 FOR I=1 TO N
1060 IF A(I,1)<0 THEN FL=1
1070 NEXT I
1080 IF FL=0 THEN 1110
1090 PRINT"THE SYSTEM IS UNSTABLE"
1100 GOTO 1120
1110 PRINT"THE SYSTEM IS STABLE"
1120 END
```

Example 11.4

Test the stability of a system whose characteristic equation is

$$D(s) = s^5 + 8s^4 + 33s^3 + 88s^2 + 122s + 60 = 0$$

Output from Program 4

```
INPUT DEGREE OF POLYNOMIAL ? 5

POLYNOMIAL COEFFICIENTS ARE ASSUMED TO BE
OF THE FORM:

A(0)*S^N+A(1)*S^(N-1)+.......A(N-1)*S+A(N)=0

INPUT COEFFICIENT A(0)
? 1
INPUT COEFFICIENT A(1)
? 8
INPUT COEFFICIENT A(2)
? 33
INPUT COEFFICIENT A(3)
? 88
```

```
INPUT COEFFICIENT A(4)
? 122
INPUT COEFFICIENT A(5)
? 60


THE ROUTH ARRAY IS:

ROW   1  :
  1                 33              122
ROW   2  :
  8                 88              60
ROW   3  :
 22                114.5
ROW   4  :
 46.36364          60
ROW   5  :
 86.02941           0
ROW   6  :
 60                 0


THE SYSTEM IS STABLE
```

11.7   Program 5: Horner's method

This program is intended for use with the Routh array program, Program 4. Often it is desirable to know a system's margin of absolute stability – that is, how fast the transients will decay after some forcing disturbance (see Section 5.9.1). Typically, this requires all the roots of the system's characteristic equation to lie to the left of a line parallel to the imaginary axis and passing through the point $s = -\delta$, where δ defines the slowest permissible rate of decay.

Translation of the imaginary axis is easily carried out, using Horner's method[15] as follows. Let the coefficients of the shifted polynomial $D(s - \delta)$ be $b_0, b_1, \ldots, b_n$, and the coefficients of $D(s)$ be $a_0, a_1, \ldots, a_n$. To find the coefficients of $D(s - \delta)$, initially let

$$b_i = a_i, \quad \text{for } i = 0, 1, \ldots, n \tag{11.11}$$

The procedure shown as a flow diagram in Fig. 11.1 is then followed, and the resulting coefficients $b_0, b_1, \ldots, b_n$ will be the coefficients of the shifted polynomial $D(s - \delta)$.

11.7.1   Listing of Program 5

```
10 REM    ***************   PROGRAM 5   ******************
20 REM    THIS PROGRAM WILL MODIFY A POLYNOMIAL F(S) TO PRODUCE
30 REM    A NEW POLYNOMIAL F(S-D). THE EFFECT IS TO GIVE A
40 REM    LINEAR TRANSLATION OF THE IMAGINARY AXIS RELATIVE TO
50 REM    THE SYSTEM'S POLES AND ZEROS.
60 REM
70 DIM A(16),B(16)
```

```
80 PRINT:PRINT
90 REM
100 REM    INPUT DEGREE OF POLYNOMIAL AND COEFFICIENTS
110 REM
120 INPUT "INPUT DEGREE OF POLYNOMIAL";ND
130 IF ND<16 THEN 170
140 PRINT"THE ORDER OF THE POLYNOMIAL MUST BE"
150 PRINT"LESS THAN 16"
160 GOTO 80
170 N= ND+1
180 IF ABS(INT(N))<>N THEN 120
190 PRINT:PRINT
200 PRINT"POLYNOMIAL COEFFICIENTS ARE ASSUMED TO BE"
210 PRINT"OF THE FORM:"
220 PRINT
230 PRINT"A(0)*S^N+A(1)*S^(N-1)+.......A(N-1)*S+A(N)=0"
240 PRINT:PRINT
250 FOR I=1 TO N
260 PRINT"INPUT COEFFICIENT A(";I-1;")"
270 INPUT A(I)
280 B(I)=A(I)
290 NEXT I
300 REM
310 REM    INPUT REQUIRED AXIS SHIFT
320 REM
330 PRINT:PRINT
340 PRINT"INPUT THE REQUIRED TRANSLATION OF IMAGINARY AXIS"
350 INPUT "IMAGINARY AXIS TO BE SHIFTED TO: ";DA
360 PRINT:PRINT
370 REM
380 REM    HORNER'S AXIS SHIFT ALGORITHM
390 REM
400 I =0
410 I=I+1
420 L=N+1-I
430 J=1
440 J=J+1
450 CT=B(J)+DA*B(J-1)
460 B(J)=CT
470 IF J<L THEN 440
480 IF I<N-1 THEN 410
490 REM
500 REM    OUTPUT THE SHIFTED POLYNOMIAL
510 REM
520 PRINT"THE SHIFTED POLYNOMIAL IS OF THE FORM:"
530 PRINT"B(0)*S^N+B(1)*S^(N-1)+.......B(N-1)*S+B(N)=0"
540 PRINT"WHERE:"
550 PRINT
560 FOR I=1 TO N
570 PRINT"COEFFICIENT B(";I-1;") =    ";B(I)
580 NEXT I
590 END
```

Example 11.5

A fifth-order polynomial $D(s)$ is given by

$$D(s) = s^5 + 8s^4 + 33s^3 + 88s^2 + 122s + 60 = 0$$

Find the coefficients of the shifted polynomial $D(s-3)$.

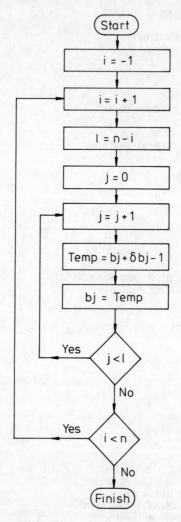

Figure 11.1 Flow diagram for Horner's method

Output from Program 5

```
INPUT DEGREE OF POLYNOMIAL 5


POLYNOMIAL COEFFICIENTS ARE ASSUMED TO BE
OF THE FORM:

A(0)*S^N+A(1)*S^(N-1)+.......A(N-1)*S+A(N)=0
```

```
INPUT COEFFICIENT A(0)
? 1
INPUT COEFFICIENT A(1)
? 8
INPUT COEFFICIENT A(2)
? 33
INPUT COEFFICIENT A(3)
? 88
INPUT COEFFICIENT A(4)
? 122
INPUT COEFFICIENT A(5)
? 60


INPUT THE REQUIRED TRANSLATION OF IMAGINARY AXIS
IMAGINARY AXIS TO BE SHIFTED TO: ? -3

THE SHIFTED POLYNOMIAL IS OF THE FORM:
B(0)*S^N+B(1)*S^(N-1)+.......B(N-1)*S+B(N)=0
WHERE:

COEFFICIENT B(0) =      1
COEFFICIENT B(1) =     -7
COEFFICIENT B(2) =     27
COEFFICIENT B(3) =    -47
COEFFICIENT B(4) =     26
COEFFICIENT B(5) =      0
```

11.8   Program 6: Root loci

The development of a robust, fast and accurate root loci program which will run on an 8 bit microcomputer has proved to be one of the more difficult programming problems. Extensive testing of many routines suggests that some sort of tradeoff must occur. If the program is fast (e.g. a full root loci plot for systems up to fifth order in under 2 minutes) and accurate (e.g. all singular points identified, and all off-axis loci drawn accurately), then there will invariably be systems for which the method fails. If the requirement is a robust and accurate program, based for example on a root solving routine, then the resulting plots tend to be slow: plots taking 20 minutes or more to draw are not uncommon. It is not surprising to find that with a fast, robust method based, say, on the angle criterion (see Section 6.3.1), the accuracy of the plot becomes questionable. The problem is further compounded when BASIC is used as the source language. In these programs compilation and execution are concurrent, and consequently response times are inevitably slow.

Apart from the direct root finding methods, there appear to be two main alternatives: the area-search and branch-following methods. Other methods do exist; these include semi-analytical and analytical methods,[16,17] which tend to be unsuitable for high-order systems, and methods requiring a high degree of user interaction (see e.g. References 18 and 19).

Area-search methods[20,21] begin with the specification of a grid covering a defined area in the s-plane. Each point in the grid is then systematically tested to find those satisfying necessary and sufficient conditions for the existence of a root locus. Obviously, the grid's mesh size is important since it determines the accuracy and resolution of the resulting locus plot as well as the amount of calculation required. In general, the main advantage of area-search methods is the direct nature of the calculations, which makes them well suited to digital computer implementation. However, when a system's poles and zeros are well spaced, the number of grid points required for reasonable accuracy becomes large, and any advantage in speed over direct methods tends to be lost.

Branch-following methods[22-24] use various schemes to track the trajectories of the locus within the s-plane. In general, a trajectory starts at an open-loop pole position, where $K = 0$, or from a singular point, where K has a known, positive (or negative) value. Some form of iterative procedure is then applied until a new point on the locus is found which is close to the original point. (Just how close depends on the particular problem, as experience – or trial and error – will show.) By continuing this procedure, the loci can be tracked to some specified terminal point such as an open-loop zero or singular point. The main advantage of these methods is that, in general, the number of iterations required to find a new point on a given locus is small, so that for certain problems they can be considerably faster than area-search methods. However, there are a number of problems. If two branches pass close together it is possible for the program to jump from one locus to another. Other problems may arise when tracking a locus close to a terminal point: a singular point could be missed, or for an open-loop zero some iterative procedures can go into a loop. Branch-following programs therefore invariably require a considerable amount of special programming in order to avoid the various problems that can arise.

For mainframe machines and minicomputers, interactive root locus plots are possibly best produced by branch-following methods, although some authors[10,25-28] suggest that the speed and versatility of these machines has essentially reduced the problem to one of finding an efficient root solving routine. With small microcomputers the problems associated with obtaining root loci plots can be resolved by arriving at some form of compromise.

The root loci routine given below is based on the modified Bairstow root solving method of Program 3. Although the routine is reasonably robust, convergence can occasionally require many iterations. To overcome this problem, an error trap has been included which will terminate the search for a pair of roots after one hundred iterations.

11.8.1   Listing of Program 6

```
10 REM     *****************    PROGRAM 6    *********************
20 REM     THIS PROGRAM USES A MODIFIED FORM OF THE ROOT SOLVING
30 REM     ROUTINE (PROGRAM 3) TO GENERATE ROOT LOCUS POINTS.
```

```
40 REM     THE PROGRAM REQUIRES THE OPEN LOOP TRANSFER FUNCTION
50 REM     IN RATIONAL POLYNOMIAL FORM AND A GAIN K (WHICH MAY BE
60 REM     VARIED). THE OPEN LOOP POLES AND ZEROS ARE FOUND. THEN FOR
70 REM     EACH VALUE OF K THE CLOSED LOOP CHARACTERISTIC POLYNOMIAL
80 REM     AND ITS ROOTS ARE CALCULATED.
90 REM
100 REM    THE ROOT SOLVING ROUTINE CONTAINS AN ERROR TRAP WHICH
110 REM    WILL DETECT CONVERGENCE PROBLEMS. IF A PROBLEM OCCURS
120 REM    A WARNING IS GIVEN, TOGETHER WITH THE CURRENT ERROR VALUE.
130 REM    AT CONVERGENCE THE ERROR IS ZERO. THE USER MAY ASSUME
140 REM    CONVERGENCE AND SOLVE FOR THE REMAINING ROOTS (THESE ROOTS
150 REM    COULD CONTAIN SIGNIFICANT ERRORS). ALTERNATIVELY, THE USER
160 REM    MAY MAKE A SMALL CHANGE IN THE GAIN K AND RECALCULATE THE
170 REM    CHARACTERISTIC ROOTS. THIS WILL USUALLY PROVIDE AN
180 REM    ACCEPTABLE SOLUTION.
190 REM
200 REM
210 DIM PN(10),PD(10),A(10),B(10),C(10)
220 PRINT:PRINT
230 REM
240 REM    INPUT DEGREE OF NUMERATOR POLYNOMIAL AND COEFFICIENTS.
250 REM
260 INPUT "INPUT DEGREE OF NUMERATOR POLYNOMIAL";NN
270 IF NN<10 THEN 310
280 PRINT"THE ORDER OF THE POLYNOMIAL MUST BE LESS"
290 PRINT"THAN 10"
300 GOTO 200
310 NT= NN+1
320 IF ABS(INT(NT))<>NT THEN 260
330 PRINT:PRINT
340 PRINT"NUMERATOR COEFFICIENTS ARE ASSUMED TO BE"
350 PRINT"OF THE FORM:"
360 PRINT
370 PRINT"A(0)*S^N+A(1)*S^(N-1)+.......A(N-1)*S+A(N)"
380 PRINT:PRINT
390 FOR I=1 TO NT
400 PRINT"INPUT COEFFICIENT A(";I-1;")"
410 INPUT PN(I)
420 NEXT I
430 PRINT:PRINT
440 REM
450 REM    INPUT DEGREE OF DENOMINATOR POLYNOMIAL AND COEFFICIENTS.
460 REM
470 INPUT "INPUT DEGREE OF DENOMINATOR POLYNOMIAL";ND
480 IF ND<10 AND NN<=ND THEN 540
490 PRINT
500 PRINT"THE ORDER OF THE DENOMINATOR MUST BE LESS"
510 PRINT"THAN OR EQUAL TO THE NUMERATOR, AND ALSO"
520 PRINT"LESS THAN 10"
530 GOTO 430
540 NB= ND+1
550 IF ABS(INT(NB))<>NB THEN 470
560 PRINT:PRINT
570 PRINT"DENOMINATOR COEFFICIENTS ARE ASSUMED TO BE"
580 PRINT"OF THE FORM:"
590 PRINT
600 PRINT"B(0)*S^N+B(1)*S^(N-1)+.......B(N-1)*S+B(N)"
610 PRINT:PRINT
620 FOR I=1 TO NB
630 PRINT"INPUT COEFFICIENT B(";I-1;")"
```

```
640 INPUT PD(I)
650 NEXT I
660 PRINT:PRINT
670 JN=ND-NN
680 REM
690 REM    FIND THE OPEN LOOP ZEROS
700 REM
710 IF NT=1 THEN 810
720 FOR I=1 TO NT
730 A(I)=PN(I)
740 NEXT I
750 N=NT
760 PRINT"THE OPEN LOOP ZERO LOCATIONS ARE:"
770 GOSUB 1130
780 REM
790 REM    FIND THE OPEN LOOP POLES
800 REM
810 FOR I=1 TO NB
820 A(I)=PD(I)
830 NEXT I
840 N=NB
850 PRINT"THE OPEN LOOP POLE LOCATIONS ARE:"
860 GOSUB 1130
870 REM
880 REM    FORM THE CLOSED LOOP CHARACTERISTIC POLYNOMIAL
890 REM
900 INPUT "INPUT THE OPEN LOOP GAIN,K = ";KG
910 FOR I=1 TO NB
920 A(I)=PD(I)
930 IF I>JN THEN A(I)=A(I)+KG*PN(I-JN)
940 NEXT I
950 REM
960 REM    FIND THE CLOSED LOOP POLES
970 REM
980 N=NB
990 PRINT"FOR A GAIN OF ";KG;"THE CLOSED LOOP POLES ARE:"
1000 GOSUB 1130
1010 REM
1020 REM    CHECK IF THE RUN IS COMPLETE
1030 REM
1040 PRINT"DO YOU WANT TO CONTINUE (Y/N)?"
1050 INPUT Z$
1060 PRINT:PRINT
1070 IF Z$="Y" THEN 900
1080 IF Z$<>"N" THEN 1030
1090 END
1100 REM
1110 REM    SUBROUTINE CONTAINING MODIFIED CODING OF PROGRAM 3
1120 REM
1130 PRINT:PRINT"REAL                IMAGINARY"
1140 PRINT:IF A(N)<>0 THEN 1160
1145 PRINT 0,0:N=N-1
1150 IF N=1 THEN 1420
1155 GOTO 1140
1160 IF N=2 THEN 1410
1170 IF A(N-2)<>0 THEN 1190
1180 P=0:Q=0:GOTO 1200
1190 P=A(N-1)/A(N-2):Q=A(N)/A(N-2)
1200 IF N=3 THEN 1360
```

```
1210 B(1)=A(1):C(1)=B(1):HI=0
1220 B(2)=A(2)-P*B(1):C(2)=B(2)-P*C(1)
1230 FOR I=3 TO N-1:B(I)=A(I)-P*B(I-1)-Q*B(I-2)
1240 C(I)=B(I)-P*C(I-1)-Q*C(I-2):NEXT I
1250 B(N)=A(N)-P*B(N-1)-Q*B(N-2)
1260 DV=C(N-2)*C(N-2)-(C(N-1)-B(N-1))*C(N-3)
1270 IF DV = 0 THEN DV =E-20
1280 P1=(B(N-1)*C(N-2)-B(N)*C(N-3))/DV
1290 Q1=(C(N-2)*B(N)-(C(N-1)-B(N-1))*B(N-1))/DV
1300 P=P+P1:Q=Q+Q1:HI=HI+1:IF HI<100 THEN 1350
1310 PRINT"WARNING ITERATION LIMIT EXCEEDED, ERROR =";P1
1320 PRINT:PRINT"DO YOU WANT THE REMAINING ROOTS (Y/N)?"
1330 INPUT Z$
1340 IF Z$="Y" THEN 1360:IF Z$="N" THEN 1420:GOTO 1320
1350 IF ABS(P1)>.00001 THEN 1220
1360 G=P*P-4*Q:PR=-P/2:IF G<0 THEN 1380
1370 PG=SQR(G)/2:PRINT PR+PG,0:PRINT PR-PG,0:GOTO 1390
1380 PG=SQR(-G)/2:PRINT PR,PG:PRINT PR,-PG
1390 N=N-2:IF N=1 THEN 1420
1400 FOR I= 1 TO N:A(I)=B(I):NEXT I:GOTO 1150
1410 PRINT -A(N)/A(N-1),0
1420 PRINT:PRINT:RETURN
```

Example 11.6

This example produces the data for a root loci plot for a system having an open-loop transfer function given by

$$G(s) = \frac{K(s^2 + 4s + 5)}{s^4 + 8s^3 + 19s^2 + 12s}$$

The root loci plot for positive values of K is shown in Fig. 11.2.

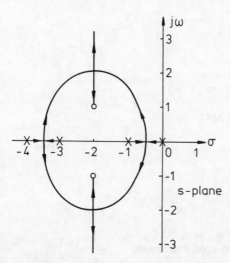

Figure 11.2 Root loci plot for $G(s) = \dfrac{K(s^2 + 4s + 5)}{s^4 + 8s^3 + 19s^2 + 12s}$

Output from Program 6

```
INPUT DEGREE OF NUMERATOR POLYNOMIAL ? 2


NUMERATOR COEFFICIENTS ARE ASSUMED TO BE
OF THE FORM:

A(0)*S^N+A(1)*S^(N-1)+.......A(N-1)*S+A(N)

INPUT COEFFICIENT A(0)
? 1
INPUT COEFFICIENT A(1)
? 4
INPUT COEFFICIENT A(2)
? 5


INPUT DEGREE OF DENOMINATOR POLYNOMIAL ? 4

DENOMINATOR COEFFICIENTS ARE ASSUMED TO BE
OF THE FORM:

B(0)*S^N+B(1)*S^(N-1)+.......B(N-1)*S+B(N)

INPUT COEFFICIENT B(0)
? 1
INPUT COEFFICIENT B(1)
? 8
INPUT COEFFICIENT B(2)
? 19
INPUT COEFFICIENT B(3)
? 12
INPUT COEFFICIENT B(4)
? 0


THE OPEN LOOP ZERO LOCATIONS ARE:

REAL           IMAGINARY

-2             1
-2             -1

THE OPEN LOOP POLE LOCATIONS ARE:

REAL           IMAGINARY

 0             0
-1             0
-3             0
-4             0


INPUT THE OPEN LOOP GAIN,K = ? .25
FOR A GAIN OF  .25 THE CLOSED LOOP POLES ARE:
```

REAL	IMAGINARY
-.1147288	0
-.9064954	0
-3.093536	0
-3.885249	0

DO YOU WANT TO CONTINUE (Y/N)?
? Y

INPUT THE OPEN LOOP GAIN,K = ? .7
FOR A GAIN OF   .7 THE CLOSED LOOP POLES ARE:

REAL	IMAGINARY
-.5306552	9.473207E-02
-.5306552	-9.473207E-02
-3.469346	9.476682E-02
-3.469346	-9.476682E-02

DO YOU WANT TO CONTINUE (Y/N)?
? Y

INPUT THE OPEN LOOP GAIN,K = ? 5
FOR A GAIN OF 5 THE CLOSED LOOP POLES ARE:

REAL	IMAGINARY
-.7752551	1.224745
-.7752551	-1.224745
-3.224745	1.224746
-3.224745	-1.224746

DO YOU WANT TO CONTINUE (Y/N)?
? Y

INPUT THE OPEN LOOP GAIN,K = ? 13.33
FOR A GAIN OF 13.33 THE CLOSED LOOP POLES ARE:

REAL	IMAGINARY
-1.999986	2.008426
-1.999986	-2.008426
-2.000014	2.072734
-2.000014	-2.072734

DO YOU WANT TO CONTINUE (Y/N)?
? Y

INPUT THE OPEN LOOP GAIN,K = ? 20
FOR A GAIN OF 20 THE CLOSED LOOP POLES ARE:

```
REAL          IMAGINARY

-2            1.349478
-2           -1.349478
-2            3.630277
-2           -3.630277

DO YOU WANT TO CONTINUE (Y/N)?
? N
```

11.9   Program 7: Frequency response routines

Program 7 consists of a main program which acts as a front end for programs 7A, 7B and 7C. These programs will between them produce data for polar (Nyquist), Bode, inverse polar (inverse Nyquist) and closed-loop frequency response plots.

Program 7 accepts transfer function data in rational polynomial form, and calculates the number of frequency points between user-specified minimum and maximum frequency values. A linear range of frequencies for the specified number of points is assumed. However, for some applications (e.g. Bode diagrams) a logarithmic frequency range is desirable; the following alterations must then be made:

Program 7: change line 740 to

```
740 DW = (WX/WM)^(1/(WP-1)) : YL = 1
```

Program 7A: change line 1380 to

```
1380 WH = WM*DW^YL : YL = YL + 1
```

Program 7B: change line 1230 to

```
1230 WH = WM*DW^YL : YL = YL + 1
```

Program 7C: change line 1460 to

```
1460 WH = WM*DW^YL : YL = YL + 1
```

The remainder of this section deals with the algorithms used to produce the data for various frequency response plots.

11.9.1   Program 7A: Polar and Bode data

Let the open-loop transfer function be represented by the rational polynomial

$$G(s) = \frac{N(s)}{D(s)} \tag{11.12}$$

where $N(s)$ and $D(s)$ are polynomials in s and are of the form

$$D(s) = a_0 s^n + a_1 s^{n-1} + \cdots + a_{n-1} s + a_n \tag{11.13}$$

$$N(s) = b_0 s^m + b_1 s^{m-1} + \cdots + b_{m-1}s + b_m \tag{11.14}$$

For a given frequency ω, the corresponding complex open-loop data is obtained by setting $s = j\omega$:

$$G(j\omega) = \frac{N(j\omega)}{D(j\omega)} \tag{11.15}$$

where

$$D(j\omega) = a_0(j\omega)^n + a_1(j\omega)^{n-1} + \cdots + a_{n-1}j\omega + a_n \tag{11.16}$$

This may be separated into real and imaginary parts:

$$\mathrm{Re}(D(j\omega)) = a_n - \omega^2 a_{n-2} + \omega^4 a_{n-4} - \omega^6 a_{n-6} + \cdots \tag{11.17}$$

$$\mathrm{Im}(D(j\omega)) = a_{n-1} - \omega^3 a_{n-3} + \omega^5 a_{n-5} - \omega^7 a_{n-7} + \cdots \tag{11.18}$$

Similar expressions may be found from $N(j\omega)$. The expressions for $D(j\omega)$ and $N(j\omega)$ are readily programmed, as shown by the coding of Program 7A.

To find the open-loop real and imaginary coordinates, the normal rules of complex algebra are applied to give

$$G(j\omega) = \frac{\mathrm{Re}(N(j\omega)) + j\mathrm{Im}(N(j\omega))}{\mathrm{Re}(D(j\omega)) + j\mathrm{Im}(D(j\omega))} \tag{11.19}$$

from which

$$\mathrm{Re}(G(j\omega)) = \frac{\mathrm{Re}(N(j\omega))\mathrm{Re}(D(j\omega)) + \mathrm{Im}(N(j\omega))\mathrm{Im}(D(j\omega))}{\mathrm{Re}(D(j\omega))^2 + \mathrm{Im}(D(j\omega))^2} \tag{11.20}$$

$$\mathrm{Im}(G(j\omega)) = \frac{\mathrm{Im}(N(j\omega))\mathrm{Re}(D(j\omega)) - \mathrm{Re}(N(j\omega))\mathrm{Im}(D(j\omega))}{\mathrm{Re}(D(j\omega))^2 + \mathrm{Im}(D(j\omega))^2} \tag{11.21}$$

If the real and imaginary parts of $G(j\omega)$ are plotted on an Argand diagram (in rectangular coordinates), then the locus of points thus obtained for $0 \leqslant \omega \leqslant +\infty$ is called the polar or Nyquist plot.

Bode data is found by converting $G(j\omega)$ into the Euler form:

$$G(j\omega) = M(\omega)e^{j\theta(\omega)} \tag{11.22}$$

as shown in Fig. 11.3. This figure shows that

$$M(\omega) = \sqrt{[\mathrm{Re}(G(j\omega))^2 + \mathrm{Im}(G(j\omega))^2]} \tag{11.23}$$

$$\theta(\omega) = \tan^{-1}\frac{\mathrm{Im}(G(j\omega))}{\mathrm{Re}(G(j\omega))} \tag{11.24}$$

There are two Bode plots, one of dB magnitude $M(\omega)$ against frequency ω, in rad/unit time, and the other of phase $\theta(\omega)$, in degrees, against frequency, again in rad/unit time. In both plots the magnitude and phase are plotted on the y-axis

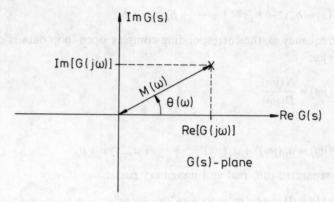

Figure 11.3 Polar coordinates of a point in the $G(s)$ plane

using a linear scale, and frequency is plotted on the x-axis using a logarithmic scale. To produce Bode data, $M(\omega)$ is converted to dB:

$$M(\omega)_{dB} = 20 \log_{10}(M(\omega)) \tag{11.25}$$

However, for most microcomputers natural logarithms are used and the expression becomes

$$M(\omega)_{dB} = 20[\log_e(M(\omega))/\log_e(10)] \tag{11.26}$$

Also, the inverse tangent function will give the angle in radians, and this must be converted to degrees. Care should be taken to ensure that the angle $\theta(\omega)$ is measured from the positive real axis in an anticlockwise direction, as shown in Fig. 11.3.

Although none of the routines contains a time delay, one could easily be accommodated. A time delay does not affect $M(\omega)$, but it does alter $\theta(\omega)$. For a given frequency the total phase shift $\theta(\omega)_T$ is given by

$$\theta(\omega)_T = \theta(\omega) - \omega\tau, \tag{11.27}$$

where $\theta(\omega)$ is the phase shift due to the transfer function, and $\omega\tau$ the phase shift due to a time delay τ. The units of $\theta(\omega)$ and $\omega\tau$ must be consistent.

11.9.2   Program 7B: Inverse polar data

The inverse polar (or inverse Nyquist) plot is a plot in rectangular coordinates of $G(j\omega)^{-1}$;

$$G(j\omega)^{-1} = \frac{D(j\omega)}{N(j\omega)} \tag{11.28}$$

The techniques for finding the real and imaginary parts of $G(j\omega)$, described in Program 7A, are equally applicable to $G(j\omega)^{-1}$.

11.9.3   Program 7C: Closed-loop frequency data

In this program closed-loop frequency response data is obtained directly from the open-loop transfer function model. Given the open-loop transfer function $G(j\omega)$, the closed-loop transfer function $T(j\omega)$ is given by

$$T(j\omega) = \frac{G(j\omega)}{1 + G(j\omega)} \tag{11.29}$$

From this expression it may be deduced that

$$\text{Re}(T(j\omega)) = \frac{\text{Re}(G(j\omega))^2 + \text{Im}(G(j\omega))^2 + \text{Re}(G(j\omega))}{[\text{Re}(G(j\omega)) + 1]^2 + \text{Im}(G(j\omega))^2} \tag{11.30}$$

$$\text{Im}(T(j\omega)) = \frac{\text{Im}(G(j\omega))}{[\text{Re}(G(j\omega)) + 1]^2 + \text{Im}(G(j\omega))^2} \tag{11.31}$$

11.9.4   Listing of main program (Program 7)

```
10 REM     *****************    PROGRAM 7    *****************
20 REM     THIS IS THE MAIN FREQUENCY RESPONSE PROGRAM. IT IS
30 REM     USED IN CONJUNCTION WITH PROGRAMS 7A, 7B AND 7C.
40 REM
50 DIM PN(10),PD(10)
60 PRINT:PRINT
70 REM
80 REM     INPUT DEGREE OF NUMERATOR POLYNOMIAL. CHECK FOR
90 REM     AN INTEGER NUMBER LESS THAN 10.
100 REM
110 INPUT "INPUT DEGREE OF NUMERATOR POLYNOMIAL";NN
120 IF NN<10 THEN 160
130 PRINT"THE ORDER OF THE POLYNOMIAL MUST BE LESS"
140 PRINT"THAN 10"
150 GOTO 60
160 NT= NN+1
170 IF ABS(INT(NT))<>NT THEN 110
180 REM
190 REM     INPUT NUMERATOR COEFFICIENTS
200 REM
210 PRINT:PRINT
220 PRINT"NUMERATOR COEFFICIENTS ARE ASSUMED TO BE"
230 PRINT"OF THE FORM:"
240 PRINT
250 PRINT"A(0)*S^N+A(1)*S^(N-1)+.......A(N-1)*S+A(N)"
260 PRINT:PRINT
270 FOR I=1 TO NT
280 PRINT"INPUT COEFFICIENT A(";I-1;")"
290 INPUT PN(NT-I+1)
300 NEXT I
310 PRINT:PRINT
320 REM
330 REM     INPUT AND CHECK DEGREE OF DENOMINATOR POLYNOMIAL.
340 REM
350 INPUT "INPUT DEGREE OF DENOMINATOR POLYNOMIAL";ND
360 IF ND<10 AND NN<=ND THEN 420
```

```
370 PRINT
380 PRINT"THE ORDER OF THE DENOMINATOR MUST BE LESS"
390 PRINT"THAN OR EQUAL TO THE NUMERATOR, AND ALSO"
400 PRINT"LESS THAN 10"
410 GOTO 310
420 NB=ND+1
430 IF ABS(INT(NB))<>NB THEN 350
440 IF ABS(INT(NB))<>NB THEN 350
450 REM
460 REM    INPUT DENOMINATOR COEFFICIENTS.
470 REM
480 PRINT:PRINT
490 PRINT"DENOMINATOR COEFFICIENTS ARE ASSUMED TO BE"
500 PRINT"OF THE FORM:"
510 PRINT
520 PRINT"B(0)*S^N+B(1)*S^(N-1)+.......B(N-1)*S+B(N)"
530 PRINT:PRINT
540 FOR I=1 TO NB
550 PRINT"INPUT COEFFICIENT B(";I-1;")"
560 INPUT PD(NB-I+1)
570 NEXT I
580 PRINT:PRINT
590 REM
600 REM    INPUT REQUIRED NUMBER OF DATA POINTS BETWEEN USER
610 REM    SPECIFIED MAXIMUM AND MINIMUM FREQUENCY VALUES.
620 REM    A LINEAR FREQUENCY RANGE IS ASSUMED BETWEEN
630 REM    THE MAXIMUM AND MINIMUM VALUES.
640 REM
650 PRINT"INPUT NUMBER OF FREQUENCY POINTS"
660 INPUT WP
670 WP=ABS(WP)
680 IF WP<0 THEN WP=2
690 PRINT:PRINT"INPUT MAXIMUM AND MINIMUM FREQUENCIES"
700 INPUT "MINIMUM FREQUENCY = ";WM
710 INPUT "MAXIMUM FREQUENCY = ";WX
720 IF WM<0 OR WX<0 THEN PRINT"FREQUENCIES MUST BE +VE":GOTO 700
730 IF WM>WX THEN WH=WM:WM=WX:WX=WH
740 DW=(WX-WM)/(WP-1)
750 FA=57.29578:WH=WM
760 REM
770 REM    THIS IS THE END OF THE MAIN SEGMENT OF PROGRAM 7
780 REM
```

11.9.5   Listing for polar and Bode data (Program 7A)

```
790 REM    ***************    PROGRAM 7A    ********************
800 REM    WHEN COMBINED WITH PROGRAM 7 THIS CODING WILL GIVE
810 REM    OPEN LOOP FREQUENCY DATA FROM WHICH A POLAR/NYQUIST
820 REM    PLOT OR BODE PLOTS MAY BE OBTAINED.
830 REM
840 REM
850 REM    PRINT THE HEADINGS FOR THE POLAR (NYQUIST) AND BODE DATA
860 REM
870 PRINT:PRINT"              OPEN LOOP FREQUENCY DATA"
880 PRINT"FREQUENCY     REAL        IMAG        GAIN        PHASE"
890 PRINT"RAD/TIME                              (DB)        (DEG)"
900 REM
```

```
910 REM     FIND THE REAL AND IMAGINARY COMPONENTS OF THE
920 REM     NUMERATOR AT THE SPECIFIED FREQUENCY.
930 REM
940 XZ=0:YZ=0:XP=0:YP=0:QR=1:QI=0
950 FOR I =1 TO NT
960 XZ=XZ+PN(I)*QR
970 YZ=YZ+PN(I)*QI
980 QZ=-(WH*QI):QI=WH*QR:QR=QZ
990 NEXT I
1000 REM
1010 REM     FIND THE REAL AND IMAGINARY COMPONENTS OF THE
1020 REM     DENOMINATOR AT THE SPECIFIED FREQUENCY.
1030 REM
1040 QR=1:QI=0
1050 FOR I = 1 TO NB
1060 XP=XP+PD(I)*QR
1070 YP=YP+PD(I)*QI
1080 QZ=-WH*QI:QI=WH*QR:QR=QZ
1090 NEXT I
1100 REM
1110 REM     CALCULATE THE OPEN LOOP REAL AND IMAGINARY
1120 REM     COORDINATES AND ALSO THE GAIN AND PHASE VALUES.
1130 REM
1140 AA=XP*XZ+YZ*YP
1150 DD=XP*XP+YP*YP
1160 IF DD=0 THEN DD=E-20
1170 AW=AA/DD
1180 BW=(YZ*XP-YP*XZ)/DD
1190 MW=SQR(AW*AW+BW*BW)
1200 IF AW=0 THEN PH=90*SGN(BW):GOTO 1250
1210 BL=ATN(BW/AW)
1220 IF AW>=0 THEN PH=BL*FA:GOTO 1250
1230 IF BW>=0 THEN PH=BL*FA+180:GOTO 1250
1240 PH=-180+BL*FA
1250 MW=20*(LOG(MW)/LOG(10))
1260 REM
1270 REM     SET THE DATA FORMAT FOR PRINTING
1280 REM
1290 WP=(INT(WH*1000+.5))/1000
1300 AW=(INT(AW*1000+.5))/1000
1310 BW=(INT(BW*1000+.5))/1000
1320 MW=(INT(MW*1000+.5))/1000
1330 PH=(INT(PH*10+.5))/10
1340 REM
1350 REM     PRINT FREQUENCY DATA, AND TEST IF RUN IS COMPLETE.
1360 REM
1370 PRINT WP,AW,BW,MW,PH
1380 WH=WH+DW
1390 IF WH<=WX THEN 940
1400 END
```

11.9.6   Listing for inverse polar data (Program 7B)

```
790 REM     ***************     PROGRAM 7B     ********************
800 REM     WHEN COMBINED WITH PROGRAM 7 THIS CODING WILL GIVE
810 REM     OPEN LOOP FREQUENCY DATA FROM WHICH AN INVERSE
820 REM     POLAR/INVERSE NYQUIST PLOT MAY BE OBTAINED.
830 REM
```

```
840 REM
850 REM     PRINT THE HEADINGS FOR THE INVERSE POLAR DATA
860 REM
870 PRINT:PRINT"FREQUENCY DATA FOR OPEN LOOP INVERSE POLAR PLOT"
880 PRINT"FREQUENCY     REAL          IMAG"
890 PRINT"RAD/TIME"
900 REM
910 REM     FIND THE REAL AND IMAGINARY COMPONENTS OF THE
920 REM     NUMERATOR AT THE SPECIFIED FREQUENCY.
930 REM
940 XZ=0:YZ=0:XP=0:YP=0:QR=1:QI=0
950 FOR I =1 TO NT
960 XZ=XZ+PN(I)*QR
970 YZ=YZ+PN(I)*QI
980 QZ=-(WH*QI):QI=WH*QR:QR=QZ
990 NEXT I
1000 REM
1010 REM     FIND THE REAL AND IMAGINARY COMPONENTS OF THE
1020 REM     DENOMINATOR AT THE SPECIFIED FREQUENCY.
1030 REM
1040 QR=1:QI=0
1050 FOR I = 1 TO NB
1060 XP=XP+PD(I)*QR
1070 YP=YP+PD(I)*QI
1080 QZ=-WH*QI:QI=WH*QR:QR=QZ
1090 NEXT I
1100 REM
1110 REM     CALCULATE THE REAL AND IMAGINARY COORDINATES OF
1120 REM     THE INVERSE FREQUENCY RESPONSE
1130 REM
1140 AA=XP*XZ+YZ*YP
1150 DD=XZ*XZ+YZ*YZ
1160 IF DD=0 THEN DD=E-20
1170 AW=AA/DD
1180 BW=(YP*XZ-XP*YZ)/DD
1190 REM
1200 REM     PRINT INVERSE DATA AND TEST IF THE RUN IS COMPLETE
1210 REM
1220 PRINT WH,AW,BW
1230 WH=WH+DW
1240 IF WH <=WX THEN 940
1250 END
```

11.9.7   Listing for closed-loop frequency data (Program 7C)

```
790 REM     ***************    PROGRAM 7C     ********************
800 REM     WHEN COMBINED WITH PROGRAM 7 THIS CODING WILL GIVE
810 REM     THE FREQUENCY RESPONSE OF THE UNITY, NEGATIVE FEEDBACK
820 REM     CLOSED LOOP SYSTEM. THE INPUT DATA IS THE OPEN LOOP
830 REM     RATIONAL POLYNOMIAL TRANSFER FUNCTION.
840 REM
850 REM
860 REM     PRINT THE HEADINGS FOR THE CLOSED LOOP FREQUENCY
870 REM     RESPONSE DATA.
880 REM
890 PRINT:PRINT"                    CLOSED LOOP FREQUENCY DATA"
900 PRINT"FREQUENCY     REAL          IMAG          GAIN          PHASE"
```

```
910 PRINT"RAD/TIME                       (DB)         (DEG)
920 REM
930 REM   FIND THE REAL AND IMAGINARY COMPONENTS OF THE
940 REM   NUMERATOR AT THE SPECIFIED FREQUENCY.
950 REM
960 XZ=0:YZ=0:XP=0:YP=0:QR=1:QI=0
970 FOR I =1 TO NT
980 XZ=XZ+PN(I)*QR
990 YZ=YZ+PN(I)*QI
1000 QZ=-(WH*QI):QI=WH*QR:QR=QZ
1010 NEXT I
1020 REM
1030 REM   FIND THE REAL AND IMAGINARY COMPONENTS OF THE
1040 REM   DENOMINATOR AT THE SPECIFIED FREQUENCY.
1050 REM
1060 QR=1:QI=0
1070 FOR I = 1 TO NB
1080 XP=XP+PD(I)*QR
1090 YP=YP+PD(I)*QI
1100 QZ=-WH*QI:QI=WH*QR:QR=QZ
1110 NEXT I
1120 REM
1130 REM   CALCULATE THE OPEN LOOP REAL AND IMAGINARY
1140 REM   COORDINATES AND ALSO THE GAIN AND PHASE VALUES.
1150 REM
1160 AA=XP*XZ+YZ*YP
1170 DD=XP*XP+YP*YP
1180 IF DD=0 THEN DD=E-20
1190 AW=AA/DD
1200 BW=(YZ*XP-YP*XZ)/DD
1210 REM
1220 REM   CALCULATE THE CLOSED LOOP REAL AND IMAGINARY
1230 REM   COORDINATES AND ALSO THE GAIN AND PHASE VALUES.
1240 REM
1250 UC=AW*AW+BW*BW+AW
1260 UD=BW*BW+(1+AW)*(1+AW)
1270 UC=UC/UD:VC=BW/UD
1280 MC=20*(LOG(SQR(UC*UC+VC*VC))/LOG(10))
1290 PC=(ATN(VC/UC))*FA
1300 IF UC>=0 THEN 1360
1310 IF VC<0 THEN PC=-(180-PC):GOTO 1360
1320 PC=180+PC
1330 REM
1340 REM   SET THE DATA FORMAT FOR PRINTING.
1350 REM
1360 WP=(INT(WH*1000+.5))/1000
1370 UC=(INT(UC*1000+.5))/1000
1380 VC=(INT(VC*1000+.5))/1000
1390 MC=(INT(MC*1000+.5))/1000
1400 PC=(INT(PC*10+.5))/10
1410 REM
1420 REM   PRINT CLOSED LOOP DATA AND TEST IF THE RUN IS COMPLETE
1430 REM
1440 PRINT WP,UC,VC,MC,PC
1450 IF WH>=WX THEN 1480
1460 WH=WH+DW
1470 GOTO 960
1480 END
```

Example 11.7

For the following open-loop transfer function $G(s)$, produce Bode, polar, inverse polar and closed-loop Bode plots:

$$G(s) = \frac{2(s+2)}{s^3 + 3s^2 + 5s + 4}$$

Output from Program 7

INPUT DEGREE OF NUMERATOR POLYNOMIAL ? 1

NUMERATOR COEFFICIENTS ARE ASSUMED TO BE
OF THE FORM:

A(0)*S^N+A(1)*S^(N-1)+.......A(N-1)*S+A(N)

INPUT COEFFICIENT A(0)
? 2
INPUT COEFFICIENT A(1)
? 4
INPUT DEGREE OF DENOMINATOR POLYNOMIAL ? 3

DENOMINATOR COEFFICIENTS ARE ASSUMED TO BE
OF THE FORM:

B(0)*S^N+B(1)*S^(N-1)+.......B(N-1)*S+B(N)

INPUT COEFFICIENT B(0)
? 1
INPUT COEFFICIENT B(1)
? 3
INPUT COEFFICIENT B(2)
? 5
INPUT COEFFICIENT B(3)
? 4

INPUT NUMBER OF FREQUENCY POINTS
? 15

INPUT MAXIMUM AND MINIMUM FREQUENCIES
MINIMUM FREQUENCY = ? .1
MAXIMUM FREQUENCY = ? 4

Output from Program 7A

FREQUENCY RAD/TIME	OPEN LOOP FREQUENCY DATA			
	REAL	IMAG	GAIN (DB)	PHASE (DEG)
.1	.998	-.075	.008	-4.3
.379	.972	-.288	.119	-16.5
.657	.904	-.518	.358	-29.8
.936	.758	-.767	.653	-45.3
1.214	.47	-.981	.731	-64.4

1.493	.066	-1.01	.103	-86.3
1.771	-.248	-.806	-1.485	-107.1
2.05	-.361	-.545	-3.691	-123.5
2.329	-.354	-.351	-6.04	-135.2
2.607	-.31	-.23	-8.28	-143.4
2.886	-.262	-.155	-10.333	-149.3
3.164	-.22	-.109	-12.197	-153.6
3.443	-.186	-.079	-13.889	-156.9
3.721	-.158	-.059	-15.433	-159.5
4	-.136	-.045	-16.848	-161.6

Output from Program 7B

```
FREQUENCY DATA FOR OPEN LOOP INVERSE POLAR PLOT
FREQUENCY        REAL              IMAG
RAD/TIME
.1               .9962468          7.493766E-02
.3785714         .9456368          .2806549
.6571429         .8328015          .4768493
.9357143         .6520102          .6597765
1.214286         .3974243          .8289509
1.492857         6.459E-02         .9861056
1.771429         -.3491717         1.133885
2.05             -.8450782         1.274924
2.328571         -1.423386         1.411421
2.607143         -2.08383          1.545036
2.885714         -2.825914         1.676947
3.164286         -3.649078         1.807958
3.442857         -4.552789         1.938599
3.721428         -5.536565         2.069208
4                -6.599998         2.2
```

Output from Program 7C

```
                CLOSED LOOP FREQUENCY DATA
FREQUENCY    REAL        IMAG        GAIN        PHASE
RAD/TIME                             (DB)        (DEG)
.1           .5          -.019       -6.01       -2.1
.379         .503        -.073       -5.871      -8.2
.657         .511        -.133       -5.547      -14.6
.936         .522        -.208       -5.003      -21.8
1.214        .529        -.314       -4.216      -30.7
1.493        .506        -.468       -3.234      -42.8
1.771        .381        -.663       -2.328      -60.1
2.05         9.4E-02     -.773       -2.173      -83.1
2.329        -.195       -.65        -3.367      -106.7
2.607        -.304       -.434       -5.517      -125
2.886        -.297       -.273       -7.886      -137.4
3.164        -.258       -.176       -10.123     -145.7
3.443        -.217       -.118       -12.143     -151.4
3.721        -.182       -.083       -13.955     -155.5
4            -.155       -.061       -15.587     -158.6
```

11.10   Program 8: Transformation of step response data into the frequency domain

The methods of transfer function identification discussed in Chapter 9 are all for the frequency domain. However, an experiment carried out on a plant or process will, in all probability, be in the time domain. This program uses the Fourier transform as the basis of a numerical method for transforming time domain data into the frequency domain.

An aperiodic function $y(t)$ in the time domain may be transformed into a corresponding function $g(j\omega)$ in the frequency domain by means of the Fourier transform:

$$g(j\omega) = \int_{-\infty}^{+\infty} y(t)e^{j\omega t}\, dt \tag{11.32}$$

However, if $y(t)$ is not an analytic function then Equation (11.32) cannot be used directly. An example is an identification experiment in which $y(t)$, the process output, is given in graphical form or as a list of experimental data. If the transient response curve is split into a number of piecewise segments and each segment approximated by a trapezoid, then for each segment

$$g_n(j\omega) = \int_{-\infty}^{+\infty} y(n\Delta t)e^{j\omega t}\, dt \tag{11.33}$$

where $y(n\Delta t)$ is the nth trapezoidal segment of the transient response (see Fig. 11.4).

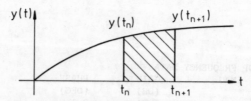

Figure 11.4 The nth segment of the time response curve

The summation of these straight line approximations of the time function gives the entire corresponding function in the frequency domain:

$$g(j\omega) = \sum g_n(j\omega) \tag{11.34}$$

By applying Equation (11.34) to a unit impulse response, with the initial conditions $t = 0$ and $y(t) = 0$, it can be shown that, splitting the transient response into K segments,

$$g(j\omega) = \sum_{n=0}^{K} \frac{Z(n\Delta t)}{\omega^2}(\cos \omega t_{n+1} - \cos \omega t_n)$$

$$-j \sum_{n=0}^{K} \frac{Z(n\Delta t)}{\omega^2}(\sin \omega t_{n+1} - \sin \omega t_n) \qquad (11.35)$$

and for a unit step response, again with the initial conditions $t = 0$ and $y(t) = 1$,

$$g(j\omega) = \sum_{n=0}^{K} \frac{Z(n\Delta t)}{\omega}(\sin \omega t_{n+1} - \sin \omega t_n)$$

$$-j \sum_{n=0}^{K} \frac{Z(n\Delta t)}{\omega}(\cos \omega t_{n+1} - \cos \omega t_n) \qquad (11.36)$$

In both cases $Z(n\Delta t)$ is determined from the nth trapezoidal segment $y(n\Delta t)$:

$$Z(n\Delta t) = \frac{y(t_{n+1}) - y(t_n)}{t_{n+1} - t_n} \qquad (11.37)$$

where $(y(t_{n+1}), t_{n+1})$ and $(y(t_n), t_n)$ are the coordinates of the end points of each line segment. Since the transient response has been split into K segments, the final response time considered would be t_{K+1}, and the corresponding output $y(t_{K+1})$.

For a given frequency the first terms in both expressions for $g(j\omega)$ give the real part of the complex frequency response, and the second gives the imaginary part. Since the case studies in Chapter 10 use step response data, the coding for Program 8 is based on the step response algorithm of Equation (11.36). As this program is intended for use with the transfer function identification routine of Program 9, it is necessary to produce a frequency response which includes the dominant corner frequencies. To this end, a considerable amount of information about the system may be obtained by inspecting the step response. The following list indicates some of the more important features (see also Section 9.4).

(i)   *Time Delays*

Most methods of transfer function identification are based on the assumption that the response is instantaneous. For a step input initiated at time $t = 0$, if there is a delay before the system begins to respond, at time $t = \tau$, identification must be applied to the input $r(t)$ and the shifted output $y(t - \tau)$: see Fig. 11.5. This extracts the time delay from the response and will produce a transfer function $G(s)$. So, if the time delay is included, the final transfer function in the s-domain is

$$e^{-\tau s}G(s) \qquad (11.38)$$

This approach works for a linear process having a time delay in the forward path. However, the extraction of any information from a response and its subsequent introduction to the model should always be undertaken with care.

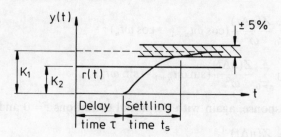

Figure 11.5 A time-delayed response; K_1 is the final steady-state gain and K_2 the input step gain

(ii)   Absolute Stability

By estimating the upper bound of the response it is possible to determine the dominant pole position. This gives an indication of the frequency range over which the transfer function identification methods will operate. For example, in Fig. 11.5 the 5 per cent settling time t_s is shown. If the dominant pole (or poles) has an absolute real value of a, then the settling time may be approximated by

$$\frac{5}{100} = e^{-at_s}$$ (11.39)

or

$$a = 3/t_s$$ (11.40)

and the first corner frequency will occur at approximately $3/t_s$ rad/unit time.

(iii)   Non-minimum Phase

The presence of non-minimum phase zeros is usually indicated by the initial time response moving in the opposite direction to that of the input. This shows that the system has at least one zero with a positive real part.

(iv)   Predominantly First-Order Process

If the step response transient leaves its steady-state position at an angle of maximum slope, then the process is predominantly first order – it may be represented by a transfer function having one more pole than zero. The process will usually have a dominant first-order pole (see Fig. 11.6).

(v)   Dominant Complex Poles

These are usually indicated by an oscillatory response.

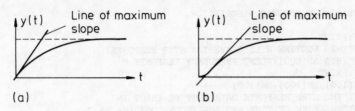

Figure 11.6 Response of system with (a) one more pole than zero and (b) two or more poles than zeros

(vi)   *Static Gain*

For a type 0 system, this is defined as the change in the steady-state output divided by the magnitude of the input step. In Fig. 11.5 the static gain is K_1/K_2.

In using the program, any time delay should be removed. This is achieved by letting the first data point record the time and amplitude of the response at the end of the time delay, and the start of the transient. For example, if a step input is initiated at time $t = 0$ and the system begins to respond at time $t = \tau$ (see Fig. 11.5), then the first data points would be τ and $y(\tau)$.

All the data is normalized such that the first time point is taken to be zero, and the initial amplitude $y(0)$ is set to zero. The final amplitude of the response is set to unity. This means that the gain of the resulting frequency response at zero frequency will be 1.

A linear range of frequencies between user-specified maximum and minimum values is assumed. If a logarithmic scale is required, then lines 800 and 870 should be replaced with

```
800 PRINT:PRINT:DW = (WX/WM)^(1/(NW-1))
```

```
870 W = WM*DW^I
```

Finally, the accuracy of the generated frequency response plot will depend on the accuracy of the input data and the spacing of the sample points. In general, the closer the spacing, the higher the frequency for which the results are valid. A useful rule of thumb is to produce an A4 size plot of the original step response. Superimpose onto this a plot of the sample points with all adjacent points connected by straight lines. If the two plots cannot be distinguished by eye, then there are sufficient points to ensure the accuracy of the generated frequency data around the dominant corner frequencies.

11.10.1   Listing of Program 8

```
10 REM ******************** PROGRAM 8 ******************
20 REM
```

```
30 REM PRINT OUT PROGRAM DISCRIPTION.
40 REM
50 PRINT:PRINT
60 PRINT"THIS ROUTINE WILL CONVERT A STEP RESPONSE"
70 PRINT"INTO AN EQUIVALENT FREQUENCY RESPONSE."
80 PRINT:PRINT
90 DIM T(100),AM(100),DA(100)
100 PRINT"THE STEP RESPONSE DATA MUST BE INPUT IN"
110 PRINT"THE FORM OF TIME AND AMPLITUDE COORDINATES."
120 PRINT:PRINT
130 PRINT"IT IS ASSUMED THAT ANY TIME DELAY IS REMOVED,"
140 PRINT"AND THAT SUFFICIENT DATA POINTS ARE AVAILABLE"
150 PRINT"TO ENSURE THAT STRAIGHT LINE CONNECTION OF THE"
160 PRINT"POINTS PRODUCES A GRAPH VISUALLY IDENTICAL TO"
170 PRINT"THE ORIGINAL RESPONSE."
180 PRINT:PRINT
190 REM
200 REM INPUT AND CHECK STEP RESPONSE DATA.
210 REM
220 PRINT"INPUT THE NUMBER OF DATA POINTS (LESS THAN 101)"
230 INPUT NN
240 IF NN>100 OR NN<=0 THEN 130
250 PRINT:PRINT
260 PRINT"AT EACH SAMPLE POINT INPUT:"
270 PRINT
280 PRINT"             TIME, AMPLITUDE"
290 FOR I=1 TO NN
300 PRINT"FOR POINT";I;:INPUT T(I),AM(I)
310 NEXT I
320 PRINT:PRINT
330 PRINT"PLEASE CHECK THE DATA"
340 PRINT"POINT          TIME          AMPLITUDE"
350 FOR I=1 TO NN
360 PRINT I,T(I),AM(I)
370 NEXT I
380 PRINT:INPUT"DO YOU WISH TO CHANGE ANY VALUE (Y/N)";Q$
390 IF Q$="N" THEN 510
400 IF Q$<>"Y" THEN 380
410 PRINT
420 INPUT"POINT NUMBER OF DATA TO BE CHANGED";I
430 PRINT"INPUT NEW TIME AND AMPLITUDE"
440 INPUT T(I),AM(I)
450 PRINT:INPUT"DO YOU WISH TO CHANGE ANY MORE VALUES(Y/N)";Q$
460 IF Q$="Y" THEN 420
470 GOTO 320
480 REM
490 REM NORMALIZE DATA.
500 REM
510 N1=NN-1:N2=NN-2
520 T1=T(1):BB=AM(1):AA=AM(NN)-BB
530 FOR I =1 TO NN
540 T(I)=T(I)-T1:AM(I)=(AM(I)-BB)/AA
550 NEXT I
560 PRINT:PRINT
570 PRINT"THE STEP STARTS AT TIME";T1
580 PRINT"AND HAS A FINAL AMPLITUDE OF";AA
590 PRINT:PRINT"THIS DATA HAS BEEN NORMALIZED SO THAT THE GAIN"
600 PRINT"WILL BE 1 AT ZERO FREQUENCY."
610 FOR I=1 TO N1
```

```
620 DA(I)=(AM(I+1)-AM(I))/(T(I+1)-T(I))
630 NEXT I
640 PRINT:PRINT
650 REM
660 REM ESTABLISH FREQUENCY RANGE AND LINEAR SPACING.
670 REM
680 INPUT"INPUT REQUIRED NUMBER OF FREQUENCY POINTS";NW
690 IF NW>1 THEN 720
700 PRINT"THE NUMBER OF FREQUENCY POINTS MUST BE GREATER THAN 1"
710 GOTO 680
720 PRINT:PRINT
730 PRINT"INPUT THE MINIMUM AND MAXIMUM FREQUENCY VALUES"
740 INPUT"MINIMUM FREQUENCY =";WM
750 INPUT"MAXIMUM FREQUENCY =";WX
760 IF WM>0 AND WX>0 THEN 790
770 PRINT"ALL FREQUENCIES MUST BE POSITIVE"
780 GOTO 720
790 IF WM=WX THEN WX=10*WM
800 PRINT:PRINT
810 REM
820 REM MAIN LOOP FOR CALCULATING FREQUENCY RESPONSE.
830 REM
840 PRINT"THE REQUIRED FREQUENCY RESPONSE IS:"
850 PRINT"FREQUENCY          REAL               IMAGINARY"
860 FOR I=0 TO NW-1
870 W=WM+I*(WX-WM)/(NW-1)
880 RE=(DA(N1)*SIN(W*T(NN))-DA(1)*SIN(W*T(1)))/W
890 IM=(DA(N1)*COS(W*T(NN))-DA(1)*COS(W*T(1)))/W
900 FOR J=1 TO N2
910 J1=J+1
920 WH=W*T(J1)
930 DF=(DA(J)-DA(J1))/W
940 DR=DF*SIN(WH):DI=DF*COS(WH)
950 RE=RE+DR:IM=IM+DI
960 NEXT J
980 PRINT W,RE,IM
990 NEXT I
1000 REM
1010 REM ADJUST FREQUENCY RANGE IF REQUIRED.
1020 REM
1030 PRINT:PRINT"DO YOU WISH TO CHANGE THE FREQUENCY RANGE(Y/N)"
1040 INPUT Q$
1050 IF Q$="Y" THEN 640
1060 IF Q$<>"N" THEN 1030
1070 END
```

Example 11.8

Step response data for a plant having an open-loop transfer function given by

$$G(s) = \frac{2}{(s+1)(s+2)}$$

was generated using Program 2B. Twenty time and amplitude points were then selected. This number of points is not enough according to the above rule of thumb, but is sufficient to demonstrate the program. For comparison purposes the frequency response of the system was calculated directly using Program 7A.

Output from Program 8

```
THIS ROUTINE WILL CONVERT A STEP RESPONSE
INTO AN EQUIVALENT FREQUENCY RESPONSE.


THE STEP RESPONSE DATA MUST BE INPUT IN
THE FORM OF TIME AND AMPLITUDE COORDINATES.


IT IS ASSUMED THAT ANY TIME DELAY IS REMOVED,
AND THAT SUFFICIENT DATA POINTS ARE AVAILABLE
TO ENSURE THAT STRAIGHT LINE CONNECTION OF THE
POINTS PRODUCES A GRAPH VISUALLY IDENTICAL TO
THE ORIGINAL RESPONSE.


INPUT THE NUMBER OF DATA POINTS (LESS THAN 101)
? 20

AT EACH SAMPLE POINT INPUT:

                 TIME, AMPLITUDE
FOR POINT 1   ?  0,    0
FOR POINT 2   ?  .1,   .0090558
FOR POINT 3   ?  .2,   .032858
FOR POINT 4   ?  .3,   .067175
FOR POINT 5   ?  .4,   .10869
FOR POINT 6   ?  .5,   .15482
FOR POINT 7   ?  .6,   .20357
FOR POINT 8   ?  .8,   .30324
FOR POINT 9   ?  1,    .39958
FOR POINT 10  ?  1.2,  .48833
FOR POINT 11  ?  1.4,  .56762
FOR POINT 12  ?  1.6,  .63697
FOR POINT 13  ?  1.8,  .69673
FOR POINT 14  ?  2,    .74765
FOR POINT 15  ?  2.2,  .79067
FOR POINT 16  ?  2.5,  .84257
FOR POINT 17  ?  2.8,  .88028
FOR POINT 18  ?  3.2,  .92014
FOR POINT 19  ?  3.9,  .95993
FOR POINT 20  ?  10,   1


PLEASE CHECK THE DATA
POINT        TIME              AMPLITUDE
1            0                 0
2            .1                .0090558
3            .2                .032858
4            .3                .067175
5            .4                .10869
6            .5                .15482
7            .6                .20357
8            .8                .30324
9            1                 .39958
10           1.2               .48833
11           1.4               .56762
12           1.6               .63697
13           1.8               .69673
14           2                 .74765
```

15	2.2	.79067
16	2.5	.84257
17	2.8	.88028
18	3.2	.92014
19	3.9	.95993
20	10	1

DO YOU WISH TO CHANGE ANY VALUE (Y/N) ? N

THE STEP STARTS AT TIME 0
AND HAS A FINAL AMPLITUDE OF 1

THIS DATA HAS BEEN NORMALIZED SO THAT THE GAIN
WILL BE 1 AT ZERO FREQUENCY.

INPUT REQUIRED NUMBER OF FREQUENCY POINTS ? 10

INPUT THE MINIMUM AND MAXIMUM FREQUENCY VALUES
MINIMUM FREQUENCY = ? .1
MAXIMUM FREQUENCY = ? 3

THE REQUIRED FREQUENCY RESPONSE IS:

FREQUENCY	REAL	IMAGINARY
.1	.9777024	-.1550075
.4222223	.7262322	-.4884481
.7444445	.4417654	-.6300301
1.066667	.148863	-.6042888
1.388889	-4.282691E-03	-.477325
1.711111	-7.338117E-02	-.362238
2.033334	-8.614131E-02	-.2871235
2.355556	-.1104824	-.2353289
2.677778	-.1187411	-.1818266
3	-.1173895	-.1304118

DO YOU WISH TO CHANGE THE FREQUENCY RANGE(Y/N)
? N

For comparison purposes, frequency data was obtained directly from the transfer function using Program 7. This data is given below.

OPEN LOOP FREQUENCY DATA

FREQUENCY RAD/TIME	REAL	IMAG	GAIN (DB)	PHASE (DEG)
.1	.983	-.148	-.054	-8.6
.422	.74	-.515	-.902	-34.8
.744	.409	-.631	-2.479	-57.1
1.067	.157	-.583	-4.387	-74.9
1.389	.008	-.48	-6.376	-89
1.711	-6.8E-02	-.377	-8.327	-100.2
2.033	-.102	-.292	-10.188	-109.3
2.356	-.114	-.226	-11.94	-116.7
2.678	-.113	-.176	-13.583	-122.8
3	-.108	-.138	-15.119	-127.9

11.11   Program 9: Transfer function identification

All the design techniques described in the text are based on a Laplace transform model of the system to be controlled. Such models may be obtained by analysis (appropriate mathematical expressions are applied to obtain a description of the dynamics of the system) or by identification (an experiment is carried out on the system, and the response examined in order to determine a model which best describes the observed dynamics). This program uses Levy's method[29] to identify a linear, low-order, time-invariant transfer function model which describes a system's dynamics for small departues from some steady-state operating point. Other methods of transfer function identification are available,[2,30,31] using both graphical and computer aided approaches.

Levy's method is a complex curve fitting technique designed for computer implementation. It uses a modified least-squares approach to fit the frequency response curve of a system to the linear expression

$$G(s) = \frac{b_0 + \sum\limits_{i=1}^{q} b_i s^i}{1 + \sum\limits_{i=1}^{r} a_i s^i}, \quad r \geqslant q \tag{11.41}$$

Replacing the operator s by $j\omega$ and separating into real and imaginary parts gives

$$G(j\omega) = \frac{\alpha + j\omega\beta}{\sigma + j\omega\tau} = \frac{N(j\omega)}{D(j\omega)} \tag{11.42}$$

It is assumed that a function $H(j\omega)$ exists which coincides exactly with the measured data; $H(j\omega)$ will then also have real and imaginary parts:

$$H(j\omega) = R(\omega) + jI(\omega) \tag{11.43}$$

At any particular value of the frequency, ω_k, the error in fitting becomes

$$\varepsilon(\omega_k) = H(j\omega_k) - G(j\omega_k) = H(j\omega_k) - \frac{N(j\omega_k)}{D(j\omega_k)} \tag{11.44}$$

The problem, then, is to minimize this error at each sampling point on the curve. The minimization could be done quite simply by summing the magnitude of the error squared, $\sum |\varepsilon(\omega_k)|^2$, and setting the partial derivatives with respect to each of the coefficients equal to zero. This corresponds to a least-squares fit, resulting in a set of linear, simultaneous algebraic equations which, in principle, could be solved for the desired coefficients of $G(j\omega)$. However, this approach can result in the optimum approximation of the transfer function having non-minimum phase poles or zeros.

Levy modified Equation (11.44) by multiplying throughout by $D(j\omega_k)$, to give

$$D(j\omega_k)\varepsilon(\omega_k) = H(j\omega_k)D(j\omega_k) - N(j\omega_k) \tag{11.45}$$

This modification is valid provided $D(j\omega_k)$ is a non-zero function. This means that the denominator of the overall process transfer function cannot contain any pure integrating elements.

Levy's basic method operates directly on Equation (11.45), although other authors have further modified this equation in order to obtain better results for noise or high-frequency data. The effect of all these modifications may be embodied in an equation of the form

$$\varepsilon'(\omega_k) = \frac{\varepsilon(\omega_k)D(j\omega_k)_L}{|D(j\omega_k)_{L-1}|^n} = \frac{H(j\omega_k)D(j\omega_k)_L - N(j\omega_k)_L}{|D(j\omega_k)_{L-1}|^n} \tag{11.46}$$

where the subscript L corresponds to the iteration number, since it has been shown[32] that an iterative procedure will increase the final accuracy.

The case where $n = 0$ corresponds to Levy's original, unmodified expression. Sanathanan and Koerner[33] proposed using $n = 1$, whilst 't Mannetje[34] developed the general form given above. The program used in this section sets $n = 1$, and thus corresponds to Sanathanan and Koerner's proposal. For the first iteration, the denominator term is set equal to unity for each frequency. This means that for the first iteration the results produced are identical to those of Levy's original procedure, Equation (11.45). On the second and subsequent iterations the denominator term in Equation (11.46) acts as a scalar weighting function associated with each frequency.

If Equation (11.46) is separated into real and imaginary parts, the Lth iteration will yield an equation of the form

$$\varepsilon'(\omega_k) = A(\omega_k)_L + jB(\omega_k)_L \tag{11.47}$$

The magnitude of the error is then

$$|\varepsilon'(\omega_k)| = \sqrt{(A^2(\omega_k)_L + jB^2(\omega_k)_L)} \tag{11.48}$$

Squaring and summing over the sampled frequencies produces a weighted error function

$$E_L = \sum_{k=0}^{N} |\varepsilon'(\omega_k)|^2 = \sum_{k=0}^{N} (A^2(\omega_k)_L + jB^2(\omega_k)_L) \tag{11.49}$$

On substituting Equations (11.42), (11.43), (11.46) and (11.47) into Equation (11.49), it follows that for the Lth iteration

$$A^2(\omega_k)_L = (\sigma_k R_k - \omega_k \tau_k I - \alpha_k)^2 D_{2n}(k) \tag{11.50}$$

$$B^2(\omega_k)_L = (\omega_k \tau_k R_k + \sigma_k I_k - \omega_k \beta_k)^2 D_{2n}(k) \tag{11.51}$$

where

$$D_{2n}(k) = |D(j\omega_k)_{L-1}|^{-2n} \tag{11.52a}$$

$$\alpha_k = b_0 - b_2\omega_k^2 + b_4\omega_k^2 - \cdots \tag{11.52b}$$

$$\beta_k = b_1 - b_3\omega_k^2 + b_5\omega_k^4 - \cdots \tag{11.52c}$$

$$\sigma_k = a_0 - a_2\omega_k^2 + a_4\omega_k^4 - \cdots \tag{11.52d}$$

$$\tau_k = a_1 - a_3\omega_k^2 + a_5\omega_k^4 - \cdots \tag{11.52e}$$

The identification problem resolves itself to that of minimizing Equation (11.49). Hence the partial derivatives of Equation (11.49) with respect to each of the coefficients a_i and b_i must be set to zero. Therefore

$$\frac{\partial E}{\partial a_i} = 0, \quad i = 1, 2, 3, \ldots, r \tag{11.53}$$

$$\frac{\partial E}{\partial b_i} = 0, \quad i = 0, 1, 2, 3, \ldots, q \tag{11.54}$$

This results in a set of linear algebraic equations which may be condensed using the following formulae:

$$\Gamma_h = \sum_{k=1}^{N} \omega_k^h D_{2n}(k) \tag{11.55a}$$

$$S_h = \sum_{k=1}^{N} \omega_k^h R_k(\omega_k) D_{2n}(k) \tag{11.55b}$$

$$T_h = \sum_{k=1}^{N} \omega_k^h I_k(\omega_k) D_{2n}(k) \tag{11.55c}$$

$$U_h = \sum_{k=1}^{N} \omega_k^h (R_k^2(\omega_k) + I_k^2(\omega_k)) D_{2n}(k) \tag{11.55d}$$

Here $R_k(\omega_k)$ and $I_k(\omega_k)$ represent the real and imaginary parts of the measured function $H(j\omega)$, respectively (Equation (11.43)). The resulting set of linear equations has been derived for a maximum size of five poles and five zeros, and is given in matrix notation as

$$Px = y \tag{11.56}$$

where

$$P = \begin{bmatrix}
\Gamma_0 & 0 & -\Gamma_2 & 0 & \Gamma_4 & 0 & T_1 & S_2 & -T_3 & -S_4 & T_5 \\
0 & -\Gamma_2 & 0 & \Gamma_4 & 0 & -\Gamma_6 & S_2 & -T_3 & -S_4 & T_5 & S_6 \\
-\Gamma_2 & 0 & \Gamma_4 & 0 & -\Gamma_6 & 0 & -T_3 & -S_4 & T_5 & S_6 & -T_7 \\
0 & \Gamma_4 & 0 & -\Gamma_6 & 0 & \Gamma_8 & -S_4 & T_5 & S_6 & -T_7 & -S_8 \\
\Gamma_4 & 0 & -\Gamma_6 & 0 & \Gamma_8 & 0 & T_5 & S_6 & -T_7 & -S_8 & T_9 \\
0 & -\Gamma_6 & 0 & \Gamma_8 & 0 & -\Gamma_{10} & S_6 & -T_7 & -S_8 & T_9 & S_{10} \\
T_1 & -S_2 & -T_3 & S_4 & T_5 & -S_6 & U_2 & 0 & -U_4 & 0 & U_6 \\
-S_2 & -T_3 & S_4 & T_5 & -S_6 & -T_7 & 0 & -U_4 & 0 & U_6 & 0 \\
-T_3 & S_4 & T_5 & -S_6 & -T_7 & S_8 & -U_4 & 0 & U_6 & 0 & -U_8 \\
S_4 & T_5 & -S_6 & -T_7 & S_8 & T_9 & 0 & U_6 & 0 & -U_8 & 0 \\
T_5 & -S_6 & -T_7 & S_8 & T_9 & -S_{10} & U_6 & 0 & -U_8 & 0 & U_{10}
\end{bmatrix} \tag{11.57}$$

and

$$x = [b_0, b_1, b_2, b_3, b_4, b_5, a_1, a_2, a_3, a_4, a_5]^T \tag{11.58a}$$

$$y = [S_0, -T_1, -S_2, T_3, S_4, -T_5, 0, -U_2, 0, U_4, 0]^T \tag{11.58b}$$

The numerical values of the unknown transfer function coefficients $b_0, b_1, \ldots$ and $a_1, a_2, \ldots$ are determined by matrix inversion once the coefficients of P and y have been evaluated.

In developing a computer program, it is useful to note that many of the elements in the Matrix P and the vector y are repeated. A skeletal matrix may be produced, together with the coefficient S_0, from which all the other coefficients may be derived. For a transfer function with five poles and five zeros, the skeletal matrix takes the form

$$
\begin{bmatrix}
\Gamma_0 & 0 & -\Gamma_2 & 0 & \Gamma_4 & 0 & T_1 & S_2 & -T_3 & -S_4 & T_5 \\
 & & & & & -\Gamma_6 & & & & & S_6 \\
 & & & & & 0 & & & & & -T_7 \\
 & & & & & \Gamma_8 & & & & & -S_8 \\
 & & & & & 0 & & & & & T_9 \\
 & & & & & -\Gamma_{10} & & & & & S_{10} \\
 & & & & & U_2 & 0 & -U_4 & 0 & U_6 \\
 & & & & & & & & & & 0 \\
 & & & & & & & & & & -U_8 \\
 & & & & & & & & & & 0 \\
 & & & & & & & & & & U_{10}
\end{bmatrix}
\tag{11.59}
$$

11.11.1   Listing of Program 9

```
10 REM *********** PROGRAM 9 ********************
20 REM THIS PROGRAM FINDS A TRANSFER FUNCTION FROM
30 REM FREQUENCY RESPONSE DATA USING A COMPLEX CURVE
40 REM FITTING TECHNIQUE PROPOSED BY E.C. LEVY.
50 REM THE MODIFICATIONS PROPOSED BY SANATHANAN AND
60 REM KOERNER TO REMOVE HIGH FREQUENCY BIAS HAVE BEEN
70 REM INCLUDED. THE METHOD WILL NOT DEAL WITH POLES
80 REM AT ZERO.
90 REM
100 REM
110 DIM P(11,11),AB(11),D2(30)
120 DIM FW(30),FR(30),FI(30),CP(11)
130 DIM IP(11),IQ(11),B(11),C(11)
140 PRINT:PRINT
150 REM
160 REM SPECIFY THE NUMERATOR AND DENOMINATOR ORDER
170 REM
180 PRINT"INPUT ORDER OF DENOMINATOR (LESS THAN 6)"
```

```
190 INPUT M:IF M>5 THEN 180
200 PRINT"INPUT ORDER OF NUMERATOR (LESS THAN 6)"
210 INPUT NN:N=NN+1:IF M>=NN THEN 240
220 PRINT"THE ORDER OF THE NUMERATOR MUST NOT BE GREATER
230 PRINT"THAN THE ORDER OF THE DENOMINATOR":PRINT:GOTO 180
240 MN=M+N:PRINT:PRINT
250 FOR I=1 TO MN
260 FOR J=1 TO MN
270 P(I,J)=0
280 NEXT J
290 CP(I)=0
300 NEXT I
310 REM
320 REM INPUT THE FREQUENCY DATA.
330 REM
340 PRINT"THE FREQUENCY RESPONSE DATA MUST BE INPUT IN"
350 PRINT"THE FORM OF A FORCING FREQUENCY, TOGETHER WITH"
360 PRINT"THE RESULTING RESPONSE IN TERMS OF A REAL AND"
370 PRINT"AN IMAGINARY PART"
380 PRINT:PRINT
390 PRINT"INPUT THE NUMBER OF FREQUENCY POINTS (LESS THAN 31)
400 INPUT NW
410 IF NW>30 OR NW<=0 THEN 380
420 PRINT:PRINT
430 PRINT"AT EACH FREQUENCY POINT INPUT:"
440 PRINT
450 PRINT"                FREQUENCY,REAL,IMAGINARY"
460 FOR I =1 TO NW
470 PRINT"FOR POINT";I;:INPUT FW(I),FR(I),FI(I)
480 D2(I)=1
490 NEXT I
500 PRINT:PRINT
510 PRINT"PLEASE CHECK THE DATA"
520 PRINT"POINT           FREQUENCY        REAL           IMAGINARY"
530 FOR I=1 TO NW
540 PRINT I,FW(I),FR(I),FI(I)
550 NEXT I
560 REM
570 REM CHECK THE FREQUENCY DATA.
580 REM
590 PRINT:INPUT"DO YOU WISH TO CHANGE ANY VALUE (Y/N)";Q$
600 IF Q$="N" THEN 710
610 IF Q$<>"Y" THEN 590
620 INPUT"POINT NUMBER OF DATA TO BE CHANGED";I
630 PRINT"INPUT NEW FREQUENCY,REAL,AND IMAGINARY"
640 INPUT FW(I),FR(I),FI(I)
650 PRINT:INPUT"DO YOU WISH TO CHANGE ANY MORE VALUES (Y/N)";Q$
660 IF Q$="Y" THEN 620
670 GOTO 500
680 REM
690 REM SET THE ARRAY SIZES AND CONSTANTS.
700 REM 710 NP=2*NN+1:N1=N+1:N2=N+2:N3=N+3
720 MP=MN+NN:MQ=MN+M-1:ZM=MN-1
730 REM
740 REM THIS IS THE MAIN LOOP SETTING THE SKELETON P MATRIX,
750 REM CP IS THE VECTOR PRODUCT y ASSOCIATED WITH Px, AND AB
760 REM IS THE VECTOR OF TRANSFER FUNCTION COEFFICIENTS x.
770 REM
780 FOR KI=1 TO NW
```

```
790 FH=FW(KI)*FW(KI)
800 P(1,1)=P(1,1)+D2(KI)
810 CP(1)=CP(1)+FR(KI)*D2(KI)
820 IF N=1 THEN 920
830 SH=D2(KI)
840 FOR I=3 TO NP STEP 2
850 IF I>N THEN 890
860 SH=-SH*FH
870 P(1,I)=P(1,I)+SH
880 GOTO 910
890 II=I-N+1:SH=SH*FH
900 P(II,N)=P(II,N)+SH
910 NEXT I
920 T1=D2(KI)*FW(KI)*FI(KI)
930 P(1,N1)=P(1,N1)+T1
940 IF M=1 THEN 1130
950 S2=-D2(KI)*FR(KI)
960 FOR J=N2 TO MP STEP 2
970 IF J>MN THEN 1050
980 S2=-S2*FH
990 P(1,J)=P(1,J)+S2
1000 JJ=J+1
1010 IF JJ>MN THEN 1080
1020 T1=-T1*FH
1030 P(1,JJ)=P(1,JJ)+T1
1040 GOTO 1120
1050 I=J-MN+1:S2=S2*FH:JJ=J+1
1060 IF I>N THEN 1120
1070 P(I,MN)=P(I,MN)+S2
1080 I=JJ+1-MN
1090 IF I>N THEN 1120
1100 T1=T1*FH
1110 P(I,MN)=P(I,MN)+T1
1120 NEXT J
1130 FX=(FI(KI)*FI(KI)+FR(KI)*FR(KI))*D2(KI)*FH
1140 P(N1,N1)=P(N1,N1)+FX
1150 IF M=1 THEN 1240
1160 FOR I= N3 TO MQ STEP 2
1170 IF I>MN THEN 1210
1180 FX=-FX*FH
1190 P(N1,I)=P(N1,I)+FX
1200 GOTO 1230
1210 FX=FX*FH:J=I-MN+1+N
1220 P(J,MN)=P(J,MN)+FX
1230 NEXT I
1240 NEXT KI
1250 REM
1260 REM THIS SECTION OF THE PROGRAM FILLS IN THE P MATRIX
1270 REM
1280 IF N<3 THEN 1370
1290 MH=-1:H=1
1300 FOR J=2 TO N
1310 H=H*MH:J1=J-1
1320 FOR L=1 TO NN
1330 K=L+1
1340 P(J,L)=P(J1,K)*H
1350 NEXT L
1360 NEXT J
1370 IF N=1 THEN 1460
```

```
1380 MH=-1:H=1
1390 FOR J=2 TO N
1400 H=H*MH:J1=J-1
1410 FOR L=N1 TO ZM
1420 K=L+1
1430 P(J,L)=P(J1,K)*H
1440 NEXT L
1450 NEXT J
1460 IF M<3 THEN 1550
1470 MH=-1:H=1
1480 FOR J=N2 TO MN
1490 H=H*MH:J1=J-1
1500 FOR L=N1 TO ZM
1510 K=L+1
1520 P(J,L)=P(J1,K)*H
1530 NEXT L
1540 NEXT J
1550 X1=1:X2=-1:II=0
1560 FOR I=1 TO N
1570 II=II+1
1580 IF II>=3 THEN II=1:X1=X1*X2
1590 JJ=0:Y1=1
1600 FOR J=N1 TO MN
1610 JJ=JJ+1
1620 IF JJ>=3 THEN JJ=1:Y1=Y1*X2
1630 P(J,I)=P(I,J)*X1*Y1
1640 NEXT J
1650 NEXT I
1660 REM
1670 REM VECTOR CP IS FOUND FROM APPROPRIATE ELEMENTS WITHIN
1680 REM THE P MATRIX.
1690 REM
1700 IF N=1 THEN 1770
1710 X1=1
1720 FOR I=1 TO NN
1730 II=I+1
1740 CP(II)=P(I,N1)*X1
1750 X1=-X1
1760 NEXT I
1770 IF M=1 THEN 1830
1780 FOR I=N2 TO MN STEP 2
1790 J=I-1
1800 CP(I)=P(J,N1)
1810 NEXT I
1820 REM
1830 REM INVERT P
1840 REM
1850 GOSUB 2410
1860 REM
1870 REM FIND AB: AB=INVERSE(P)*CP
1880 REM
1890 FOR I= 1 TO MN
1900 AB(I)=0
1910 FOR J= 1 TO MN
1920 AB(I)=AB(I)+P(I,J)*CP(J)
1930 NEXT J
1940 NEXT I
1950 REM
1960 REM PRINT OUT THE TRANSFER FUNCTION COEFFICIENTS.
```

```
1970 REM
1980 PRINT:PRINT
1990 PRINT"THE TRANSFER FUNCTION IS OF THE FORM:"
2000 PRINT:PRINT"A(0)+A(1)*S+A(2)*S^2+....."
2010 PRINT"_____"
2020 PRINT"B(0)+B(1)*S+B(2)*S^2+....."
2030 PRINT:PRINT"WHERE:"
2040 PRINT
2050 FOR I=1 TO N
2060 PRINT"A(";I-1;")= ";AB(I)
2070 NEXT I
2080 PRINT:PRINT"B( 0 )= 1"
2090 FOR I=N+1 TO MN
2100 PRINT"B(";I-N;")= ";AB(I)
2110 NEXT I
2120 REM
2130 REM REMOVE BIAS IF REQUIRED.
2140 REM
2150 PRINT:PRINT
2160 INPUT"REFINE THE ESTIMATE (Y/N)";Q$
2170 IF Q$="N" THEN 2370
2180 IF Q$<>"Y" THEN 2160
2190 REM
2200 REM SET D2 AND RE-ENTER DATA
2210 REM
2220 FOR J=1 TO NW
2230 W=FW(J):RE=FR(J):IM=FI(J)
2240 SR=0:SI=W:DR=1:DI=0
2250 FOR I= N+1 TO MN
2260 DR=DR+AB(I)*SR
2270 DI=DI+AB(I)*SI
2280 HS=SR:SR=-SI*W:SI=HS*W
2290 NEXT I
2300 D2(J)=1/(DR*DR+DI*DI)
2310 NEXT J
2320 FOR I=1 TO MN
2330 FOR J=1 TO MN
2340 P(I,J)=0:NEXT J
2350 CP(I)=0:NEXT I
2360 GOTO 780
2370 END
2380 REM
2390 REM THIS SUBROUTINE FINDS THE INVERSE OF P
2400 REM
2410 FOR K=1 TO MN
2420 PVOT=0
2430 FOR I=K TO MN
2440 FOR J=K TO MN
2450 IF (ABS(P(I,J))-ABS(PVOT)<=0) THEN 2470
2460 PVOT=P(I,J):IP(K)=I:IQ(K)=J
2470 NEXT J
2480 NEXT I
2490 IF (ABS(PVOT)=0) THEN 2870
2500 II=IP(K):PP=IQ(K)
2510 IF (II-K=0) THEN 2550
2520 FOR J=1 TO MN
2530 Z=P(II,J):P(II,J)=P(K,J):P(K,J)=Z
2540 NEXT J
2550 IF (PP-K=0) THEN 2590
```

```
2560 FOR I=1 TO MN
2570 Z=P(I,PP):P(I,PP)=P(I,K):P(I,K)=Z
2580 NEXT I
2590 FOR J=1 TO MN
2600 IF (J-K=0) THEN 2620
2610 GOTO 2640
2620 B(J)=1/PVOT:C(J)=1
2630 GOTO 2660
2640 B(J)=-1*P(K,J)/PVOT
2650 C(J)=P(J,K)
2660 P(K,J)=0:P(J,K)=0
2670 NEXT J
2680 FOR I= 1 TO MN
2690 FOR J= 1 TO MN
2700 P(I,J)=P(I,J)+C(I)*B(J)
2710 NEXT J
2720 NEXT I
2730 NEXT K
2740 MX=MN+1
2750 FOR L= 1 TO MN
2760 K=MX-L:II=IP(K):PP=IQ(K)
2770 IF (II-K=0) THEN 2810
2780 FOR I= 1 TO MN
2790 Z=P(I,II):P(I,II)=P(I,K):P(I,K)=Z
2800 NEXT I
2810 IF (PP-K=0) THEN 2850
2820 FOR J= 1 TO MN
2830 Z=P(PP,J):P(PP,J)=P(K,J):P(K,J)=Z
2840 NEXT J
2850 NEXT L
2860 RETURN
2870 PRINT"ERROR IN MATRIX INVERSION"
2880 RETURN
```

Example 11.9

The example used to demonstrate this program is taken from Levy's original paper,[29] and is derived for a dynamic system having the transfer function

$$F(s) = \frac{1-s}{1 + 0.1s + 0.01s^2}$$

Output from Program 9

```
INPUT ORDER OF DENOMINATOR (LESS THAN 6)
? 2
INPUT ORDER OF NUMERATOR (LESS THAN 6)
? 1


THE FREQUENCY RESPONSE DATA MUST BE INPUT IN
THE FORM OF A FORCING FREQUENCY, TOGETHER WITH
THE RESULTING RESPONSE IN TERMS OF A REAL AND
AN IMAGINARY PART
```

INPUT THE NUMBER OF FREQUENCY POINTS (LESS THAN 31)
? 13

AT EACH FREQUENCY POINT INPUT:

              FREQUENCY,REAL,IMAGINARY
FOR POINT 1 ? .1,   1, -.113
FOR POINT 2 ? .2,   1, -.22
FOR POINT 3 ? .5,   .975, -.55
FOR POINT 4 ? .7,   .963, -.78
FOR POINT 5 ? 1,   .905, -1.12
FOR POINT 6 ? 2,   .588, -2.2
FOR POINT 7 ? 4,   -.925, -4.34
FOR POINT 8 ? 7,   -5.87, -5.69
FOR POINT 9 ? 10,   -10, -1.05
FOR POINT 10 ? 20, -3.31,   4.46
FOR POINT 11 ? 40, -.724,   2.44
FOR POINT 12 ? 70, -.227,   1.43
FOR POINT 13 ? 100, -.113,   .993

PLEASE CHECK THE DATA

POINT	FREQUENCY	REAL	IMAGINARY
1	.1	1	-.113
2	.2	1	-.22
3	.5	.975	-.55
4	.7	.963	-.78
5	1	.905	-1.12
6	2	.588	-2.2
7	4	-.925	-4.34
8	7	-5.87	-5.69
9	10	-10	-1.05
10	20	-3.31	4.46
11	40	-.724	2.44
12	70	-.227	1.43
13	100	-.113	.993

DO YOU WISH TO CHANGE ANY VALUE (Y/N) ? N

THE TRANSFER FUNCTION IS OF THE FORM:

$$A(0)+A(1)*S+A(2)*S^2+.....$$
$$\overline{}$$
$$B(0)+B(1)*S+B(2)*S^2+.....$$

WHERE:

A(0)= 1.01213
A(1)= -.995576

B(0)= 1
B(1)= .100207
B(2)= 9.990139E-03

REFINE THE ESTIMATE (Y/N) ? Y

THE TRANSFER FUNCTION IS OF THE FORM:

A(0)+A(1)*S+A(2)*S^2+.....
──────────────────────────
B(0)+B(1)*S+B(2)*S^2+.....

WHERE:

A(0)= 1.000726
A(1)= -1.001689

B(0)= 1
B(1)= .100186
B(2)= 9.980949E-03

REFINE THE ESTIMATE (Y/N) ? Y

THE TRANSFER FUNCTION IS OF THE FORM:

A(0)+A(1)*S+A(2)*S^2+.....
──────────────────────────
B(0)+B(1)*S+B(2)*S^2+.....

WHERE:

A(0)= 1.000725
A(1)= -1.001683

B(0)= 1
B(1)= .1001854
B(2)= 9.980961E-03

REFINE THE ESTIMATE (Y/N) ? N

REFERENCES

1. Chai A S 1974 Modified Merson's integration algorithm which saves two evaluations at each step. *Simulation* **22**(3): 90–2
2. Chen C, Haas I J 1968 *Elements of Control Systems Analysis: Classical and Modern Approaches*. Prentice-Hall, Englewood Cliffs, NJ
3. Vernon J B 1967 *Linear Vibrations and Control Systems Theory, with Computer Applications*. Wiley, New York
4. Watkins B O 1971 A partial fraction algorithm. *IEEE Transactions on Automatic Control* **AC-16**: 489–91
5. Pottle C 1964 On the partial fraction algorithm. *IEEE Transactions on Automatic poles by digital computer. IEEE Transactions on Circuit Theory* **CT-11**: 161–2
6. N-Nagy F L, Al-Tikriti M N 1972 General matrix formulation for dynamic simulation of linear time-invariant systems. *Proceedings of the Institute of Electrical Engineers* **119**(4): 493–6
7. Hovanessian S A, Pipes L A 1969 *Digital Computer Methods in Engineering*. McGraw-Hill, New York
8. Lin S N 1941 Method of successive approximations of evaluating the real and complex roots of cubic and higher order equations. *Journal of Mathematics and Physics* **20**(3): 231–42
9. Bairstow L 1914 Investigations relating to the stability of the aeroplane. *Reports and Memoranda, Advisory Committee for Aeronautics* (154)
10. Melsa J L 1970 *Computer Programs for Computational Assistance in the Study of Linear Control Theory*. McGraw-Hill, New York
11. Borrie A J 1986 *Modern Control Systems: A Manual of Design Methods*. Prentice-Hall International, Hemel Hempstead
12. Schwarz R J, Friedland B 1965 *Linear Systems*. McGraw-Hill, New York (Electrical and Electronic Engineering Series)
13. Shamash Y 1980 Comments on the Routh–Hurwitz criterion. *IEEE Transactions on Automatic Control* **AC-25**(2): 132–3
14. Yeung K S 1983 Routh–Hurwitz test under vanishing leading array elements. *IEEE Transactions on Automatic Control* **AC-28**(1) 104–6
15. Truxal J G 1955 *Automatic Control Systems Synthesis*. McGraw-Hill, New York
16. Krishnan V 1966 Semi-analytic approach to root locus. *IEEE Transactions on Automatic Control* **AC-11**(1): 102–8
17. Power H M, Simpson R J 1978 *Introduction to Dynamics and Control*. McGraw-Hill UK, Maidenhead

18. Harden R C, Simons F O Jr 1980 Root-locus algorithms and routines adapted to hand-held HP-67 computers. In Proceedings of the 12th IEEE Southeastern Conference, Virginia Beach, VA, pp 285–9

19. Mee D H 1970 A computer graphics demonstration for teaching root locus. *International Journal of Electrical Engineering Education* **13**: 41–7

20. Chen C F, Hsu C 1966 The determination of root loci using Routh's algorithm. *Journal of the Franklin Institute* **281**(2): 114–21

21. Hennci P 1974 *Applied and Computational Analysis.* Wiley, New York, vol 1

22. Ash R H, Ash G R 1968 Numerical computation of root loci using the Newton–Raphson technique. *IEEE Transactions on Automatic Control* **AC-13**(5): 576–82

23. Pan T C, Chao K S 1978 A computer-aided root-locus method. *IEEE Transactions on Automatic Control* **AC-23**(5): 856–60

24. Wang L C, Han K W, Thaler G J 1978 A new algorithm for plotting root loci (with applications). In Proceedings of the IEEE Conference on Circuits, Systems, and Computers, Asilomar, CA, pp 110–16

25. N-Nagy F L, Al-Tikriti M N 1972 Root-locus plotting by digital computer using operational-array technique. *Proceedings of the Institute of Electrical Engineers* **119**: 243–6

26. Richard S, Sims F, Naim A 1979 An interactive computer-aided design program using Bode, Nyquist and root locus. In Proceedings of the 11th IEEE Southeastern Conference, pp 311–13

27. Thomas R J 1976 An efficient root-locus program for feedback control system design. *IEEE Transactions on Education* **E-19**: 76–8

28. Vernon J B 1976 *Linear Vibration Theory, Computer Applications.* Wiley, New York

29. Levy E C 1959 Complex curve fitting. *IRE Transactions on Automatic Control* **AC-4**: 37–44

30. Jones N B 1967 On fitting a rational function to a polynomial. *IEEE Transactions on Automatic Control* **AC-12**: 795–6

31. Lawrence P J, Rogers C J 1979 Sequential transfer-function synthesis from measured data. *Proceedings of the Institute of Electrical Engineers* **126**(1): 104–6

32. Kardashov A A, Kaniushin L V 1958 Determination of system parameters from experimental frequency characteristics. *Automation and Remote Control* **19**: 327–38

33. Sanathanan C K, Koerner J 1963 Transfer function synthesis as a ratio of two complex polynomials. *IEEE Transactions on Automatic Control* **AC-8**: 56–58

34. 't Mannetje J J 1973 Transfer function identification using a complex curve-fitting technique. *Journal of Mechanical Engineering Science* **15**(5): 339–45

ANSWERS TO PROBLEMS

1.1.   The disturbance inputs are the hot oil temperature and flow rate, and the inlet cooling water temperature. The manipulable input is the inlet cooling water volumetric flow rate.

   The measurable outputs could be any of the measured quantities on the pipework or heat exchanger.

   With one manipulable input one output may be controlled. (It is possible to control several outputs with a single input using state space techniques, but this book is concerned only with single input/single output control methods.) The controlled output could be the temperature of the cool oil, the temperature of the water outlet or the flow rate of the water outlet.

   The simplest control scheme is to feedback the outlet oil temperature measurement and compare it with a desired temperature setting. This error signal is adjusted by means of a forward path compensator, the output of which is used to drive a flow control valve on the cooling water inlet pipe. Such a scheme is shown in Fig. 1.3.

1.2.   The control scheme required is shown in Fig. A.1.

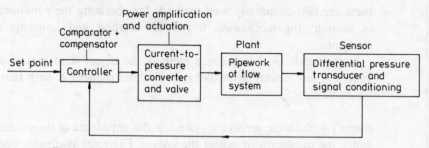

Figure A.1

1.3.   There is no unique solution, the degree of complexity of the resulting model depending on the assumptions made and the information required. A typical model for vertical motion may take the form shown in Fig. A.2.

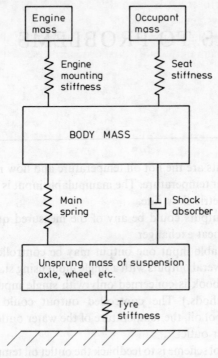

Figure A.2

1.4.   None of the mechanical compensators shown in Fig. 1.23 contains a mass, so Newton's second law is inappropriate. For this type of system there are two commonly used methods for obtaining the equations of motion: the mechanical impedance method and Lagrange's equations.

In the mechanical impedance method, the object is to replace a system of mechanical elements by an idealized element Z, such that

$$f = Zx$$

where f is the force across the system, Z the impedance of the system and x the displacement across the system. There are two rules, one for mechanical elements in series (for example, the two springs and the dashpot in the lag compensator) and one for mechanical elements in parallel (for example the spring k_1 and dashpot b_1 in the lag–lead compensator).

Mechanical series laws

From Newton's third law (to every action there is an equal and opposite reaction), the force across each element equals the total force applied to the series system.

From the geometric constraints, the total displacement of the series system is equal to the sum of the individual displacements across each element.

Mechanical parallel laws

Again from Newton's third law, the sum of the forces across each element must equal the total force applied to the parallel system.

From the geometric constraints, the displacement of the parallel system is equal to the displacement of each element.

For the three mechanical systems shown in Fig. 1.23 the equations of motion are derived below from Lagrange's equation. The mechanical impedence method is demonstrated for the lag and lag–lead compensators.

(a) Lag compensator

Let the displacement between dashpot b and spring k_2 be $x(t)$. The potential energy and power function terms for this system are, respectively,

$$V = \frac{1}{2}k_1(u - y)^2 + \frac{1}{2}k_2x^2$$

$$P = \frac{1}{2}b\left(\frac{dy}{dt} - \frac{dx}{dt}\right)^2$$

Applying Lagrange's equation for the coordinates y and x respectively gives

$$-k_1(u - y) + b\left(\frac{dy}{dt} - \frac{dx}{dt}\right) = 0$$

$$k_2x - b\left(\frac{dy}{dt} - \frac{dx}{dt}\right) = 0$$

These are two simultaneous linear differential equations, and may therefore be used to eliminate $x(t)$ and hence establish the relationship between $y(t)$ and $u(t)$.

Using the mechanical impedance method, the first series law

indicates that

$$f = f_{\text{spring } k_1} = f_{\text{dashpot}} = f_{\text{spring } k_2}$$

and the second series law that

$$u = (u - y) + (y - x) + x$$

Now,

$$f_{\text{spring } k_1} = k_1(u - y)$$

$$f_{\text{dashpot}} = b\left(\frac{dy}{dt} - \frac{dx}{dt}\right)$$

$$f_{\text{spring } k_2} = k_2 x$$

Solving the above three equations for $u - y$, $y - x$ and x, and substituting the results into the second series law will, after some manipulation, produce an equation of the form

$$f = Zu$$

The required relationship between $y(t)$ and $u(t)$ is obtained by noting that

$$f = k_1(u - y)$$

At this point the reader may wonder how the dashpot equation is solved to give the relationship between f and $y - x$. Fortunately, such problems are easily dealt with using Laplace transforms, which are described in Chapter 2.

(b)   Lead compensator

$$V = \frac{1}{2}k_1(u - y)^2 + \frac{1}{2}k_2 y^2$$

$$P = \frac{1}{2}b\left(\frac{du}{dt} - \frac{dy}{dt}\right)^2$$

Applying Lagrange's equation for the coordinate $y(t)$ produces

$$-k_1(u - y) + k_2 y - b\left(\frac{du}{dt} - \frac{dy}{dt}\right) = 0$$

which is the required solution.

(c)   Lag–lead compensator

Let $x(t)$ be the displacement between the dashpot b_2 and the spring

k_2. Then

$$V = \frac{1}{2}k_1(u-y)^2 + \frac{1}{2}k_2 x^2$$

$$P = \frac{1}{2}b_1\left(\frac{du}{dt} - \frac{dy}{dt}\right)^2 + \frac{1}{2}b_2\left(\frac{dy}{dt} - \frac{dx}{dt}\right)^2$$

Applying Lagrange's equations for the coordinates $y(t)$ and $x(t)$ gives

$$-k_1(u-y) - b_1\left(\frac{du}{dt} - \frac{dy}{dt}\right) + b_2\left(\frac{dy}{dt} - \frac{dx}{dt}\right) = 0$$

and

$$k_2 x - b_2\left(\frac{dy}{dt} - \frac{dx}{dt}\right) = 0$$

which may be solved to produce the required equations of motion.

Using the mechanical impedence method, first consider the parallel elements b_1 and k_1. From the first parallel mechanical law

$$f = f_{\text{dashpot } b_1} + f_{\text{spring } k_1}$$

and from the second parallel law

$$(u-y)_{\text{system}} = (u-y)_{\text{dashpot } b_1} = (u-y)_{\text{spring } k_1}$$

From the force equation and the displacement relationship, the impedence Z_1 of the parallel elements may be found such that

$$f = Z_1(u-y)$$

The method of solution is now identical to that for the lag compensator.

1.5.   All three parts may be solved using either Kirchhoff's laws or Lagrange's equation. (Note that the mechanical impedence method, described in the solution to Problem 1.4, is essentially the application of Kirchhoff's laws to a mechanical system.)

With a series electrical circuit the total voltage drop is the sum of the individual voltage drops across each element, and the same current flows through each element. With a parallel electrical circuit the voltage drop across each element is the same, and the total current flowing into the circuit is the sum of the currents flowing through each element.

As in Problem 1.4, solutions are obtained for each compensator using Lagrange's equation. In addition, the lead and lag compensators are solved using Kirchhoff's voltage law.

(a)   *Lag compensator*

$$V = \frac{1}{2C} q^2$$

$$P = \frac{1}{2} R_1 \left(\frac{dq}{dt}\right)^2 + \frac{1}{2} R_2 \left(\frac{dq}{dt}\right)^2$$

and the external force for the coordinate q is the input voltage $v_i(t)$. From Lagrange's equation,

$$\frac{1}{C} q + R_1 \frac{dq}{dt} + R_2 \frac{dq}{dt} = v_i(t)$$

To obtain the required solution, note that from Equation (1.12)

$$\frac{dq}{dt} = i$$

Also, the current flow through the resistor R_1 is given by

$$i = \frac{1}{R_1}(v_i - v_o)$$

The alternative solution is obtained by applying Kirchhoff's voltage law. Let the voltage drop across the capacitor be v, then

$$v_i = (v_i - v_o) + (v_o - v) + v$$

The equations for each element are

$$R_1 i = (v_i - v_o)$$

$$R_2 i = (v_o - v)$$

$$\frac{1}{C} \int i\, dt = v$$

These equations are used to find a solution of the form

$$i = Zv_i$$

To find the desired relationship, i is replaced by

$$i = \frac{1}{R_1}(v_i - v_o)$$

(b)   *Lead compensator*

Let i_1 and q_1 be the current flow and charge, respectively, through the resistor R_1. Hence

$$V = \frac{1}{2C}(q - q_1)^2$$

$$P = \frac{1}{2}R_1\left(\frac{dq_1}{dt}\right)^2 + \frac{1}{2}R_2\left(\frac{dq}{dt}\right)^2$$

For the charge q the external force is $v_i(t)$. Using Lagrange's equation for the generalized coordinate q gives

$$\frac{1}{C}(q - q_1) + R_2\frac{dq}{dt} = v_i(t)$$

and for the coordinate q_1

$$-\frac{1}{C}(q - q_1) + R_1\frac{dq_1}{dt} = 0$$

which are the required system equations. To obtain the relationship between $v_o(t)$ and $v_i(t)$, eliminate q_1 and substitute for

$$i = \frac{dq}{dt} = \frac{v_o}{R_2}$$

Use Kirchhoff's laws, and let $i(t)$ be the circulating current. For the parallel resistor and capacitor

$$i = i_{\text{capacitor}} + i_{\text{resistor } R_1}$$

where

$$i_{\text{capacitor}} = C\frac{d}{dt}(v_i - v_o)$$

$$i_{\text{resistor } R_1} = \frac{1}{R_1}(v_i - v_o)$$

These equations may be solved to obtain a relationship between $i(t)$ and $v_i(t) - v_o(t)$. The circuit equation may be obtained from

$$v_i = (v_i - v_o) + v_o$$

by substituting the expression for the parallel elements together with the relationship

$$v_o = R_2 i$$

This produces an equation relating $i(t)$ to $v_i(t)$. The required solution is obtained by making a final substitution for $i(t)$, as given above.

(c)   *Lag–lead compensator*

Applying Lagrange's equation for coordinates q (the total charge) and q_1 (the charge through the resistor R_1), respectively, produces

$$\frac{1}{C_2}q + \frac{1}{C_1}(q - q_1) + R_2\frac{dq}{dt} = v_i(t)$$

and

$$-\frac{1}{C_1}(q - q_1) + R_1\frac{dq_1}{dt} = 0$$

By eliminating q_1 the equation relating $q(t)$ to $v_i(t)$ is obtained.

To find the required input–output relationship, let $v(t)$ be the voltage drop across capacitor C_2. Then

$$v_o = (v_o - v) + v$$

and

$$i = \frac{1}{R_2}(v_o - v) = C\frac{dv}{dt} = \frac{dq}{dt}$$

from which the relationship between $v_o(t)$ and $q(t)$ may be established, and hence the required relationship between $v_o(t)$ and $v_i(t)$.

1.6. The force–current and displacement–voltage analogies are used. These are valid only if there are no masses, or inductances, in the mechanical and electrical systems, respectively.

1.7.

$$\frac{dT_2}{dt} = \frac{1}{C_2}\frac{T_2 - T_1}{q_{12}} - \frac{T_2 - T_3}{q_{23}} - \frac{T_2 - T_a}{q_{2a}}$$

$$\frac{dT_3}{dt} = \frac{1}{C_3}\frac{T_3 - T_1}{q_{13}} + \frac{T_2 - T_3}{q_{23}} - \frac{T_3 - T_a}{q_{3a}}$$

1.8. The equivalent mechanical system is shown in Fig. A.3.

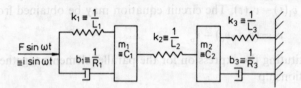

Figure A.3

1.9.

$$L_1\frac{di}{dt} + L(x)\frac{di}{dt} + i\frac{dL}{dt}(x) + Ri = e(t)$$

$$m\frac{d^2x}{dt^2} + b\frac{dx}{dt} + kx - \frac{1}{2}\frac{dL}{dx}(x)\left(\frac{di}{dt}\right)^2 = 0$$

2.1.
$$M\ddot{y}_1 + B\dot{y}_1 + Ky_1 - U\dot{y}_2 = 0$$
$$L\ddot{y}_2 + R\dot{y}_2 + U\dot{y}_1 = e(t)$$

2.2. Solutions given.

2.3.
$$y_1(t) = \frac{1}{2} - \frac{1}{2}e^{-2t}$$

$$y_2(t) = \frac{1}{4}(10t^2e^{-2t} - 10te^{-2t} + 5e^{-2t} - 5e^{-4t})$$

$$y_3(t) = \frac{1}{\sqrt{2}}e^{-t}\sin\sqrt{(2t)}$$

2.4.
$$f_1(t) = \frac{1}{75}(2 + 25e^{-3t} - 27e^{-5t} + 60te^{-5t})$$

$$f_1(0) = 0 \quad\text{and}\quad f_1(\infty) = \frac{2}{75}$$

$$f_2(t) = 0.8e^{-t} - 0.8e^{-2t}(\cos 2t + 0.4\sin 2t)$$
$$f_2(0) = 0 \quad\text{and}\quad f_2(\infty) = 0$$

2.5. Since Program 2 deals only with distinct poles and zeros, only $y_1(t)$ and $y_3(t)$ can be found in Problem 2.3, and $f_2(t)$ in Problem 2.4.

2.6.

(a) $$x(t) = \frac{1}{2} - \frac{3}{2}e^{-2t} + 2e^{-t}$$

(b) $$x(t) = 7(e^{-2t} - te^{-2t} - e^{-3t})$$

3.1. The required block diagrams are shown in Fig. A.4.

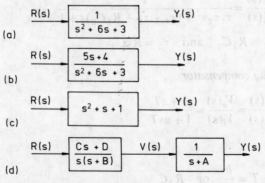

Figure A.4

3.2.
$$\frac{d^3\theta_o}{dt^3} + (0.9 + 0.7K)\frac{d^2\theta_o}{dt^2} + (1.18 + 0.42K)\frac{d\theta_o}{dt} + 0.68\theta_o$$

$$= 0.7\frac{d\theta_i}{dt} + 0.42\theta_i$$

3.3.
$$C_1(s) = -\frac{1}{4}\frac{(s+2)}{(s+1)} \quad \text{and} \quad C_2(s) = -\frac{1}{4}\frac{(s+2)}{(s+1)}$$

$$\frac{Y_1(s)}{R_1(s)} = \frac{K_1(15s^2 + 28s + 12)}{8s^3 + (32 + 15K_1)s^2 + (40 + 28K_1)s + (16 + 12K_1)}$$

$$\frac{Y_2(s)}{R_2(s)} = \frac{(3s^2 + 4s)K_2}{4s^3 + (16 + 3K_2)s^2 + (20 + 4K_2)s + 8}$$

3.4.
$$\frac{Y(s)}{R(s)} = \frac{K}{s^2 + 4s + (K - 5)}$$

3.5.   The block diagram of the network is shown in Fig. A.5.

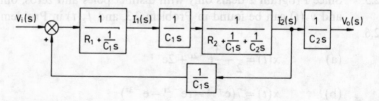

Figure A.5

$$\frac{V_o(s)}{V_i(s)} = \frac{1}{\tau_1\tau_2 s^2 + (\tau_1 + \tau_2 + R_1C_2)s + 1}$$

where $\tau_1 = R_1C_1$ and $\tau_2 = R_2C_2$

3.6.
(a)   *Lag compensator*

$$\frac{Y(s)}{U(s)} = \frac{V_o(s)}{V_i(s)} = \frac{1 + sT}{1 + \alpha sT}$$

where

$$T = \frac{b}{k_2} \quad \text{or} \quad R_2C$$

and

$$\alpha = \frac{k_1 + k_2}{k_1} \quad \text{or} \quad \frac{R_2 + R_1}{R_2}$$

(b)   *Lead compensator*

$$\frac{Y(s)}{U(s)} = \frac{V_o(s)}{V_i(s)} = \frac{\alpha(1 + sT)}{1 + \alpha sT}$$

where

$$T = \frac{b}{k_1} \quad \text{or} \quad R_1 C$$

and

$$\alpha = \frac{k_1}{k_1 + k_2} \quad \text{or} \quad \frac{R_2}{R_2 + R_1}$$

(c)   *Lag–lead compensator*

For the electrical network it is required to find

$$\frac{V_o(s)}{V_i(s)} = \frac{R_2 + 1/(sC_2)}{R_2 + 1/(sC_2) + (R_1/(sC_1))/(R_1 + 1/(sC_1))}$$

Let

$$T_1 = R_1 C_1, \qquad T_2 = R_2 C_2, \qquad \alpha = \frac{R_2}{R_1 + R_2}$$

The equation then reduces to

$$\frac{(1 + T_1 s)(1 + T_2 s)}{(1 + T_1 s)(1 + T_2 s) + sR_1 C_2}$$

Expanding the denominator of the above expression produces

$$1 + (T_1 + T_2 + R_1 C_2)s + T_1 T_2 s^2$$

This may be compared with the denominator of the ideal lag–lead compensator, see section 8.4.3, which may be expanded to give

$$1 + (\alpha T_1 + T_2/\alpha)s + T_1 T_2 s^2$$

If

$$T_1 + T_2 + R_1 C_2 \simeq \alpha T_1 + T_2/\alpha$$

then the electrical network may be considered as providing lag–lead compensation. Substituting for α in T_2/α, indicates that

$$T_1 + T_2 \simeq \alpha T_1 + T_2$$

which is valid providing $T_2 \gg T_1$ and $\alpha \ll 1$.

3.7.

$$\frac{H_2(s)}{Q_i(s)} = \frac{R_2}{(1 + sT_1)(1 + sT_2)}$$

where

$$T_1 = R_1 C_1 \quad \text{and} \quad T_2 = R_2 C_2$$

4.1.   (i)

$$F(s) = \frac{s + 3}{(s + 1)(s + 2)}$$

$$f(t) = 2e^{-2t} - e^{-3t}$$

The response starts at 1 and decays exponentially to zero. As time increases, the rate of decay approaches e^{-2t}.

(ii)

$$F(s) = \frac{39}{(s + 4)(s^2 + 2s + 5)}$$

$$f(t) = 3e^{-4t} - 3e^{-t}(\cos 2t - 1.5 \sin 2t)$$

The response starts at zero and decays, after oscillating, to zero. The damped frequency is approximately 2 rad/s (i.e. it has a period of π) and the damping ratio is approximately 0.447.

(iii)

$$F(s) = \frac{4(s + 3)}{(s + 2)(s^2 + 2s + 2)}$$

$$f(t) = 2e^{-2t} - e^{-t}(2 \cos t - 6 \sin t)$$

The response starts at zero and decays to zero. The damped frequency is approximately 1 rad/s (period 2π) and the damping ratio is approximately 0.707.

(iv)

$$F(s) = \frac{4(s^2 + 9)}{s(s + 1)(s^2 + 4)}$$

$$f(t) = 9 - 8e^{-t} - \cos 2t - 2 \sin 2t$$

The response starts at zero and then settles to a continuous oscillation around the value $f(t) = 9$.

All the above solutions should be checked using Program 2 of Chapter 11.

4.2.
$$F(0, 0) = 1, \qquad F(0, -4) = (7 + j6)/8$$

$$F(-2, 2) = (8 + j10)/20$$

4.3. To use the graphical technique, both $F_1(s)$ and $F_2(s)$ must first be transformed into the form

$$F_1(s) = \frac{10(0.1 + s)}{1 + s}$$

$$F_2(s) = \frac{10 + s}{10(1 + s)}$$

The plot of $F_1(j\omega)$ is a semicircle of radius 4.5, centred at $(5.5, 0)$. The plot of $F_2(j\omega)$ is a semicircle of radius 0.45, centred at $(0.55, 0)$. Both plots are sketched in Fig. A.6.

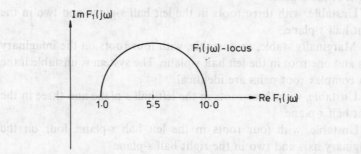

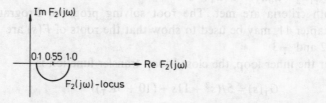

Figure A.6

4.4. (a) $\omega_d = 4.49$ rad/s, $\omega_n = 5.74$ rad/s, $\zeta = 0.62$ and $\zeta\omega_n = 3.56$. The amplitude at peak overshoot is 1.082, and the steady-state error is zero.
(b) $\omega_d = 1.91$ rad/s, $\omega_n = 1.94$ rad/s, $\zeta = 0.172$ and $\zeta\omega_n = 0.334$. The amplitude at peak overshoot is 1.577.

(c) $\omega_d = 0.870\,\text{rad/s}, \omega_n = 0.931\,\text{rad/s}, \zeta = 0.358$ and $\zeta\omega_n = 0.333$. The subsidence ratio is 11.1:1.

The response for all three cases may be produced using Program 1 or 2 of Chapter 11.

4.5. The closed-loop characteristic equation is $1 + KG(s) = 0$. Thus $G(s_0) = -1/K$, and taking the modulus and argument of both sides of this equation gives the results indicated.

4.6. For Design A, $t_r = 0.362$, $t_p = 0.641$, $t_s = 3$ and the percentage overshoot is 52.6%.

For Design B, $K = 0.12$, $t_r = 0.389$, $t_p = 0.631$, $t_s = 1.2$ and the percentage overshoot is 18%.

System B has the best overall performance.

5.1. (a) Stable, with three roots in the left half s-plane.

(b) Marginally stable, with one root in the left half s-plane, two complex conjugate roots on the imaginary axis and one root at the origin of the s-plane.

(c) Unstable, with three roots in the left half s-plane and two in the right half s-plane.

(d) Marginally stable, or unstable, with four roots on the imaginary axis and one root in the left half s-plane. The system is unstable if the two complex root pairs are identical.

(e) Unstable, with three roots in the left half s-plane and three in the right half s-plane.

(f) Unstable, with four roots in the left half s-plane, four on the imaginary axis and two in the right half s-plane.

5.2. $a_3(a_1 a_2 - a_0 a_3) - a_1^2 a_4 > 0$

5.3. Both criteria are met. The root solving program, Program 3 in Chapter 11, may be used to show that the roots of $F(s)$ are $-1 \pm j$, -2 and -3.

5.4. For the inner loop, the closed-loop transfer function is

$$G_T(s) = 5/(s^2 + 11s + (10 + 5K_T))$$

For critical damping of the loop $K_T = 4.05$. With this value of K_T the closed-loop system transfer function is

$$Y(s)/R(s) = 20(s + 0.1)K_B/(s^5 + 15s^4 + 74s^3 + 121s^2$$

$$+ 20K_B s + 2K_B)$$

(a) $0 < K_B < 25.3$

(b) $K_B = 25.3$ and $\omega = 7.62\,\text{rad/unit time}$.

5.5.   The system's closed-loop characteristic equation is given by

$$s^4 + 9s^3 + (88 + 80T_d)s^2 + 160s + 80/T_i = 0$$

To satisfy the necessary condition for stability, see section 5.6,

$$T_i > 0 \quad \text{and} \quad T_d > -1.1$$

Using the Routh array to test for stability, then from row 3:

$$T_d > -79/90$$

and from row 4:

$$T_d > 9/160T_i - 79/90$$

In the above inequality, as T_i become infinite,

$$T_d > -79/90$$

which is identical to the condition on row 3. Hence for asymptotic stability

$$T_i > 0 \quad \text{and} \quad T_d > -79/90$$

5.6.   Under pure proportional control the closed-loop characteristic equation of the model is

$$s^4 + 6s^3 + 11s^2 + 6s + K = 0$$

From which it will be found that

$$K_c = 10 \quad \text{and} \quad T = 2\pi$$

Hence the PID settings are

$$K = 6, \quad T_i = \pi \quad \text{and} \quad T_d = \pi/4$$

The closed-loop characteristic equation of the model and PID controller is

$$s^5 + 6s^4 + 11s^3 + (6 + 1.5\pi)s^2 + 6s + 6/\pi = 0$$

Routh's array may be used to show stability.

6.1.   The characteristic equation and roots are:
(a) $s^2 + 2s + K = 0$, $\quad s = -1 \pm \sqrt{(1 - K)}$
(b) $s^2 + 4s + (3 + K) = 0$, $\quad s = -2 \pm \sqrt{(1 - K)}$
(c) $s^2 + (2 + K)s + (1 + K) = 0$, $\quad s = -1$ or $-(1 + K)$
The root-loci plots for these systems are shown in Fig. A.7. Note that in system (c) the open-loop transfer function has pole–zero cancellation; with cancellation the system becomes $K/(s + 1)$, for

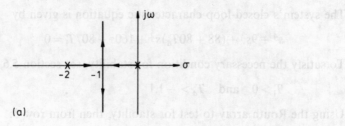

(a)

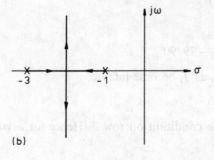

(b)

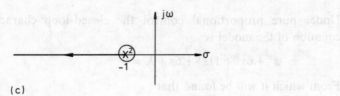

(c)

Figure A.7

which the closed-loop characteristic equation is $s + (1 + K) = 0$, and the dominant pole is at $-(1 + K)$.

6.2.　The root loci plot, shown in Fig. A.8, may be checked by solving for the roots of the closed-loop characteristic equation directly. For a damping ratio of 0.707, $K = 10$ and $t_s = 3/4$ time units.

6.3.　The plots are shown in Fig. A.9, and should be checked using Program 6 of Chapter 11.

6.4.　The root loci plot is shown in Fig. A.10 and the required solutions are:
(a) $K = 74$ and 21.5.
(b) $K = 45.584$.

(c) $$\frac{45.584}{s^3 + 9s^2 + 18s + 45.584} \quad \text{or} \quad \frac{45.584}{(s + 7.4)(s^2 + 1.6s + 6.16)}$$

(d) $1 - 0.126e^{-7.4t} - e^{-0.8t}(0.875 \cos 2.35t + 0.693 \sin 2.35t)$

6.5.　$K = 223.2$. The root loci plot is shown in Fig. A.11.

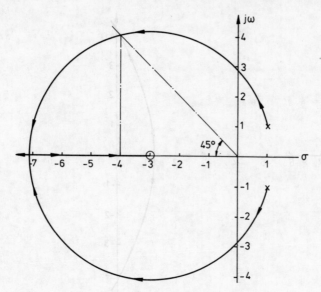

Figure A.8

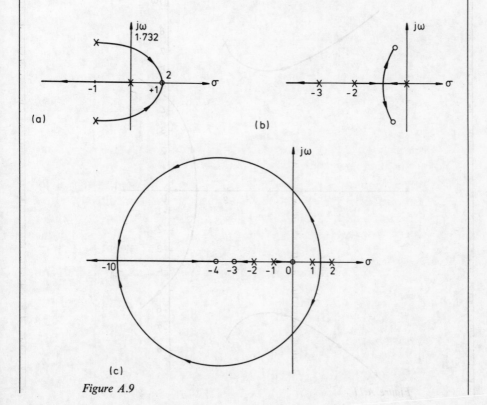

(a)

(b)

(c)

Figure A.9

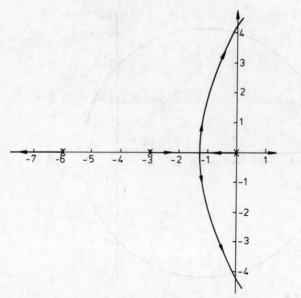

Figure A.10

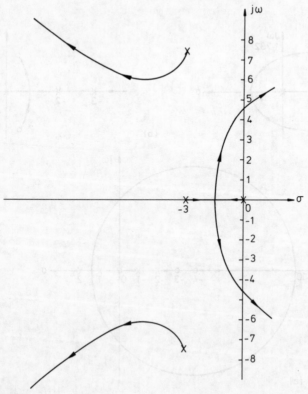

Figure A.11

6.6.   The open-loop transfer function should be first written in the normal form, as

$$G(s) = \frac{K(s + 0.1)}{s^2(s + 0.2)}, \quad K = 0.0225$$

This gives the root loci plot shown in Fig. A.12.

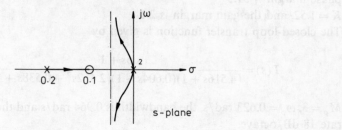

Figure A.12

The modulus criterion is then used to solve for the real pole when the gain is 0.0225, and gives $s = -0.15$. By factoring the closed-loop characteristic equation the other poles are found to be at $s = -0.025 \pm j0.12$.

7.1.   (a) $M(\omega) = \dfrac{1}{\omega\sqrt{(1 + \omega^2)}}$

$$\phi(\omega) = \frac{3\pi}{2} + \tan^{-1}(\omega)$$

(b) $M(\omega) = \dfrac{K}{\omega\sqrt{(\omega^2\delta_1^2 + 1)}\sqrt{(\omega^2\delta_2^2 + 1)}}$

$$\phi(\omega) = \frac{3\pi}{2} - \tan^{-1}(\omega\delta_1) - \tan^{-1}(\omega\delta_2)$$

(c) $M(\omega) = \dfrac{10\sqrt{(1 + 4\omega^2)}}{\omega\sqrt{(1 + 4 \times 10^{-4}\omega^2)}\sqrt{(1 + 0.25\omega^2)}\sqrt{(1 + \omega^2)}}$

$$\phi(\omega) = \tan^{-1}(2\omega) + \frac{3\pi}{2} - \tan^{-1}(0.02\omega) - \tan^{-1}(0.5\omega)$$

$$- \tan^{-1}\omega$$

7.2.

$$M(\omega) = \frac{\sqrt{((0.2\omega)^2 + 1)}}{\sqrt{(\omega^2 + 1)}\sqrt{((0.5\omega)^2 + 1)}\sqrt{((0.25\omega)^2 + 1)}}$$

$$\phi(\omega) = \tan^{-1}(0.2\omega) - \tan^{-1}(\omega) - \tan^{-1}(0.5\omega)$$
$$- \tan^{-1}(0.25\omega)$$

which gives $M(2) = 0.305$ and $\phi(2) = -113.2°$.

7.3.   The gain margin is 0.28 (11.1 dB) and the phase margin $-31.5°$. With a transfer function gain of 1, the gain margin is 2.8 (8.9 dB) and the phase margin $+31.5°$.

7.4. $K = 1.52$, and the gain margin is 2.1.

7.5.   The closed-loop transfer function is given by

$$T(s) = \frac{5s + 1}{(4.516s + 1)(0.604s + 1)(2.412s^2 + 0.538s + 1)}$$

$M_p = 3$, $\omega_p = 0.623$ rad/s, the bandwidth is 0.964 rad/s and the cutoff rate 18 dB/octave.

7.6.   The results should be checked by comparing points with the Bode plots obtained for Problem 7.5.

7.7.   With the lag compensator included, a loop gain of $K = 5$ is required for a 20° phase margin. From Table 7.1, the steady-state ramp error of a type 1 system is $1/(KG_cG(0))$. Hence the percentage reduction in error is

$$\frac{1/1.52 - 1/5}{1/1.52} \times 100\% = 69.6\%$$

7.8.   (a) 54.4°.

(b) The increase in phase margin frequency is $0.708 - 0.587 = 0.121$ rad/s. The required gain is $K = 4.2$. Note that the lead compensator must be transformed to Bode form:

$$G_c(s) = \frac{1}{5} \frac{(10s + 1)}{(2s + 1)}$$

7.9.   It is normal to fix K at some nominal value, say $K = 2.5$. This produces the Nyquist plot, shown in Fig. A.13. For this system $N = 1$ and $Z = P - N = 0$, so the closed-loop system is stable and will be stable for $1.25 < K \leqslant +\infty$.

7.10.   (a) $K = 0.6$.

(b) The Nyquist diagram is shown in Fig. A.14. For this system $N = 0$ and $Z = P - N = 0$, so it is stable.

(c) The phase margin is approximately 30°.

7.11.   (a) $K = 139$.

(b) Gain margin $= 2.05$.

(c) Phase margin $= 100°$.

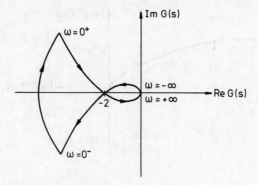

Figure A.13

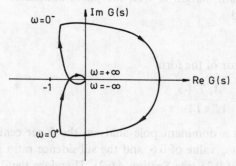

Figure A.14

7.12. (a) $\sin \psi = \dfrac{ac}{Oc} = \dfrac{1}{M}$

(b) $(\vec{Oa})^2 = (\vec{Oc})^2 - (\vec{ca})^2$ and $\vec{ba} = \vec{Ob} \sin \psi$. As $(\vec{Ob})^2 = (\vec{Oa})^2 - (\vec{ba})^2$, substituting for $\vec{Oa}$ and $\vec{ba}$ yields the desired result $\vec{Ob} = 1$.

7.13. $K = 10.4$, gain margin $= 4.59$ and phase margin $= 42.2°$.

7.14. $M_p = 1.5$ when the gain is adjusted by 0.5. Then $\omega_p = 3.5$ rad/s and $\omega_d = 3.79$ rad/s. Note that Equation (7.34) uses M_p to find ζ. From Equation (7.35), ω_n may be found. The value of ω_p is the frequency at the point of tangency of the $G(j\omega)$ locus and the M_p circle. The damped frequency for a second-order dominant system is given approximately by $\omega_d = \omega_n \sqrt{(1 - \zeta^2)}$ (see Example 4.1).

7.15. Using the inverse polar plot, draw a line at angle ψ, where $\tan \psi = 1/M$. By trial and error find a circle of radius r which just touches the $G^{-1}(j\omega)$ locus. The radius r is then the perpendicular distance from the real axis to the line at angle ψ (see Fig. A.15).

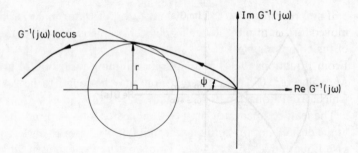

Figure A.15

$K = 11.7$, the gain margin is 4.11, the phase margin 39° and the bandwidth 1.9 rad/s.

8.1. A lead compensator of the form

$$G_c(s) = \frac{1/T + s}{1/(\alpha T) + s}$$

is required. Using a dominant pole analysis, the 5 per cent settling time indicates a $\zeta\omega_n$ value of 0.6, and the subsidence ratio indicates a damping ratio of 0.25 (see Section 4.6.2). Therefore the dominant closed-loop poles should be at $-0.6 \pm j2.32$.

Suitable values of T and α are found by trial and error. A typical procedure is to set T (guess a zero position), and use the angle criterion (see Section 6.3.1) to locate the compensator pole. Check that $0.1 \leqslant \alpha < 1$ (that is, that the compensator is a lead device). Produce the root loci diagram, select the loop gain to give the required closed-loop poles and check these poles for dominance. Produce a closed-loop step response and verify that the performance specifications are met. The procedure is clearly iterative, and the programs in Chapter 11 should be used.

Section 8.4.2 suggests using the compensator zero to cancel the pole at $s = -10$ (that is, to set $T = 0.1$). If this value of T is used the angle criterion indicates a pole contribution of $-51°$, thus producing lag compensation. In this case some alternative must be sought.

There are a number of possible solutions. One is to set $T = 1$, which gives $\alpha = 0.25$ and a compensator pole at -4. With $K = 90.3$ the closed-loop poles are at $-0.6 \pm j2.32$, -1.43 and -11.01. Since the pole at -1.43 is close to the zero at -1, the response should be second-order dominant. A step response of this system produces a 5 per cent settling time of about 5.6 s and a subsidence ratio of 5.6:1.

If improvements are required the compensator pole should be moved further into the left half s-plane, thus increasing the dominance of the complex poles.

8.2. From Equations (7.34) and (7.35), a damping ratio of 0.25 gives an M_p value of 2.07. Also, the ω_n value may be found to be 2.4 rad/s which, from Equation (7.35), makes $\omega_p = 2.245$ rad/s.

The lead compensator may be designed as follows. Find the phase at the desired ω_p and estimate the amount of phase advance required. Use Equation (8.16) to find α. Knowing the frequency at which the maximum phase advance is required, use Equation (8.17) to find T. Produce a frequency response plot and adjust the gain to give the required M_p value.

In practice, the procedure will be iterative and further adjustments to the phase advance will normally be required before an acceptable solution is found.

Using the programs of Chapter 11, the phase at 2.245 rad/s is found to be $-192.7°$. From the 2.07 M circle a total phase advance of 38° seems appropriate. Equation (8.16) gives $\alpha = 0.238$, and Equation (8.17) gives $T = 1/1.095$. This result appears close to that obtained for Problem 8.1; however, with $\omega_p = 2.245$, $M_p = 2.3$ which is slightly high, and for this M_p the gain is $K = 110$.

A further adjustment to the phase advance could be made and the procedure repeated.

9.1. The method used in Example 9.1 will yield the transfer function

$$G(s) = \frac{1}{s(s+1)(0.1s+1)}$$

9.2. Levy's method cannot be used directly since the Bode plots indicate that there is a pole at the origin of the s-plane. The frequency contribution of this pole may be removed from the data by subtracting its gain and phase contribution (see Table A.1).

Levy's method (Program 9 in Chapter 11) may now be applied. Another transformation is required to change the amplitude and phase data into real and imaginary components (see Table A.2).

9.3. The exact frequency data (Program 7A) and that obtained from the step response (using Program 8) should be similar. The accuracy will improve with increasing number of step response data points. However, satisfactory results may be obtained with 75 equally spaced time and amplitude points between 0 and 6 s.

Using Levy's method (Program 9) and the exact frequency data (12 logarithmically scaled points between 0.2 and 8 rad/s) and a

Table A.1

| Frequency ω (rad/s) | Contribution of real pole | | Modified system | |
	Amplitude ratio (dB)	Phase	Amplitude ratio (dB)	Phase
0.2	13.98	$-90°$	-0.17	$-12.5°$
0.369	8.65	$-90°$	-0.56	$-22.4°$
0.682	3.32	$-90°$	-1.68	$-38.2°$
1.260	-2.00	$-90°$	-4.20	$-58.7°$
2.327	-7.34	$-90°$	-8.30	$-79.8°$
4.298	-12.66	$-90°$	-13.63	$-100.2°$
7.937	-17.99	$-90°$	-20.18	$-121.3°$
14.659	-23.32	$-90°$	-28.32	$-141.8°$
27.073	-28.65	$-90°$	-37.86	$-157.6°$
50.0	-33.98	$-90°$	-48.13	$-167.5°$

Table A.2

Frequency ω (rad/s)	Real component	Imaginary component
0.2	0.957	-0.211
0.369	0.867	-0.357
0.682	0.648	-0.510
1.260	0.32	-0.527
2.327	0.068	-0.379
4.298	-0.037	-0.205
7.937	-0.051	-0.084
14.659	-0.030	-0.024
27.073	-0.012	-0.005
50.0	-0.004	-0.001

transfer function structure with numerator 1 and denominator 3, produces the following pole–zero locations: a zero at $s = -9.348$, and three poles at $s = -1.112$, -1.537 and -7.856.

With a transfer function structure of numerator 0 and denominator 2, the pole locations are at $s = -1.26 \pm j0.15$.

Over the range of frequencies both models will produce frequency response data similar to that of the original model. This example demonstrates the use of Programs 7 and 9 for model reduction. (Further details on the use of these programs for model reduction may be found in S Thompson 1985 'Model reduction for closed-loop performance'. Transactions of the Institute of Measurement and Control, 7(3): 127–132.)

INDEX